Lecture Notes in Operations Research

Lecture Notes in Operations Research is an interdisciplinary book series which provides a platform for the cutting-edge research and developments in both operations research and operations management field. The purview of this series is global, encompassing all nations and areas of the world.

It comprises for instance, mathematical optimization, mathematical modeling, statistical analysis, queueing theory and other stochastic-process models, Markov decision processes, econometric methods, data envelopment analysis, decision analysis, supply chain management, transportation logistics, process design, operations strategy, facilities planning, production planning and inventory control.

LNOR publishes edited conference proceedings, contributed volumes that present firsthand information on the latest research results and pioneering innovations as well as new perspectives on classical fields. The target audience of LNOR consists of students, researchers as well as industry professionals.

This series is indexed by Research Papers in Economics (RePEc), zbMATH and has been accepted by Scopus.

Fouad El Ouardighi
Editor

Essays on Pollution Control in Economics and Management Science

An Interdisciplinary View

Forewords by Rick Van Der Ploeg and Franz Wirl

Springer

Editor
Fouad El Ouardighi
ESSEC Business School
Cergy Pontoise Cedex, France

ISSN 2731-040X ISSN 2731-0418 (electronic)
Lecture Notes in Operations Research
ISBN 978-3-031-78226-8 ISBN 978-3-031-78227-5 (eBook)
https://doi.org/10.1007/978-3-031-78227-5

This Springer imprint is published by the registered company Springer Nature Switzerland AG
The registered company address is: Gewerbestrasse 11, 6330 Cham, Switzerland

"While there is no shortage among articles and books about the economics of pollution (in particular about CO_2 emissions leading to global warming with far reaching consequences), the book by Professor El Ouardighi has some unique selling propositions. First it emphasizes the dynamics of pollution accumulation (mostly but not only relevant to global warming). Second it addresses (in some chapters) the (mathematically) difficult issue of complex policies with an emphasis on thresholds, which is highly relevant for policy making. Third strategic aspects of pollution policies are addressed not only at the level of economies (i.e., states) but also for firms (supply chain). And this adds the fourth point, namely, the inclusion of the analysis of management strategies.

Given the book's focused yet broad coverage, I suggest the book to readers who are interested in an analytical treatment of pollution problems from the point of view of economics and management. In particular for researchers who look for a kick start when entering this field (e.g., Ph.D. students), because the reader is exposed to the inner workings of the models rather than an informal description, which the many existing surveys fail."

—*Franz Wirl,* Professor Emeritus of Industry, Energy, and Environment, University of Vienna, Austria

"Pollution of the environment varying from greenhouse gas or nitrogen emissions to deforestation and plastics and heavy metals waste is destroying biodiversity and ultimately threatens sustainable growth. Many of these pollutants are transboundary so game theory approaches are crucial. This volume by a top expert in the field gives a comprehensive overview of pollution control in economics and in management science. Political issues such as myopia of policymakers and management issues such as supply chains, inventory control and recycling materials are treated. This volume is essential reading for anyone who wants to get up to date on pollution control."

—*Rick Van Der Ploeg,* Professor of Economics, University of Oxford, UK

I dedicate this book to my beloved father.

Acknowledgement

This book is a compilation of articles on pollution control models and policies in economics and management science published in various scientific journals in the last 10 years. It was an honor for me to collaborate with my dearest friends and highly esteemed colleagues on this topic. I would like to thank each of them wholeheartedly for sharing their knowledge and expertise with me. I believe that the re-edition of these papers may provide the readership with a comprehensive and unified perspective on pollution control issue. I am grateful to the Springer Nature editors for publishing the book, especially to Ms. Jialin Yan for her support for this project, and Ms. Sneha Arunagiri for her coordination efforts. I also thank Ananya Thakur for her research assistance and Ms. Karen Sherman for the English editing.

Introduction to Pollution Control Models in Economics and Management Science

Introduction

Carbon pollution accumulation is steadily rising (Fig. 1) despite repeated strident calls from an increasing number of scientists for national governments to adopt an urgent, disruptive environmental policy (Lenton et al., 2008, 2019; Penuelas, 2023). However, such a policy must take into account the growing recognition that a drastic reduction of emissions is neither possible in the short run, nor sufficient in the long run.

A drastic reduction of emissions is not possible in the short run because the decoupling between economic growth and annual emissions occurs at a relatively sluggish pace in various countries (Wang & Su, 2020). The main reasons for this slowness include both the exorbitant cost of technological transition to more environmentally friendly economic processes (Goering, 2022) and the difficulty of imposing socially fair carbon sobriety on the populations concerned (Maitland et al., 2022). Nor is the drastic reduction sufficient in the long run because of the vulnerable state of the pollution sinks and the need for rapid actions to enhance their resilience (Malhi et al., 2020).

As illustrated in Fig. 2, the main pollution sinks that are still effective are the atmosphere (about 45%), the woodlands, the soil and the forested areas (about 30%), and the oceans (about 25%).

Conversely, the main pollution sources are fossil carbon and land use change (Fig. 3). One important source related to land use change is deforestation, due to the rapidly increased oxidation of soil organic matter engendered by tree extraction (Harris et al., 2012; Houghton et al., 2012; Canadell et al., 2007).

Regarding the terrestrial carbon sink, its efficiency has been shown to have decreasing marginal returns (Fig. 4). Moreover, while the global forest sink has endured almost undiminished for three decades until 2019 (Pan et al., 2024), the most recent investigations revealed a large decline in the efficiency of the land carbon sink in 2023 (Ke et al., 2024).

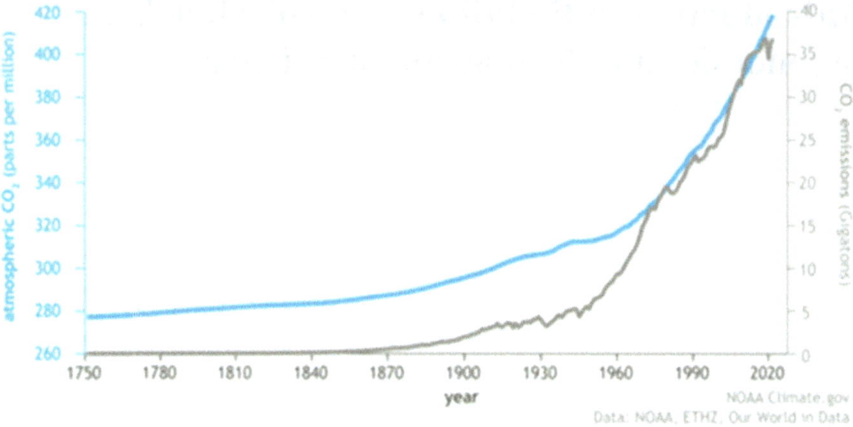

Fig. 1 Global atmospheric carbon dioxide compared to annual emissions between 1751 and 2022 (Lindsay, 2024)

One important observation related to this evolution lies with the increasing tendency of terrestrial carbon sinks to switch to carbon sources. A local illustration of this pattern is provided by the case of French forests (Fig. 5), where wide regions have become net carbon emitters.

A more global illustration of the increasing tendency of terrestrial carbon sinks to switch to carbon sources is provided by the case of tropical forests (Fig. 6). In this figure, the red areas correspond to net carbon emitter regions, while green areas are for net carbon absorber regions.

As for the ocean carbon sink, two important consequences of carbon emissions are ocean acidification (Jiang et al., 2019) and ocean warming (Storto & Yuang, 2024). It is well-known that warming decreases the absorption efficiency of the ocean sink (IPCC, 2014). Recent investigations suggest that ocean warming weakens ocean circulation, which might lead to a release of carbon from the deep ocean into the atmosphere (Lauderdale, 2024). As a result, oceans might also switch from carbon sinks into carbon sources.

The abovementioned studies are all consistent with the long-established prediction that the biosphere is expected to switch from a carbon sink to a source within this century (e.g., Cox et al., 2000; Cramer et al., 2001; Joos et al., 2001; Lenton et al., 2006; Canadell et al., 2007; Piao et al., 2008; Baccini et al., 2017). Such an extreme scenario would give rise to uncontrollable consequences due to an increase of exceptional magnitude in atmospheric carbon. For instance, the stock of carbon stored in the permafrost, forests, wetlands, and oceans that would potentially be released into the atmosphere due to global warming is estimated at more than 3700 Gt (Giga-tons).

An important related issue lies with the hysteretic if not irreversible nature of this evolution, which results from an inertia in the destructive impact of the pollution

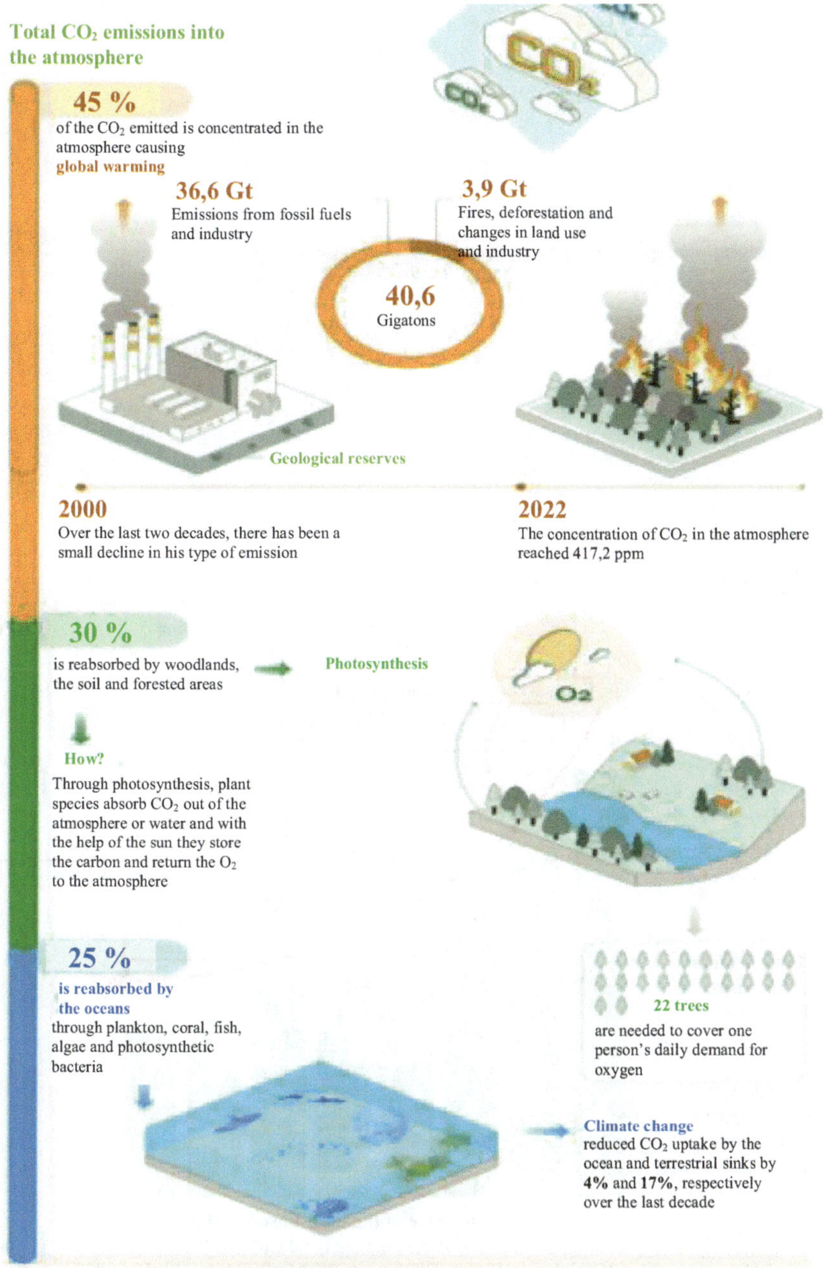

Fig. 2 Anthropic carbon emissions and natural absorption (Global Carbon Project, 2022)

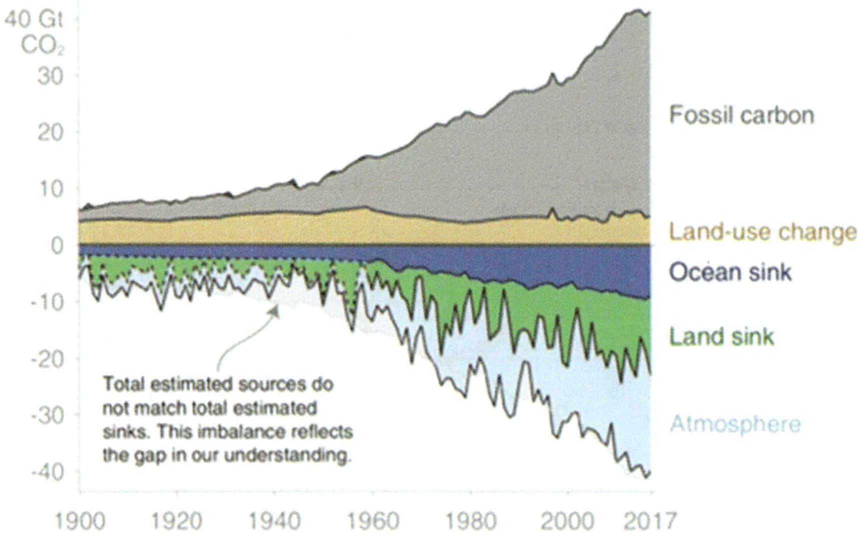

Fig. 3 Balance of pollution sinks and pollution sources (Global Carbon Project, 2022)

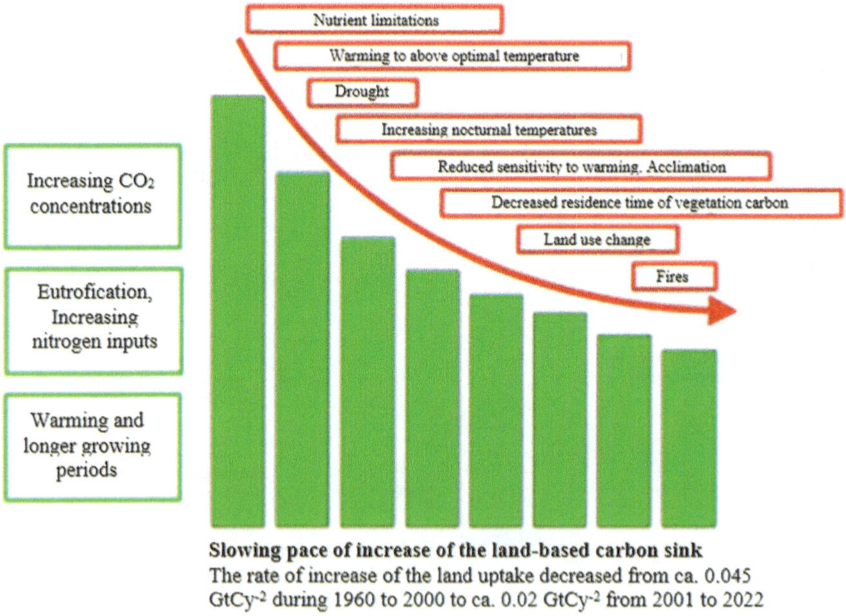

Fig. 4 Decreasing absorption efficiency of the terrestrial carbon sink (Penuelas, 2023)

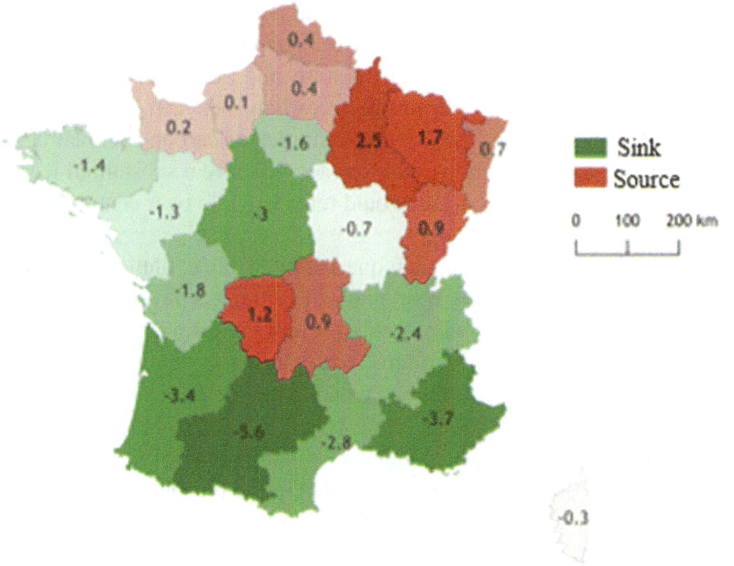

Fig. 5 Forest carbon sinks and carbon sources in France in millions of tons (CITEPA, 2024)

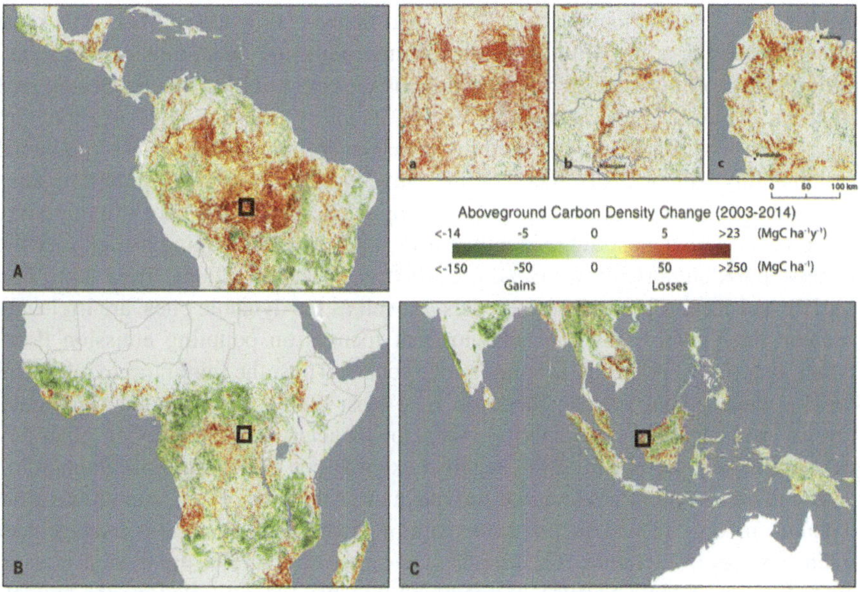

Fig. 6 Geography of carbon density change in the South American, African, and Asian tropical forests (**a**) to (**c**) in MgC $ha^{-1}year^{-1}$ (Megatons of carbon produced by hectare per year) from 2003 to 2014 (Baccini et al., 2017)

stock on the biosphere. This inertia implies that a large artificial removal of carbon from the atmosphere would not result in the simultaneous return to preindustrial climate conditions, as surface air temperature and sea level response exhibit a substantial time lag relative to atmospheric carbon, between 100 and 1000 years (Zickfeld et al., 2013). Even after several centuries of carbon removal deployment from the atmosphere, past carbon emissions would leave a substantial legacy in the marine environment. That is, oceans would remain acidic over centuries (Mathesius et al., 2015).

Therefore, it is vital to prevent such quasi-irreversible evolution. Over more than half a century, the environmental economics literature has developed a considerable body of knowledge to characterize optimal levers that prevent environmental irreversibilities. To this end, this literature mostly rests on continuous-time dynamic models with infinite planning horizons, which have the merit of accounting for the future generations' best interests (Jørgensen et al., 2010). More than ever, these economic models face the critical challenge of finding an economically sustainable path toward a *stable* state that prevents a switch of the biosphere from a carbon sink into a source, along with the negative implications of this transition for terrestrial and marine life.

Besides this scientific challenge for the environmental economics literature, a significant contribution is also required from the research in management science. During the last two decades, an important stream of publications on environmental sustainability has developed in various areas of management, e.g., marketing (Kotler, 2011), operations (Jaehn, 2016), and finance (Jing et al., 2023). The most closely related area to pollution control is environmental operations management, which seeks to infer managerial prescriptions on how to reduce the environmental footprint of production, transportation, and inventory activities. Because they are mainly process-improving oriented, the management approaches of environmental sustainability cannot easily focus on the prevention of a switch of pollution sinks into pollution sources. However, dynamic approaches in this area are still quite rare (Xiao et al., 2019); most of the literature uses static decision-making models (see the reviews by Kleindorfer et al. (2005), Corbett and Klassen (2006), Tang and Zhou (2012), and Jaehn (2016). These models, as well as the dynamic ones, are irrelevant because they involve myopic behavior that focuses on polluting emission flows rather than on the pollution accumulation process. Thus, the minimum requirement for the management science literature is at least opting for a dynamic approach toward environmental sustainability issues to characterize optimal policies that rely on a proper internalization of the negative externalities of pollution accumulation.

In this introductory chapter, we provide a critical review of the continuous-time dynamic models related to pollution control in the economics and management science pieces of literature. The emphasis is on the structure of the models considered. We argue that, because of their one-dimensional formulation, the models generally lack both relevance and flexibility to take up the challenge of switch-prevention of pollution sinks into pollution sources, especially in the context of multiple noncooperative polluters. However, such models would be useful in the handling of sustainability issues by the management science literature, as they would enable a shift from a flow- toward a stock-oriented approach of pollution control.

The Dynamics of Pollution Accumulation

The history of dynamic pollution control models is closely related to the modeling of environmental absorption efficiency. These models belong to six main categories described below.

Consider one country that produces an output of a given product that provides a social utility. The output generates polluting emissions at time t, $t \in [0, \infty[$, denoted by $e(t) \geq 0$, which is a control variable. The instantaneous emissions accumulate over time as a pollution stock, $P(t) \geq 0$, which acts as a state variable, that engenders negative externalities. If we let $\delta > 0$ denote the marginal environmental absorption efficiency, the pollution accumulation dynamics are alternatively described by the following model structures.

The six main categories are the following:

- *Linear environmental absorption efficiency* (LAE):

$$\dot{P}(t) = e(t) - \delta P(t), \quad P(0) = P_0 \geq 0$$

This simplistic model structure has been commonly used in the literature in diverse contexts, e.g., Keeler et al. (1972), Nordhaus (1991), van der Ploeg and Withagen (1991), Dockner and Long (1993), Ulph and Ulph (1994), Caputo and Wilen (1995), Farzin (1996), Benchekroun and Long (1998), Stockey (1998), Bertinelli et al. (2014), Li (2014), Benchekroun and Martin-Herran (2016), Fünfgelt and Schulze (2016), Dragone et al. (2022), Li and Chen (2021), and Martin-Herran and Rubio (2023). The main reason for its frequent use lies in the linear specification of the state variable and the mathematical tractability it enables. Given the limitations imposed by the biosphere in terms of pollution absorption, this model is nevertheless clearly invalid for the treatment of global pollution issues. However, it can still be used to resolve local pollution management issues if the decision problem accounts for the cost of the negative externalities engendered by pollution accumulation (Kogan & El Ouardighi, 2024).

- *Nonlinear environmental absorption efficiency* (NLAE):

$$\dot{P}(t) = e(t) - \delta P(t) + h(P(t)), \qquad P(0) = P_0 \geq 0$$

where $h(P(t)) \geq 0$, $h_P > 0$. This model structure initially appeared in the study by Forster (1975). Because of the mathematical complexity it induced, however, it remained unused for decades, until it was reintroduced with the specification

$h(P(t)) \equiv \frac{P(t)^2}{1+P(t)^2}$ by Schaeffer (1997), to describe the shallow lake ecology. It was then adopted under various specifications by Mäler et al. (2003), Kossioris et al. (2008), Dockner and Wagener (2014), Moser et al. (2014), and Kogan and El Ouardighi (2024). This model structure is notoriously complex to handle mathematically, and is thus generally resolved by numerical means. Its main merit lies with the nonlinear (i.e., hysteretic) and eventually non-convex (i.e., irreversible) feedback it involves, which can result in pollution accumulation even if the source of pollution vanishes.

- *Environmental absorption efficiency with switching regimes* (SR):

$$\begin{cases} \dot{P}(t)=e(t)-\delta P(t), \quad P(0)=P_0\geq 0 \text{ and } \begin{cases} P(T)=\bar{P}, \text{if } T<+\infty \\ \lim_{t\to\infty} P(t)\leq \bar{P}, \text{if } T=+\infty \end{cases} & \text{(Period I)} \\ \dot{P}(t)=e(t), \quad P(T)=P_T & \text{(Period II)} \end{cases}$$

This model structure was introduced by Tahvonen and Withagen (1996) and used in the studies by Chevé (2000), Prieur (2009), Boucekkine et al. (2013a, 2013b, 2023), Wagener and de Zeeuw (2021), and Gromov et al. (2023). In this structure, the planning horizon is decomposed into two time intervals: Period I, during which there is linear environmental absorption efficiency; and Period II, wherein environmental absorption efficiency is exhausted.

- *Time-dependent environmental absorption efficiency* (TDAE):

$$\begin{cases} \dot{P}(t)=e(t)-A(t)P(t), & P(0)=P_0\geq 0 \\ \dot{A}(t)=w(t)A(t)^{\beta}-\gamma P(t), & A(0)=A_0>0 \end{cases}$$

where $A(t) \geq 0$ denotes the environmental absorption efficiency (state variable) at time t, $t \in [0, \infty[$, $w(t) \geq 0$ the instantaneous restoration efforts (control variable), $\gamma > 0$ the marginal destructive impact of pollution on the absorption efficiency, and $1 \geq \beta \geq 0$ a scale parameter. This two-dimensional model structure was proposed by El Ouardighi et al. (2014, 2016a), and developed by El Ouardighi et al. (2018, 2020a), and El Ouardighi et al. (2020b) to account for environmental absorption efficiency as a state variable in its own right, rather than an instantaneous function of the pollution stock.

- *Stochastic environmental absorption efficiency* (SAE):

$$\mathrm{d}P(t)=(e(t)-\delta P(t))\mathrm{d}t-\sigma P(t)\mathrm{d}W(t), \quad P(0)=P_0\geq 0$$

where $\delta P(t)$ and $\sigma P(t)$ represent the mean and the volatility of the environmental absorption efficiency, respectively, and $W(t)$ is a standard Wiener process. This stochastic formulation was recently used by Kogan et al. (2017) to account for the impact of variability in environmental absorption efficiency over an infinite time horizon.

– *No environmental absorption efficiency* (NAE):

$$\dot{P}(t) = e(t), \quad P(0) = P_0 \geq 0$$

This reduction model is used in the studies by El Ouardighi et al. (2016b, 2021) for a finite time horizon only. It is more suitable for solving management issues related to local pollution than for addressing economic concerns about global pollution.

Under LAE, NLAE, SR, and NAE, the country's discounted objective is defined as:

$$\underset{e(t) \geq 0}{\text{Max}}\, J = \int_0^{\infty} e^{-rt}[f(e(t)) - c(P(t))]\mathrm{d}t$$

where f denotes an increasing concave utility function of emissions and c an increasing convex cost function of the negative externalities of pollution, $r > 0$ being the discounting rate.

Under SAE, the problem writes:

$$\underset{e(t) \geq 0}{\text{Max}}\; J = E\left\{\int_0^{\infty} e^{-rt}[f(e(t)) - c(P(t))]\mathrm{d}t\right\}$$

Under TDEAE, the country's discounted objective is:

$$\underset{e(t),\, w(t) \geq 0}{\text{Max}}\; J = \int_0^{\infty} e^{-rt}[f(e(t)) - c(P(t)) - d(w(t))]\mathrm{d}t$$

where d is an increasing convex cost function of the restoration effort.

While the SR formulation considers the case of irreversible exhaustion of the pollution sinks, the NLAE and TDAE models extend to the case of a switch of pollution sinks into pollution sources. The structure of these models is more consistent with the abovementioned studies in environmental sciences. In economic terms, the difference between a mere exhaustion of the pollution sinks and a switch of pollution sinks into pollution sources is huge. That is, with the SR formulation, the tradeoff between the pollution cost and the benefit from instantaneous polluting emissions can make irreversible exhaustion of pollution sinks an economically viable option (Tahvonen & Withagen, 1996; Boucekkine et al., 2023). In contrast, with the NLEA and TDEA models, the pollution stock, being out of control, would grow in such a disproportionate way that no tradeoff could reasonably justify the economic viability of such a switch. We conclude that the SR formulation is well-suited to the economics of local pollution problems while the NLAE and TDAE models are more suitable for the economics of global pollution issues.

Transient and Long Run-Behavior with a Single Polluter

In the case of a single polluter, the most remarkable solution patterns are derived from the NLAE and TDAE model structures. For instance, in the context of a single technology of production, multiple, history-dependent steady states emerge from an NLAE model structure with a hysteretic effect (Fig. 7).

The phase diagram below shows three steady states, two being locally stable, namely $\widehat{P}_1$ and $\widehat{P}_3$, and one being unstable, namely $\widehat{P}_2$. The history-dependency property is characterized by the so-called Skiba threshold (Grass et al., 2008), $\overline{P}$, which divides the phase diagram into two basins of attraction, each one converging to the corresponding stable steady state. That is, for a relatively low (high) initial pollution stock, i.e., $P_0 < (>) \overline{P}$, the transient path converges toward the most favorable (unfavorable) outcome, $\widehat{P}_1$ ($\widehat{P}_3$). If the initial pollution stock is exactly equal to $\overline{P}$, then the polluter is indifferent between $\widehat{P}_1$ and $\widehat{P}_3$.

Another example of an interesting solution pattern is provided by the TDAE model structure (Fig. 8). An oscillatory transient path to the steady state can be observed that is due here to a greater discounting rate.

In Fig. 8a, a lower discounting rate gives rise to a monotonic transient path that ends up with less pollution and both greater environmental absorption capacity and production activity level. In contrast, in Fig. 8b, a greater discounting rate, i.e., a more impatient polluter, exhibits a spiraling behavior in which the pollution sinks switch several times to a pollution source, and reciprocally. Though the spiraling

Fig. 7 Phase portrait diagram in the state-control space (Moser et al., 2014)

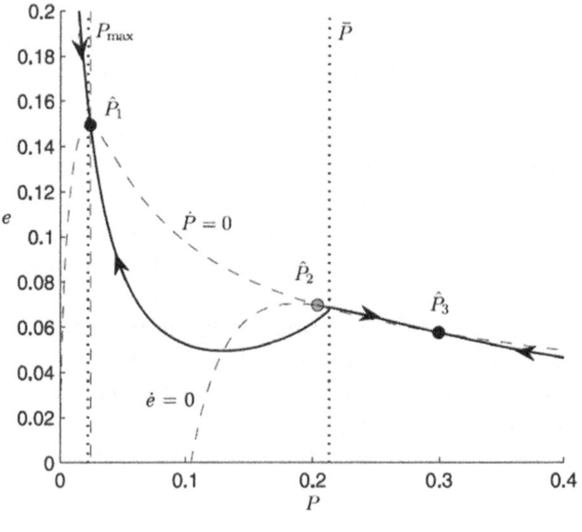

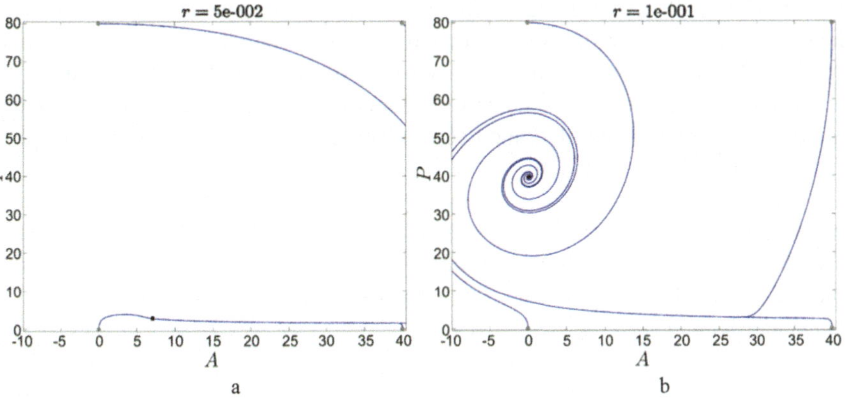

Fig. 8 Phase plane in the state-state space (El Ouardighi et al., 2014). (**a**) Lower discounting rate. (**b**) Higher discounting rate

transient behavior is reversible, it ends up in a much less favorable steady state than the case with a lower discounting rate. Most of all, this pattern raises the issue of the practical management of such versatile evolution.

The Transboundary Pollution Game

In a transboundary pollution game, the players, i.e., countries, are supposed to play noncooperatively. In dynamic noncooperative games, players have the choice between two solution concepts (Başar & Olsder, 1979; Dockner et al., 2000):

- Either they pre-commit to the execution of a given plan of action throughout the game, i.e., they decide by the clock, and thus use open-loop equilibrium strategies,
- Or their actions at each instant of time are contingent upon the state of the system, i.e., they decide by the stock, and thus apply Markov-perfect equilibrium strategies.

The two solution concepts differ in terms of how they adjust for changing values of the pollution stock. In an open-loop equilibrium, the players' strategies are weakly time consistent, which allows for the continuation of the players' equilibrium actions at any time only if the equilibrium path is followed. The Markov-perfect equilibrium has the desirable property of subgame perfectness because current actions depend on the pollution stock and time, and not only on time, as open-loop equilibrium does.

In practice, voluntary "contributions" to global emissions reduction, supposed to act as commitments, are agreed to by the countries attending the successive International Conferences on the Climate since the implementation of the Kyoto Protocol in 2005. However, despite massive commitments (195 countries) at the Paris

Conference in 2015, a growing number of countries (e.g., Sweden; cf. Hivert, 2024) have since opted for a more contingent approach (i.e., more flexible emissions policy with regular updates of the targets to reach, depending on the current state of the environment). Indeed, the absence of commitment at an international level may affect the pollution accumulation process and the eventual transition of the biosphere from a carbon sink to a source.

A transboundary pollution game usually involves symmetric players with simultaneous noncooperative moves, i.e., Nash equilibrium strategies. The players thus seek to maximize their individual utility over an infinite time horizon. Assuming a one-dimensional model, country i's noncooperative game problem, $i = 1 \ldots n$, is generally defined as:

$$\underset{e_i(t)\geq 0}{\text{Max}}\, J^i = \int_0^{\infty} e^{-rt}[f(e_i(t)) - c(P(t))]\mathrm{d}t$$

such that one of the following constraints applies:

$$\dot{P}(t) = \sum_{i=1}^{n} e_i(t) - \delta P(t), \quad P(0) = P_0 \geq 0$$

$$\dot{P}(t) = \sum_{i=1}^{n} e_i(t) - \delta P(t) + h(P(t), \qquad P(0) = P_0 \geq 0$$

$$\begin{cases} \dot{P}(t) = \sum_{i=1}^{n} e_i(t) - \delta P(t), P(0) = P_0 \geq 0 \text{and} \begin{cases} P(T) = \bar{P}, \text{if}\, T < +\infty \\ \lim\limits_{t\to\infty} P(t) \leq \bar{P}, \text{if}\, T = +\infty \end{cases} & (\text{Period I}) \\ \dot{P}(t) = \sum_{i=1}^{n} e_i(t), \quad P(T) = P_T & (\text{Period II}) \end{cases}$$

$$\mathrm{d}P(t) = \left(\sum_{i=1}^{n} e_i(t) - \delta P(t)\right)\mathrm{d}t - \sigma P(t)\mathrm{d}W(t), \quad P(0) = P_0 \geq 0$$

In a two-dimensional model, country i's objective function is defined as:

$$\underset{e_i(t),\, w_i(t)\geq 0}{\text{Max}}\, J^i = \int_0^{\infty} e^{-rt}[f(e_i(t)) - c(P(t)) - d(w_i(t))]\mathrm{d}t$$

such that:

$$\dot{P}(t) = \sum_{i=1}^{n} e_i(t) - A(t)P(t), \qquad P(0) = P_0 \geq 0$$

$$\dot{A}(t) = \sum_{i=1}^{n} w_i(t)A(t)^{\beta} - \gamma P(t), \qquad \mathrm{A}(0) = A_0 > 0$$

Most often, Markov-perfect Nash equilibrium (i.e., MPNE) strategies are less environmentally friendly than open-loop Nash equilibrium i.e., OLNE) strategies (Jørgensen et al., 2010; Long, 2010; Benchekroun & Long, 2011), unless:

- Softly discounted, nonlinear MPNE strategies are used in an LAE or NLEA setup (Dockner & Long, 1993; Rubio & Casino, 2002; Rowat, 2006; Dockner & Wagener, 2014).
- There is regime switching, in which case (linear) MPNE strategies might significantly lower the probability of crossing the irreversibility thresholds (Boucekkine et al., 2023).

An important milestone in the analysis of transboundary pollution games was achieved by Dockner and Long (1993). Using a simple derivation procedure borrowed from Tsutsui and Mino (1990) in an LAE pollution accumulation model,[1] Dockner and Long (1993) showed that if the polluters are sufficiently patient, the use of nonlinear MPNE strategies can be an almost perfect substitute for fully coordinating environmental policies.[2] Although the problem of selecting among infinitely many pairs of nonlinear MPNE strategies requires some pre-play communication, the authors presume that the most efficient pair of equilibrium strategies will be selected. This presumption is confirmed by De Frutos and Martin-Herran (2018) for the realistic case of a high initial pollution state.

An extension of nonlinear MPNE strategies to an NLAE pollution model was provided by Kossioris et al. (2008) with the Schaeffer specification, i.e., the shallow lake model. Using numerical methods, the authors suggest that it can generally not be concluded that nonlinear MPNE strategies enable a near-optimal policy in an NLAE pollution model (Fig. 9). However, Dockner and Wagener (2014) claim that this result is due to the fact that local rather than global Nash equilibria were derived.

While nonlinear strategies support the pre-eminence of the concept of subgame perfectness over open-loop strategies, we posit that they are of limited use in dynamic models of pollution control. The reasons are threefold.

- First, nonlinear MPNE strategies only work in one-dimensional models, which means that they confine the notion of environmental absorption efficiency to an instantaneous function of the pollution stock. Therefore, they do not allow for the consideration of restoration effort as a policy instrument. This is an important omission. Restoration of ecosystems has become an essential lever of environmental policies (Bhatia et al., 2023) as well as an important research topic (Shen et al., 2023). Though wrongly confounded with geoengineering (and rejected for this reason) by some economists, the restoration effort has been recently validated (Dasgupta, 2021) as a major strategy for the revitalization of ecosystems. Taking into account the effort to restore ecosystems requires one to define environmental absorption efficiency as a state variable in its own right. Any artifice that

[1] This procedure consists of differentiating the Hamilton-Jacobi-Bellman equation with respect to the (scalar) state variable, and then deriving a closed-form solution of the ensuing differential equation for the value function gradient.

[2] According to Wirl (2007), this result is not due to nonlinear strategies but rather to the non-decreasing elasticity of the marginal utility function of the players. In contrast, Rubio and Casino (2002) pointed out that the result obtained by Dockner and Long (1993) does not hold if the initial state is lower than the optimal steady state.

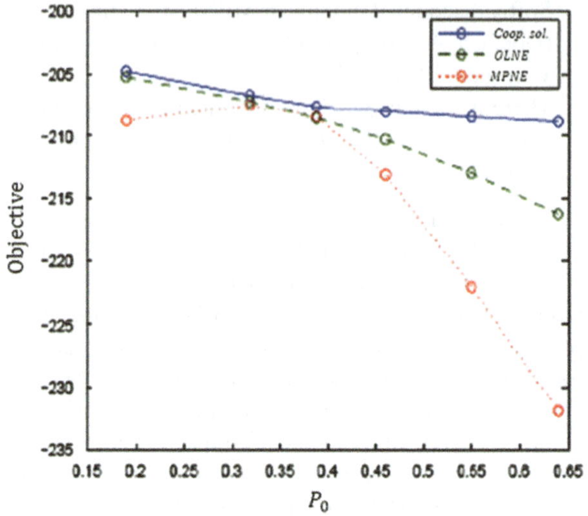

Fig. 9 Initial pollution stock and overall utility in an NLAE model with nonlinear MPNE strategies (Kossioris et al., 2008)

circumvents this requirement will face considerable mathematical difficulty and result in biased policy prescriptions. Another important omission due to the one-dimensional model structure condition for the use of nonlinear MPNE strategies lies in the fact that it precludes any extension toward other issues closely related to pollution control, namely deforestation, renewable natural resource harvesting, etc. More reductionist models are likely to generate misleading environmental policies. Given the stakes at hand, that is, the possibility of a switch of the biosphere into a pollution source, there is a need for more flexible models that include systemic interactions that are likely to alter the evolution of the biosphere. Obviously, flexible environmental models are not amenable to the use of nonlinear MPNE strategies.

– Second, nonlinear MPNE strategies do not neutralize history-dependency. This phenomenon is far from being uncommon in noncooperatively controlled nonlinear dynamical systems (e.g., Karp & Seguin, 2013). An illustration in the setup of the NLAE structure with Schaeffer's specification (i.e., the shallow lake model) is notably provided by Dockner and Wagener (2014), who show that with nonlinear MPNE strategies, a discontinuity occurs that involves history-dependency (Fig. 10). Depending on whether the initial pollution stock is below or above P_i, the steady state can be either lightly, P_1, or heavily polluted, P_2.

From the broader perspective of prevention of an irreversible switching of the biosphere to a pollution source, the neutralization of the switch-inducing history-dependency phenomenon is highly desirable, though implausible. A more realistic approach would consist of maximally pushing back the threshold at which

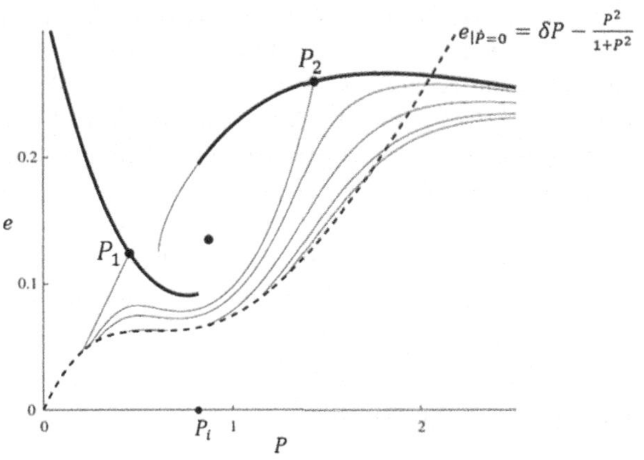

Fig. 10 Local (*thin*) and global (*thick*) nonlinear MPNE strategies in the symmetric two-player shallow lake game with discontinuous global Nash equilibrium strategy at P_i (Dockner & Wagener, 2014)

such discontinuity could occur.[3] In this regard, a policy based on a combination of carbon tax,[4] emissions abatement and the restoration of ecosystems might be switch-preventing or at least switch-delaying. More specifically, it is unlikely that the incorporation of these decision levers into nonlinear MPNE strategies is mathematically tractable.

– Third, nonlinear MPNE strategies are *supposed* to operate over an infinite time horizon, which precludes their use in the resolution of sustainable management issues. The existing dynamic models in this area frequently, though not exclusively, assume a finite time horizon (El Ouardighi et al. 2016b, 2021; Xiao et al., 2019). In this context, there is no way in which MPNE equilibria could be implemented in noncooperative dynamic game models of sustainable management to support a cooperative solution.

Despite its supposed Pareto-superiority over OLNE strategies and its mathematical convenience, the concept of nonlinear MPNE strategies has therefore reached the limits of its utility in the study of pollution control models. From the broader perspective of non-cooperative dynamic game theory, the debate about the Pareto-superiority of this concept is anyway likely to end up soon with the recent questioning of its mathematical validity by Bernhard (2024), on the basis of a disturbing proof that nonlinear MPNE strategies do not even exist for linear-quadratic differential games. The implications of this belated result are like an earthquake for the proponents of these strategies in environmental economics because they invalidate three decades of investigation on the subject.[5]

[3]Given the current context, this approach is consistent with the *ad minima* sustainability requirement criterion (somehow inspired by Wald, 1945) advocated by Léandri and Tidball (2019).

[4]A recent important contribution on the carbon tax is the study by van den Bremer and van der Ploeg (2021).

[5]To paraphrase Gaston Bachelard (1938), Bernhard (2024)'s result comes as the *reformation of an illusion* on the laborious path of resolution of the dynamic tragedy of the commons.

For all these reasons, the discarding of nonlinear MPNE strategies should then definitely sound the death knell for one-dimensional pollution control models in environmental economics. Because these models lead to a restrictive approach to pollution control problems, they should be limited to addressing sustainable management issues, wherein abatement rather than restoration efforts are to be leveraged effectively. Regarding the economic issues related to pollution control, more flexible models should be favored to enable the inclusion of deforestation and renewable natural resource exploitation activities. These approaches have been adopted in the following essays on pollution control in economics and management science.

Contents of the Book

The book is divided into two parts. The first part deals with the problem of pollution control from an economic viewpoint while the second one adopts a management science viewpoint.

In the first part, the pollution accumulation model with time-dependent environmental absorption efficiency (TDAE) is investigated in various directions. The first essay, co-authored with H. Benchekroun and D. Grass, introduces the TDAE model structure where a single polluter controls both polluting emissions and restoration efforts. The second essay, with the same co-authors, extends this model to include self-regenerating absorption efficiency. The third essay of the first part, co-authored with K. Kogan, G. Gnecco, and M. Sanguineti, extends the TDAE model to the transboundary pollution problem. In the fourth essay, co-written with E. Khmelnitsky and M. Léandri, two features are added to the single polluter model, that is, a deforestation activity and a production activity; and state-dependent restoration efforts.

The second part of the book is composed of four essays, which are all based on dynamic modeling. The first essay, co-authored with J.E. Sim and B. Kim, determines an optimal abatement policy in a one manufacturer-one retailer supply chain that generates pollution accumulation. The next essay, with the same co-authors, extends the problem to the case of two supply chains under both vertical and horizontal competition and the consideration of strategy types. The third essay, co-authored with A. Bensoussan, characterizes an optimal environmental management policy in the context of (s, S) inventory control. Finally, the last essay, co-written with R. Boucekkine and K. Kogan, considers the problem of an optimal recycling policy in an aggregate circular economy where waste is generated either by production or consumption activities.

ESSEC Business School, Cergy Pontoise Cedex, France

Fouad El Ouardighi

References

Baccini, A., Walker, W., Carvalho, L., Farina, M., Sulla-Menashe, D., & Houghton, R. A. (2017). Tropical forests are a net carbon source based on aboveground measurements of gain and loss. *Science, 358*(6360), 230–234.

Bachelard, G. (1938). *Etudes*. Librairie philosophique J. Vrin.

Başar, T., & Olsder, G.-J. (1979). *Dynamic noncooperative game theory*. SIAM Society for Industrial & Applied Mathematics.

Benchekroun, H., & Long, N. V. (1998). Efficiency inducing taxation for polluting oligopolists. *Journal of Public Economics, 70*(2), 325–342.

Benchekroun, H., & Martin-Herran, G. (2016). The impact of foresight in a transboundary pollution game. *European Journal of Operational Research, 251*(1), 300–309.

Benchekroun, H., & Chaudhuri, A. R. (2014). Transboundary pollution and clean technologies. *Resource and Energy Economics, 36*(2), 601–619.

Benchekroun, H., & Long, N. V. (2011). Game theory: Static and dynamic games. In A. A. Batabyal & P. Nijkamp (Eds.), *Research tools in natural resource and environmental economics*. World Scientific Publishing.

Bernhard, P. (2024). There is no known nonlinear Markov perfect equilibrium strategies for the infinite horizon linear quadratic differential game. *Journal of Economic Theory, 222*(105927). https://doi.org/10.1016/j.jet.2024.105927

Bertinelli, L., Camacho, C., & Zou, B. (2014). Carbon capture and storage and transboundary pollution: A differential game approach. *European Journal of Operational Research, 237*(2), 721–728.

Bhatia, U., Dubey, S., Gouhier, T. C., & Ganguli, A. R. (2023). Network-based restoration strategies maximize ecosystem recovery. *Communications Biology, 6*(1256).

Boucekkine, R., Pommeret, A., & Prieur, F. (2013a). Optimal regime switching and threshold effects. *Journal of Economic Dynamics and Control, 37*(12), 2979–2997.

Boucekkine, R., Pommeret, A., & Prieur, F. (2013b). Technological vs ecological switch and the environmental Kuznets curve. *American Journal of Agricultural Economics, 95*(2), 252–260.

Boucekkine, R., Ruan, W., & Zou, B. (2023). The irreversible pollution game. *Journal of Environmental Economics and Management, 120*(102841).

Canadell, J. G., Le Quéré, C., Raupach, M. R., Field, C. B., Buitenhuis, E. T., Ciais, P., Conway, T. J., Gillett, N. P., Houghton, R. A., & Marland, G. (2007). Contributions to accelerating atmospheric CO_2 growth from economic activity, carbon intensity, and efficiency of natural sinks. *Proceedings of the National Academy of Sciences, 104*(47), 18866–18870.

Caputo, M., & Wilen, J. (1995). Optimal cleanup of hazardous wastes. *International Economic Review, 36*(1), 217–243.

Chevé, M. (2000). Irreversibility of pollution accumulation. *Environmental and Resource Economics, 16*(1), 93–104.

CITEPA. (2024). *Rapport Secten: Émissions de Gaz à Effet de Serre et de Polluants Atmosphériques en France 1990-2023.*

Corbett, C. J., & Klassen, R. D. (2006). Extending the horizons: Environmental excellence as key to improving operations. *Manufacturing and Service Operations Management, 8*(1), 5–22.

Cox, P. M., Betts, R. A., Jones, C., Spall, S. A., & Totterdell, I. (2000). Acceleration of global warming due to carbon-cycle feedbacks in a coupled climate model. *Nature, 408*, 184–187.

Cramer, W., Bondeau, A., Woodward, F. I., Prentice, I. C., Betts, R. A., Brovkin, V., Cox, P. M., Fisher, V., Foley, J. A., Friend, A. D., Kucharik, C., Lomas, M. R., Ramankutty, N., Sitch, S., Smith, B., White, A., & Young-Molling, C. (2001). Global response of terrestrial ecosystem structure and function to CO_2 and climate change: Results from six dynamic global vegetation models. *Global Change Biology, 7*(4), 357–373.

Dasgupta, P. (2021). *The economics of biodiversity: The Dasgupta review*. HM Treasury.

De Frutos, J., & Martin-Herran, G. (2018). Selection of a Markov perfect Nash Equilibrium in a class of differential games. *Dynamic Games and Applications, 8*(3), 620–636.

Dockner, E., Jørgensen, S., Long, N. V., & Sorger, G. (2000). *Differential games in economics and management science*. Cambridge University Press.

Dockner, E. J., & Long, N. V. (1993). International pollution control: Cooperative versus non-cooperative strategies. *Journal of Environmental Economics and Management, 25*(1), 13–29.

Dockner, E. J., & Wagener, F. (*2014*). Markov perfect Nash equilibria in models with a single capital stock. *Economic Theory, 56*(3), 585–625.

Dragone, D., Lambertini, L., & Palestini, A. (2022). Emission taxation, green innovation, and inverted-U aggregate R&D Efforts in a linear state differential game. *Research in Economics, 76*(1), 62–68.

El Ouardighi, F., Benchekroun, H., & Grass, D. (2014). Controlling pollution and environmental absorption capacity. *Annals of Operations Research, 220*(1), 111–134.

El Ouardighi, F., Benchekroun, H., & Grass, D. (2016a). Self-regenerating environmental absorption efficiency and the *Soylent Green* scenario. *Annals of Operations Research, 238*(1), 179–198.

El Ouardighi, F., Sim, J. E., & Kim, B. (2016b). Pollution accumulation and abatement policy in a supply chain. *European Journal of Operational Research, 248*(3), 982–996.

El Ouardighi, F., Kogan, K., Gnecco, G., & Sanguineti, M. (2018). Commitment-based equilibrium environmental strategies under time-dependent absorption efficiency. *Group Decision and Negotiation, 27*(2), 235–249.

El Ouardighi, F., Khmeniltsky, E., & Léandri, M. (2020a). Production-based pollution versus deforestation: Optimal policy with state-independent and -dependent environmental absorption efficiency restoration process. *Annals of Operations Research, 292*, 1–26.

El Ouardighi, F., Kogan, K., Gnecco, G., & Sanguineti, M. (2020b). Transboundary pollution control and environmental absorption efficiency management. *Annals of Operations Research, 287*(2), 653–681.
El Ouardighi, F., Sim, J., & Kim, B. (2021). Pollution accumulation and abatement policies in two supply chains under vertical and horizontal competition and strategy types. *Omega, 98*(102108), 2–19.
Farzin, Y. H. (1996). Optimal pricing of environmental and natural resource use with stock externalities. *Journal of Public Economics, 62*(2), 31–57.
Forster, B. (1975). Optimal pollution control with a nonconstant exponential rate of decay. *Journal of Environmental Economics and Management, 2*(1), 1–6.
Fünfgelt, J., & Schulze, G. G. (2016). Endogenous environmental policy for small open economies with transboundary pollution. *Economic Modelling, 57*, 294–310.
Global Carbon Project. (2022). *The Global Carbon Budget 2022*. www.globalcarbonproject.org/carbonbudget/archive/2022/GCP_CarbonBudget_2022.pdf
Goering, L. (2022, January 27). It's economic and social barriers, not technology, slowing the green transition. *World Economic Forum*.
Grass, D., Caulkins, J. P., Feichtinger, G., Tragler, G., & Behrens, D. A. (2008). *Optimal control of nonlinear processes with applications in drugs, corruption, and terror*. Springer.
Gromov, D., Shigoka, T., & Bondarev, A. (2023). Optimality and sustainability of hybrid limit cycles in the pollution control problem with regime shifts. *Environment, Development and Sustainability, 26*, 10701–10718.
Harris, N. L., Brown, S., Hagen, S. C., Saatchi, S. S., Petrova, S., Salas, W., Hansen, M. C., Potapov, P. V., & Lotsch, A. (2012). Baseline map of carbon emissions from deforestation in tropical regions. *Science, 336*(6088), 1573–1576.
Hivert, A.-F. (2024, January 27). Sweden is moving backward on climate policy. *Le Monde*.
Houghton, R. A., House, J. I., Pongratz, J., Van der Werf, G. R., DeFries, R. S., Hansen, M. C., Le Quéré, C., & Ramankutty, N. (2012). Carbon emissions from land use and land-cover change. *Biogeosciences, 9*, 5125–5142.
IPCC. (2014). *AR5 Synthesis Report: Climate Change 2014*. https://www.ipcc.ch/report/ar5/syr/
Jaehn, F. (2016). Sustainable operations. *European Journal of Operational Research, 253*(2), 243–264.
Jiang, L. Q., Carter, B. R., Feely, R. A., Lauvset, S. K., & Olsen, A. (2019). Surface ocean pH and buffer capacity: Past, present and future. *Scientific Reports, 9*(18624).
Jing, C., Keasey, K., Lim, I., & Xu, B. (2023). Analyst coverage and corporate environmental policies. *Journal of Financial and Quantitative Analysis*, 1–34. https://doi.org/10.1017/S0022109023000340
Joos, F., Prentice, I. C., Sitch, S., Meyer, R., Hooss, G., Plattner, G.-K., Gerber, S., & Hasselmann, K. (2001). Global warming feedbacks on terrestrial carbon uptake under the Intergovernmental Panel on Climate Change (IPCC) emission scenarios. *Global Biogeochemical Cycles, 15*(4), 891–907.

Jørgensen, S., Martín-Herrán, G., & Zaccour, G. (2010). Dynamic games in the economics and management of pollution. *Environmental Modelling and Assessment, 15*(6), 433–467.

Karp, L., & Seguin, C. (2013). *Nonlinear differential games with Skiba points.* Mimeo.

Ke, P., Ciais, P., Sitch, S., Li, W., Bastos, A., Liu, Z., Xu, Y., Gui, X., Bian, J., Goll, D., Xi, Y., Li, W., O'Sullivan, M., Souza, J., Friedlingstein, P., Chevallier, F., 2024. Low latency carbon budget analysis reveals a large decline of the land carbon sink in 2023. arXiv:2407.12447

Keeler, E., Spence, M., & Zeckhauser, R. (1972). The optimal control of pollution. *Journal of Economic Theory, 4*(1), 19–34.

Kleindorfer, P. R., Singhal, K., & Van Wassenhove, L. N. (2005). Sustainable operations management. *Production and Operations Management, 14*(4), 482–492.

Kogan, K., Venturi, B., & Shnaiderman, M. (2017). The effect of uncertainty on production-inventory policies with environmental considerations. *IEEE Transactions on Automatic Control, 62*(9), 4862–4868.

Kogan, K., & El Ouardighi, F. (2024). Environmental sustainability under production competition with costly adjustments: An appraisal of firms' behaviour with regard to pollution dynamics. *Omega, 129*(103163).

Kossioris, G., Plexousakis, M., Xepapadeas, A., de Zeeuw, A. J., & Mäler, K.-G. (2008). Feedback Nash equilibria for non-linear differential games in pollution control. *Journal of Economic Dynamics and Control, 32*(4), 1312–1331.

Kotler, P. (2011). Reinventing marketing to manage the environmental imperative. *Journal of Marketing, 75*(4), 132–135.

Lauderdale, J. M. (2024). Ocean iron cycle feedbacks decouple atmospheric CO_2 from meridional overturning circulation changes. *Nature Communications, 15*(5532).

Léandri, M., & Tidball, M. (2019). Assessing the sustainability of optimal pollution paths in a world with inertia. *Environmental Modeling & Assessment, 24*(224), 249–263.

Lenton, T. M., Williamson, M. S., Edwards, N. R., Marsh, R., Price, A. R., Ridgwell, A. J., Shepherd, J. G., Cox, S. J., & The GENIE team. (2006). Millennial timescale carbon cycle and climate change in an efficient Earth system model. *Climate Dynamics, 26*(7/8), 687–711.

Lenton, T. M., Held, H., Kriegler, E., Hall, J. W., Lucht, W., Rahmstorf, S., & Schellnhuber, H. J. (2008). Tipping elements in the earth's climate system. *Proceedings of the National Academy of Sciences, 105*(6), 1786–1793.

Lenton, T. M., Rockström, J., Gaffney, O., Rahmstorf, S., Richardson, K., Steffen, W., & Schellnhuber, H. J. (2019). Climate tipping points – too risky to bet against. *Nature, 575*(7784), 592–595.

Li, S. (2014). A differential game of transboundary industrial pollution with emission permits trading. *Journal of Optimization Theory and Applications, 163*(2), 642–659.

Li, L., & Chen, W. (2021). The impact of subsidies in a transboundary pollution game with myopic players. *Omega, 103*, 102383.

Lindsay, R. (2024, April 9). *Climate change: Atmospheric carbon dioxide*. https://www.climate.gov/news-features/understanding-climate/climate-change-atmospheric-carbon-dioxide

Long, N. V. (2010). *A survey of dynamic games in economics*. World Scientific.

Maitland, A., Lawson, M., Stroot, H., Poidatz, A., Khalfan, A., & Dabi, N. (2022). *Carbon billionaires: The investment emissions of the world's richest people*. Oxfam International.

Mäler, K.-G., Xepapadeas, A., & de Zeeuw, A. J. (2003). The economics of shallow lakes. *Environmental and Resource Economics, 26*(4), 603–624.

Malhi, Y., Franklin, J., Seddon, N., Solan, M., Turner, M. G., Field, C. B., & Knowlton, N. (2020). Climate change and ecosystems: Threats, opportunities and solutions. *Philosophical Transactions of the Royal Society B, 375*(1794), 1–8.

Martín-Herrán, G., & Rubio, S. (2023). Efficiency-inducing policy for polluting oligopolists. *Dynamic Games and Applications*, 1–28. https://doi.org/10.1007/s13235-023-00534-7

Mathesius, S., Hofmann, M., Caldeira, K., et al. (2015). Long-term response of oceans to CO_2 removal from the atmosphere. *Nature Climate Change, 5*, 1107–1113.

Moser, E., Seidl, A., & Feichtinger, G. (2014). History-dependence in production-pollution-trade-off models: A multi-stage approach. *Annals of Operation Research, 222*, 457–481.

Nordhaus, W. D. (1991). To slow or not to slow: The economics of the greenhouse effect. *The Economic Journal, 101*(407), 920–937.

Pan, Y., Birdsey, R. A., Phillips, O. L., Houghton, R. A., Fang, J., Kauppi, P. E., Keith, H., Kurz, W. A., Ito, A., Lewis, S. L., Nabuurs, G.-J., Shvidenko, A., Hashimoto, S., Lerink, B., Schepaschenko, D., Castanho, A., & Murdiyarso, D. (2024). The enduring world forest carbon sink. *Nature, 631*, 563–569.

Penuelas, J. (2023). Decreasing efficiency and slowdown of the increase in terrestrial carbon-sink activity. *One Earth, 6*(6), 591–594.

Piao, S. L., Ciais, P., Friedlingstein, P., Peylin, P., Reichstein, M., Luyssaert, S., Margolis, H., Fang, J. Y., Barr, A., Chen, A. P., Grelle, A., Hollinger, D. Y., Laurila, T., Lindroth, A., Richardson, A. D., & Vesala, T. (2008). Net carbon dioxide losses of northern ecosystems in response to autumn warming. *Nature, 451*, 49–52.

Prieur, F. (2009). The environmental Kuznets curve in a world of irreversibility. *Economic Theory, 40*(1), 57–90.

Rowat, C. (2006). Nonlinear strategies in a linear quadratic differential game. *Journal of Economic Dynamics and Control, 31*(10), 3179–3202.

Rubio, S., & Casino, B. (2002). A note on cooperative versus noncooperative strategies in international pollution control. *Resource and Energy Economics, 24*(3), 251–261.

Schaeffer, M. (1997). *The ecology of shallow lakes*. Chapman Hall.

Shen, Z., Tian, Y., Yao, Y., Jiang, W., Dong, J., Huang, X., Wu, X., Farooq, T. H., & Yan, W. (2023). Ecological restoration research progress and prospects: A bibliometric analysis. *Ecological Indicators, 155*(110968).

Stokey, N. (1998). Are there limits to growth? *International Economic Review, 39*(1), 1–31.

Storto, A., & Yang, C. (2024). Acceleration of the ocean warming from 1961 to 2022 unveiled by large-ensemble reanalyses. *Nature Communications, 15*(545).

Tahvonen, O., & Withagen, C. (1996). Optimality of irreversible pollution accumulation. *Journal of Economic Dynamics and Control, 20*(9), 1775–1795.

Tang, C. S., & Zhou, S. (2012). Research advances in environmentally and socially sustainable operations. *European Journal of Operational Research, 223*(3), 585–594.

Tsutsui, S., & Mino, K. (1990). Nonlinear strategies in dynamic duopolistic competition with sticky prices. *Journal of Economic Theory, 52*(1), 136–161.

Ulph, A., & Ulph, D. (1994). The optimal time path of a carbon tax. *Oxford Economic Papers, 46*(5), 857–868.

van den Bremer, T., & van der Ploeg, F. (2021). The risk-adjusted carbon price. *American Economic Review, 111*(9), 2782–2810.

van der Ploeg, F., & Withagen, C. (1991). Pollution control and the Ramsey problem. *Environmental and Resource Economics, 1*(2), 215–236.

Wagener, F., & de Zeeuw, A. (2021). Stable partial cooperation in managing systems with tipping points. *Journal of Environmental Economics and Management, 109*(102499).

Wald, A. (1945). Statistical decision functions which minimize the maximum risk. *The Annals of Mathematics, 46*(2), 265–280.

Wang, Q., & Su, M. (2020). Drivers of decoupling economic growth from carbon emission – an empirical analysis of 192 countries using decoupling model and decomposition method. *Environmental Impact Assessment Review, 81*(106356).

Wirl, F. (2007). Do multiple Nash equilibria in Markov strategies mitigate the tragedy of the commons? *Journal of Economic Dynamics and Control, 31*(10), 3723–3740.

Xiao, W., Gaimon, C., Subramanian, R., & Biehl, M. (2019). Investment in environmental process improvement. *Production and Operations Management, 28*(2), 407–420.

Zickfeld, K., et al. (2013). Long-term climate change commitment and reversibility: An EMIC intercomparison. *Journal of Climate, 26*, 5782–5809.

Contents

Contributors

Hassan Benchekroun McGill University, Montréal, QC, Canada

Alain Bensoussan Jindal School of Management, University of Texas at Dallas, Dallas, TX, USA

Raouf Boucekkine Aix-Marseille University (AMSE), Marseille, France

Giorgio Gnecco IMT – School for Advanced Studies, Lucca, Italy

Dieter Grass Vienna University of Technology, Vienna, Austria

Evgeni Khmelnitsky Tel Aviv University, Ramat Aviv, Israel

Bowon Kim KAIST Business School, Seoul, Korea

Konstantin Kogan Bar-Ilan University, Ramat-Gan, Israel

Marc Leandri Université Paris Saclay, Guyancourt, France

Fouad El Ouardighi ESSEC Business School, Cergy Pontoise, France

Marcello Sanguineti University of Genova, Genova, Italy

Jeongeun Sim Sauder School of Business, University of British Columbia, Vancouver, BC, Canada

Part I
Pollution Control in Economics

Chapter 1
Controlling Pollution and Environmental Absorption Capacity

Fouad El Ouardighi, Hassan Benchekroun, and Dieter Grass

Abstract This pollution accumulation model shows that the environmental absorption capacity is impacted by economic activity. The resulting optimal control problem has two interrelated state variables: the stock of pollution and the absorption capacity of the environment. The stock of pollution decreases with environmental absorption capacity and increases with the rate of current emissions, which is controlled by a production level as well as an emissions reduction effort. However, the environmental absorption capacity is positively affected by an absorption development effort, and negatively impacted by the stock of pollution. Under specific conditions, it is shown that an optimal path, which can be either monotonic or following transient oscillations, leads to a (nontrivial) saddle point characterized by a positive environmental absorption capacity.

Keywords Pollution · Environmental absorption capacity · Production rate · Emissions reduction effort · Absorption development effort

---

This research was financially supported by the Centre for Research of ESSEC Business School (CERESSEC), the Canadian Social Sciences and Humanities Research Council (SSHRC), the Fonds Québécois de la Recherche sur la Société et la Culture (FQRSC), and the Austrian Science Fund (FWF) under grant N° P 23084-N13 (A MATLAB Package for Analyzing Optimal Control Problems). The authors acknowledge very helpful comments from two referees, and Guest Editor, David K. Yeung. They also thank Andreas Novak, Gustav Feichtinger, and Georges Zaccour for constructive suggestions on an early draft presented at the *Eleventh Workshop on Optimal Control, Dynamic Games and Nonlinear Dynamics*, June 2010, Amsterdam, The Netherlands. The usual disclaimer applies.

F. El Ouardighi (✉)
ESSEC Business School, Cergy Pontoise Cedex, France
e-mail: elouardighi@essec.edu

H. Benchekroun
McGill University, Montreál, QC, Canada

D. Grass
Vienna University of Technology, Vienna, Austria

F. El Ouardighi (ed.), *Essays on Pollution Control in Economics and Management Science*, Lecture Notes in Operations Research,
https://doi.org/10.1007/978-3-031-78227-5_1

Introduction

For quite some time, a considerable stream of economic literature has pointed out the bounded absorption capacity of pollution by the environment and the necessity for global economic activity to comply with environmental limits (e.g., Georgescu-Roegen, 1971; Costanza & Daly, 1992). Despite these recurrent warnings, the carbon emissions trend has been constantly rising over the years, and increased from a yearly rate of 1.1% during the 90s to 3% between 2000 and 2004 (Raupach et al., 2007), aggravating the global warming phenomenon.

Along with increased activity and deterioration of the global economy's carbon intensity, one main driver of this recent acceleration in carbon emissions is the decreased efficiency of natural sinks (Canadell et al., 2007). Over the last 50 years, the efficiency of natural carbon sinks has decreased by 5%, and will continue to fall. One illustration of this trend is the oceanic carbon sink, the decreased efficiency of which has contributed substantially to the increasing growth rate of atmospheric carbon for the 2000–2006 period (Canadell et al., 2007).

This trend should quickly become worse with global warming as the terrestrial biosphere is expected to switch from a carbon sink to a source within this century (Cox et al., 2000; Schaphoff et al., 2006; Lewis et al., 2011). Such an extreme scenario would give rise to uncontrollable consequences due to an increase of exceptional magnitude of atmospheric carbon. For instance, the stock of carbon stored in the permafrost, forests, wetlands, and oceans that would potentially be released into the atmosphere due to global warming is estimated at more than 3700 Gt.

These general observations have critical theoretical implications. Broadly speaking, the pollution stock reduces the environmental absorption capacity, which may progressively switch from a pollution sink to a source. Since a reduction in the environmental absorption capacity increases in return the pollution stock, a "mere" reduction in the current emissions rate seems insufficient to counteract the positive feedback between pollution and environmental absorption capacity. It is thus essential that an additional policy instrument be considered for directly controlling changes in the environmental absorption capacity, to prevent it from switching from a sink to a source. It is exactly these fundamentals that we want to implement in our representation of pollution dynamics.

The first goal of this paper is to identify the optimal paths of production and efforts spent on reducing emissions that are compatible with nonexplosive damages from pollution. Whether such pair paths are chosen depends on the trade-off between revenues drawn from the production rate, on the one hand, and the cost of emissions reduction efforts plus the net cost of negative externalities from pollution, on the other hand.

As shown in Table 1.1, two kinds of strategies for reducing current emissions may be envisioned, that is, strategies based either on substitutability or on complementarity between the production rate and the technological effort level for reducing emissions.

Table 1.1 Economic consequences of possible emissions reduction strategies

		Technological effort for emissions reduction	
		High	Low
Production rate	High	*High production revenues* *Moderate pollution costs* *High emissions reduction costs*	*High production revenues* *High pollution costs* *Low emissions reduction costs*
	Low	*Low production revenues* *Low pollution costs* *High emissions reduction costs*	*Low production revenues* *Moderate pollution costs* *Low emissions reduction costs*

Substitutability-based strategies, which combine low (resp., high) production revenues and pollution costs and high (resp., low) emissions reduction costs, do not actually result from any trade-off. They merely reflect a situation where either environmental or economic sustainability, that is, either the environmental or economic ability to meet the needs of the present and future generations (Pezzey, 1992), prevails while the other fails, which is likely to be hardly manageable in the long run.

Conversely, *complementarity-based* strategies are part of the trade-off. The first one is consistent with the idea that large technological efforts for the reduction of emissions may reconcile a high level of production activity with the preservation of the environment (World Bank, 1992; Grossman & Krueger, 1995), while the second one is more in line with the general premise that mass production activity is destructive for the environment, and therefore not viable in the long run, including from an economic viewpoint (Georgescu-Roegen, 1971; Daly, 1997). The first type of strategy widely reflects the industrialist viewpoint,[1] and will be referred to hereafter as the *industrialist strategy*. The second strategy reflects an ecologically-oriented analysis of the impact of mass production on the environment, i.e., ecological economics (Harris, 2002), and will be designated as the *ecologist strategy*. We seek to introduce a simple analytical framework to evaluate both the economic and environmental viability of these two types of strategies.

The second goal of this paper is to determine how the development of the environmental absorption capacity can contribute to better control the stock of pollution. As Dasgupta and Mäler (2004) wrote, "in the popular literature, the morals that would appear to have been drawn from the finding are [...] that resource degradation is reversible: degrade all you want now, Earth can be relied upon to rejuvenate it later should you require it." Following this principle, the analysis of optimal pollution accumulation rested, until recently, on the assumption of positive, constant environmental absorption capacity. But, as noted by Wirl (1999), constant environmental absorption capacity "attributes implicitly an advantage to large

[1]For instance, "economic growth is good for the environment, because countries need to put poverty behind them in order to care" (Editorial, *The Independent*, December 4, 1999); or "trade improves the environment, because it raises incomes, and the richer people are, the more willing they are to devote resources to cleaning up their living space" (*The Economist*, December 4, 1999, p. 17).

pollution stocks [...] because a large stock of pollution implies in this case large cleaning."

With increasing evidence of "nature's non-convexities," typically exemplified in the phosphorus discharge into a shallow lake (Holling, 1973), the representation of the environmental absorption capacity as an instantaneous nonlinear function of the stock of pollution has become prevalent (Tahvonen & Salo, 1996; Dasgupta & Mäler, 2003). This change in perspective indeed allowed more accurate handling of potential irreversible environmental degradation. However, the extension of the nonlinear specification to complementary essential considerations, e.g., investment in the environmental absorption capacity, inherent self-restoring capabilities of the environmental absorption capacity, and generalization to a noncooperative game framework, would add substantial mathematical complexity to current difficulties related to non-concave representation.

On these restrictions, we propose to model the environmental absorption capacity as a state variable. Besides the destructive influence of the pollution stock, we assume that the environmental absorption capacity is susceptible to be developed through a specific policy instrument. Such additional control of the environmental absorption capacity should make it possible to counteract the fatal interdependence between pollution and environmental absorption capacity, as described above. In addition, it should alleviate the exclusive, weighty constraint traditionally devolved upon production activity, and then prevent the requirement of an unmanageable zero production terminal state.

This paper is organized as follows: The next section presents a two-state-variable model in which a social planner can control the stock of pollution by reducing the rate of current emissions and/or by developing the environmental absorption capacity. The social planner is supposed to maximize over an infinite time horizon the discounted social utility resulting from the difference between production revenues, the cost of pollution, and the respective costs of efforts to reduce emissions and to improve the environmental absorption capacity. In the third section, the optimal emissions reduction and environmental absorption capacity development strategy is identified. In the fourth section, steady-state values associated with the optimal emissions reduction strategy are computed, and their stability is analyzed. In the fifth section, the transition paths associated with the optimal strategy are determined with numerical means, and a sensitivity analysis of the parameter values is conducted. Sixth section concludes this paper.

Model Formulation

As noted by Wirl (1999), "dynamic considerations of environmental policies (...) are by large restricted to one state variable control models." According to the most employed formulation, the pollution stock increases with the rate of current emissions and depreciates at a decay rate reflecting the environmental absorption capacity. Let $P(t) \geq 0$ denote the pollution stock at time t, its evolution is given by:

$$\dot{P}(t) = e(t) - \kappa(P(t)) \quad P(0) = P_0 > 0 \tag{1.1}$$

where $e(t) \geq 0$ is the rate of current emissions, and $\kappa(.) \geq 0$ is the environmental absorption rate as a function of $P(t)$. Most often, the key specification of $\kappa(.)$ is that of a linear environmental absorption rate, i.e., $\kappa(P(t)) \equiv \rho P(t),\ \rho > 0$. This specification was introduced in Keeler et al. (1972) and extensively used in the existing literature, notably to determine an optimal economic growth path compatible with environmental preservation (Nordhaus, 1991; Stockey, 1998), an optimal cleanup policy of a pollution stock (van der Ploeg & Withagen, 1991; Caputo & Wilen, 1995), and an optimal exploitation of a natural resource whose consumption inflicts negative externalities on the environment (Ulph & Ulph, 1994; Farzin, 1996).

Although a linear environmental absorption rate is mathematically more tractable, only a few natural processes, such as radioactive decay, can be characterized in this way. Further, the linear specification implies that the environmental absorption capacity is increasing with respect to the pollution stock, and disregards the empirical evidence that high pollution stock levels may destroy the environmental absorption capacity (Holling, 1973).

A significant refinement of this model assumed that the environmental absorption function is nonlinear, concave with low pollution stock, and convex when pollution is higher, with a switching point in between. More precisely, there exists $\overline{P} > 0$ such that $\kappa(P(t)) > 0$, for all $P(t) \in \left[0, \overline{P}\right[$ and $\kappa(P(t)) = 0$ for all $P(t) \geq \overline{P}$. This formulation was suggested by Forster (1975) and notably used by Tahvonen and Withagen (1996), and Chevé (2000). Interestingly, this formulation gives rise to alternative multiple equilibria, typically optimal current emissions paths leading either to reversible or irreversible pollution equilibria. The trajectories obtained may be discontinuous and non-monotonic in the case of irreversible pollution (Tahvonen & Salo, 1996; Tahvonen & Withagen, 1996).

Another extension of the linear version of the model in Eq. 1.1 to include a costly adjustment process for reducing the current emissions level results in even more complex trajectories, i.e., limit cycles, including in the simpler case of reversible pollution (Wirl, 2000).[2]

It is well known that the assumption of the nonlinear environmental absorption function generates challenging technicalities, notably related to the non-differentiability absorption function at the critical pollution level beyond which the environmental absorption capacity is zero. Further, the environmental absorption function is represented as an instantaneous process, with initial conditions similar to those of the pollution stock. As a result, this representation limits the

[2] An alternative formulation assumes that the evolution of environmental quality is similar to that of a renewable resource. Along with extensions where the environmental regeneration function depends on other variables, such as the stock of manufactured capital, the stock of natural resources, or labor input, this formulation was used by Krautkraemer (1985), Mäler (1991), and Tsur and Zemel (1996), to derive an optimal strategy to protect the environment. The extension of this model to include an adjustment process for reducing the current emissions level gives dampened oscillations and limit cycles as possible optimal strategies (Wirl, 1999).

possibility of considering complementary influences, among which are activities aiming to develop the environmental absorption capacity, and inherent self-restoration capabilities of the environmental absorption capacity. These considerations, if they were also taken into account, would add significant complexity to current difficulties related to the non-concave representation.

The limitations associated with the pollution accumulation model in Eq. 1.1 justify that the environmental absorption capacity be treated as a state variable rather than just an instantaneous response function.[3] This paper presents the evolution of pollution accumulation and environmental absorption capacity as separate but interlinked dynamic processes. In this respect, we extend the pollution accumulation model in Eq. 1.1, introducing two distinguishing features. First, the rate of current emissions is controlled twofold, that is, through a production rate and an emissions reduction technological effort. Second, the environmental absorption capacity is defined as a stock variable.

To do so, we consider an economy with an instantaneous production activity. The production rate, denoted by $u(t) \geq 0$, determines the current emissions rate, $e(t) \geq 0$. Emissions can be reduced thanks to a technological reduction effort, the level of which is denoted by $v(t) \geq 0$. The reduction effort may consist of *de-carbonization* initiatives, that is, limiting energy consumption, using renewable energies, etc. The current emissions rate is then given as:

$$e(t) = \alpha u(t)(1 - \beta v(t)) \tag{1.2}$$

where α, $\beta > 0$. According to Eq. 1.2, there are no emissions when production activity is nil. Further, reduction effort efficiency is such that, for a given positive output, the current emissions rate reaches a maximum value in the case where the reduction effort is inexistent, but decreases proportionally with a marginal increase in the emissions reduction effort. Given thermodynamic restrictions to the extent to which emissions can be reduced, the reduction effort should be bounded, that is, $v \leq 1/\beta$. A similar formulation to Eq. 1.2 can be found in van der Ploeg and Withagen (1991).

To describe the evolution of the pollution stock, we use the following formulation:

$$\dot{P}(t) = e(t) - \delta A(t)P(t) \quad P(0) = P_0 > 0 \tag{1.3}$$

According to Eq. 1.3, the stock of pollution increases with $e(t)$ as defined in Eq. 1.2, and decreases linearly with the environmental absorption capacity $A(t)$ at the rate $\delta A(t)$, $\delta > 0$.

The environmental absorption capacity obeys the following transition equation:

[3]A similar idea is found in Léandri (2009), where the pollution accumulation is nevertheless disregarded.

$$\dot{A}(t) = w(t) - \gamma P(t) \quad A(0) = A_0 > 0 \tag{1.4}$$

where $\gamma > 0$. According to Eq. 1.4, the environmental absorption capacity is decreasing with the stock of pollution (Holling, 1973; Schaeffer, 1997). For tractability, we assume that the destructive impact of pollution on the rate of change of the environmental absorption capacity is linear. It will be shown below that linearity does not imply any loss of generality compared with the nonlinear environmental absorption function in Eq. 1.1.

The natural absorption capacity in Eq. 1.4 can also be restored and/or developed by an absorption development activity, the instantaneous effort level of which is denoted by $w(t) \geq 0$. The rationale for this effort relies on the premise that investing in natural capital is economically beneficial (Carpenter & Cottingham, 1997; Costanza et al., 1997; Costanza, 1998). The numerous programs and initiatives implemented at national and international levels (e.g., European Centre for River Restoration, United National Environment Programme) toward environmental restoration and development illustrate the importance of such policy option. It may consist of maintaining and/or developing wetlands, promoting reforestation, restoring rivers, erosion reduction, combating desertification, expanding marine protected areas, promoting biodiversity conservation, carbon sequestration, etc.

Finally, we assume that the inherent capabilities of self-restoration of the environmental absorption capacity are negligible in Eq. 1.4. Yet, we believe that extending our model to the environmental self-restoration capabilities assumption is an important topic for future research.

Although the initial value of A is strictly positive in Eq. 1.4, an insufficient absorption development effort may cause it to decrease so that it becomes strictly negative. In this case, the natural absorption capacity turns to a source, that is, a natural rejection capacity. Since a return from source to sink would be difficult, switching from a positive to a negative value of A would then correspond to an irreversible situation causing the pollution stock to increase over time independently of the current emissions level.

To illustrate this evolution, we consider the differential system given by Eqs. 1.3–1.4 and assume that there is no human interference (that is, no emissions and no absorption development activity) over an infinite horizon, i.e., $e(t) = w(t) = 0$, for all $t \geq 0$. Compared with the real-world situation, in which emissions are still largely irreducible and the environmental absorption capacity is massively degraded by deforestation (Canadell et al., 2007), assuming that there is no human interference reflects a utopic situation. In this case, the pollution-absorption dynamics is represented by the planar dynamical system:

$$\dot{P}(t) = -\delta A(t)P(t) \qquad P(0) = P_0 > 0$$
$$\dot{A}(t) = -\gamma P(t) \qquad A(0) = A_0 > 0$$

At the steady state, $P^\infty = 0$, which yields $A^\infty \gtreqless 0$. The corresponding Jacobian matrix is:

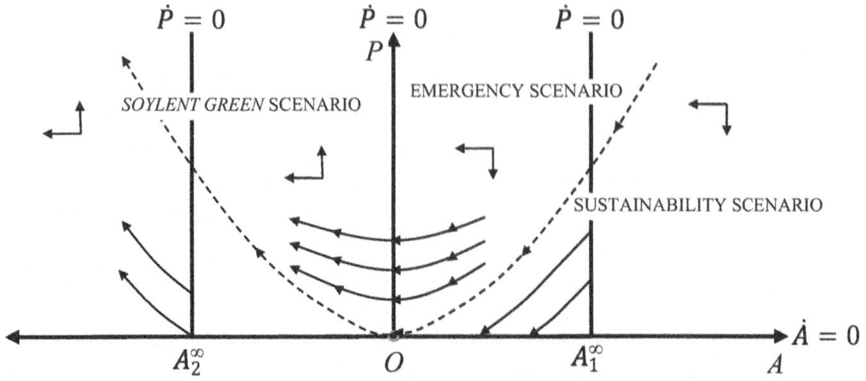

Fig. 1.1 Pollution-absorption dynamics without human interference

$$\widehat{J} = \begin{bmatrix} -\delta A^\infty & -\delta P^\infty \\ -\gamma & 0 \end{bmatrix}$$

For $A^\infty > 0$ (resp., $A^\infty < 0$), $\widehat{J}$ has one zero eigenvalue, and one eigenvalue, $-\delta A^\infty$, which is either negative or positive. This gives rise to a line of either stable or unstable equilibria. For $A^\infty = 0$, $\widehat{J}$ has two zero eigenvalues, which reflect a non-hyperbolic steady state (e.g., Grass et al., 2008).

The phase diagram below (Fig. 1.1) shows that the horizontal axis $\dot{A} = 0$ is a line of possible steady-state points, where points on the positive axis are sinks, with stable equilibria (e.g., A_1^∞), and the negative axis is a source with diverging equilibria (e.g., A_2^∞). Origin O separates two branches of the saddle-node bifurcation curve.

The phase diagram displays similar properties to those of the nonlinear environmental absorption function in Eq. 1.1, i.e., the environmental absorption capacity suffers from *hysteresis* as well as an *irreversibility effect*, since the system is unable to return to where it had been in the beginning. Also, the phase diagram depicts the possibility of switching from a pollution sink to a source.

Though the usual partition of the state space into a reversible and an irreversible region is easily obtained (e.g., Tahvonen and Withagen (1996)), the phase diagram can be alternatively divided into three regions, that is:

- The *Sustainability scenario*, in which any initial pollution stock on the right side of the dashed line in the positive quadrant can be totally eliminated while preserving a positive environmental absorption capacity. In this sense, paths in this scenario should end in a sustainable steady state that ensures environmental resilience, conceived of as stability of the ecological-economic system that does not threaten sustainability (Common & Perrings, 1992). Multiple degrees of sustainability can occur along the positive horizontal axis, where the highest sustainable states depict the pre-industrial period, characterized by a natural steady state without current emissions and thus no damage related to the pollution stock.

- The *Emergency scenario*, between the dashed line in the positive quadrant and the P-axis, in which any pollution path, although declining, inexorably tends to an exhaustion of the environmental absorption capacity. Paths in this region may provide early warning signals thus averting an imminent pessimistic evolution, notably the critical slowing down of the pollution absorption process, which is reflected in the rapidly increasing negative eigenvalue to zero (Scheffer et al., 2009). A final alternative to prevent breaching the threshold beyond which natural systems will lose resilience and suffer catastrophic collapse would consist of an urgent and massive effort to develop environmental absorption capacity, which would then eventually allow for a shift toward the sustainability region.[4]
- The *Soylent Green* scenario, on the left side of the $\dot{P} = 0$ vertical axis, which corresponds to a lasting pollution stock in an exhausted absorption capacity environment, irreversibly leads to a self-sustained evolution of the pollution path due to an increasing rejection of the past absorbed pollution by the environment. In this scenario, irreversibility stems from the impossibility for the environmental absorption capacity to recover a positive state, and then for the social planner to avoid an infinitely increasing disutility of pollution.[5]

One merit of the formulation proposed in the differential system Eqs. 1.3–1.4 lies in its relative "simplicity," that is, the potentially irreversible degradation of the environmental absorption capacity can be taken into account while avoiding some of the cumbersome technicalities implied by the assumption of the non-convex absorption function. There is however a difference in the way the decay of pollution is modeled in this paper compared to the existing literature using a nonlinear environmental absorption function. Tahvonen and Withagen (1996), for instance, assume that once a threshold of the pollution stock is reached, the natural rate of decay of pollution becomes zero. In our formulation, once the pollution stock reaches a certain threshold, the absorption capacity of the environment becomes negative and pollution grows even in the absence of anthropogenic emissions. Nature adds to the stock of carbon. This is in line with scientific evidence that the biosphere may

[4] An estimate of the disruption threshold is given by the cumulative emissions level of CO_2, that is, 1000 Gt, above which there is a 50% probability of global mean warming exceeding 2 °C (Board on Atmospheric Sciences and Climate, 2011). This limit of 2 °C adopted by the European Union "stands a strong chance of provoking drought and storm responses that could challenge civilized society, leading potentially to the conflict and suffering that go with failed states and mass migrations. Global warming of 2°C would leave the Earth warmer than it has been in millions of years, a disruption of climate conditions that have been stable for longer than the history of human agriculture. Given the drought that already afflicts Australia, the crumbling of the sea ice in the Arctic, and the increasing storm damage after only 0.8°C of warming so far, calling 2°C a danger limit…[may even seem] pretty cavalier" (Allen et al., 2009).

[5] This pessimistic scenario has been labeled in reference to the American film *Soylent Green* directed by Richard Fleisher in 1973, in which mankind can do nothing except miserably endure a worldwide ecological catastrophe.

Table 1.2 Various configurations of the environmental absorption capacity

		Existence of environmental absorption capacity development activity	
		No	Yes
Pollution impact on the environmental absorption capacity	Neutral	*Production-pollution dynamics with constant absorption capacity*	*Production-pollution dynamics with developing absorption capacity*
	Positive	*Production-pollution dynamics with exhaustible absorption capacity*	*Production-pollution dynamics with controllable absorption capacity*

switch from a carbon sink to a source (e.g., Cox et al. (2004)). Also, in terms of the problems addressed, there is a difference between the existing literature using a nonlinear environmental absorption function and this paper. While the existing literature considers only pollution emissions as a policy instrument (e.g., Tahvonen and Withagen (1996)), in our paper the policy maker has two instruments available, i.e., pollution emissions and the efforts to build up the natural absorption capacity.

The formulation in Eqs. 1.3–1.4 encompasses the pollution accumulation model with constant absorption capacity as a special case, i.e., when the impact of pollution on the environmental absorption capacity is neutral ($\gamma = 0$), and there is no environmental absorption capacity development activity ($w(t) = 0, \forall\, t$). The table below (Table 1.2) depicts the various configurations encompassed by the differential system Eqs. 1.3–1.4.

Another configuration that can be obtained as a special case of this model is the *developing absorption capacity* configuration, i.e., absorption capacity insensitive to pollution ($\gamma = 0$) and effective absorption capacity development activity ($w(t) > 0$ for all $t \geq 0$).

A far more realistic configuration is the *exhaustible absorption capacity* configuration, which combines the negative impact of pollution on the absorption capacity ($\gamma > 0$) and no development activity of absorption capacity ($w(t) = 0$ for all $t \geq 0$). This configuration may be seen as a soft reflection of the real-world situation, which is actually more in line with a reduction of the environmental absorption capacity due to, e.g., deforestation, that is, $w(t) < 0$. According to the Stern review (Stern, 2006), the exhaustible absorption capacity configuration is preferable to the current situation because "curbing deforestation is a highly cost-effective way to reduce emissions." This configuration remains nevertheless consistent with the fact that a large enough pollution stock may destroy the environmental absorption capacity.

Finally, a more general configuration is the *controllable absorption capacity* configuration, where the absorption capacity development activity ($w(t) > 0, \forall\, t$) can be leveraged to overwhelm the negative impact of pollution on the absorption capacity ($\gamma > 0$).[6]

[6] As the environmental absorption capacity can be enhanced by human intervention rather than self-restoration capabilities, it is controllable rather than renewable.

We now turn to the definition of the social planner's objective function. The economy's instantaneous net utility function is supposed to be a separable additive in its arguments. That is, the economy's revenue function is defined as an increasing concave function of the production rate. For simplicity, we use $a \ln u(t)$, $a > 0$ (e.g., Chevé, 2000).

At each period, the economy incurs a social cost of pollution given as an increasing convex function of the pollution stock, that is, $cP(t)^2/2$, $c > 0$. The reduction effort generates an increasing quadratic cost denoted by $dv(t)^2/2$, $d > 0$. Finally, the absorption capacity development effort generates an increasing quadratic cost denoted by $fw(t)^2/2, f > 0$.[7]

Denoting by $r > 0$ the social planner's discounting rate, and assuming that the planning horizon is infinite, the social planner's unconstrained optimal control problem then is:

$$\underset{u(.),v(.),w(.)}{Max} \quad W = \int_0^{+\infty} e^{-rt}\left\{a \ln u(t) - \frac{cP(t)^2}{2} - \frac{dv(t)^2}{2} - \frac{fw(t)^2}{2}\right\} dt$$

subject to Eqs. 1.2–1.3, and $v(t) \geq 0$, $w(t) \geq 0$, $\forall t$.

Optimality Conditions

The current-value Hamiltonian is (Grass et al., 2008):

$$H = a \ln u - \frac{cP^2}{2} - \frac{dv^2}{2} - \frac{fw^2}{2} + \lambda_1[\alpha u(1-\beta v) - \delta AP] + \lambda_2(w - \gamma P) \tag{1.5}$$

where $\lambda_j \equiv \lambda_j(t)$ are costate variables, $j = 1, 2$.

Necessary conditions for optimality write:

$$H_u = \frac{a}{u} + \lambda_1 \alpha(1-\beta v) = 0 \Rightarrow u = -\frac{a}{\lambda_1 \alpha(1-\beta v)} \tag{1.6}$$

$$H_v = -dv - \lambda_1 \alpha \beta u = 0 \Rightarrow v = -\frac{\lambda_1 \alpha \beta u}{d} \tag{1.7}$$

$$H_w = -fw + \lambda_2 = 0 \Rightarrow w = \frac{\lambda_2}{f} \tag{1.8}$$

We will focus on the case of an interior solution. In theory, it is possible to clean up all emissions, i.e., $v = 1/\beta$ and enjoy an infinitely large rate of production.

[7] Note that the limit conditions characterizing the functions composing the objective criterion are expected to prevent boundary solutions from occurring.

However, this is an unrealistic outcome that we do not investigate. Given a feasible value of the production rate, i.e., $u \geq 0$, a feasible emissions reduction effort value in Eq. 1.7, i.e., $v \geq 0$, requires $\lambda_1 \leq 0$. Since the pollution stock has a negative marginal influence on the social planner's objective criterion, the implicit price for the pollution stock should be nonpositive, i.e., $\lambda_1 \leq 0$, and can be interpreted as a disincentive to pollute more. However, for the production rate to have a feasible value in Eq. 1.6, i.e., $u \geq 0$, it is both required that $\lambda_1 < 0$ and $v < 1/\beta$. If these conditions are fulfilled, a strictly positive production rate should be obtained over the whole planning horizon, which in return implies a strictly positive value of the emissions reduction effort. Finally, a feasible value for the absorption capacity development effort, i.e., $w \geq 0$, in Eq. 1.8 requires a nonnegative value of the implicit price of the absorption capacity rate, i.e., $\lambda_2 \geq 0$, which can be interpreted as an incentive to develop the environmental absorption capacity.

The costate equations are given by:

$$\dot{\lambda}_1 = r\lambda_1 - H_P = cP + (r + \delta A)\lambda_1 + \gamma\lambda_2 \tag{1.9}$$

$$\dot{\lambda}_2 = r\lambda_2 - H_A = r\lambda_2 + \lambda_1\delta P \tag{1.10}$$

Substituting the expression of v in Eq. 1.7 into that of u in Eq. 1.6 gives:

$$(\alpha\beta\lambda_1)^2u^2 + (\alpha d\lambda_1)u + ad = 0$$

which can be solved with:

$$u_i = -\frac{h_i}{\lambda_1} > 0|_{d \geq 4a\beta^2} \tag{1.11}$$

where $h_i \equiv \left[d \pm \sqrt{d(d - 4a\beta^2)}\right]/2\alpha\beta^2$, $i = 1, 2$.

Given a sufficiently large cost coefficient of the emissions reduction effort, that is, $d \geq 4a\beta^2$, the production rate should have a strictly positive value over the whole planning horizon. In the remainder of the paper, we shall assume that $d \geq 4a\beta^2$. Note that, for $d \geq 4a\beta^2$, $u_2 \leq u_1$.

Plugging the RHS of Eq. 1.11 into the expression of v in Eq. 1.7 gives, respectively:

$$v_i = \frac{\alpha\beta h_i}{d} > 0|_{d \geq 4a\beta^2} \tag{1.12}$$

$i = 1, 2$. Provided $d \geq 4a\beta^2$, it can be shown that $v_2 \leq v_1 < 1/\beta$. Note also that $\dot{v}_1 = \dot{v}_2 = 0$, i.e., whatever the production rate, the emissions reduction effort should remain constant over time.

Substituting the RHS of Eq. 1.11 and Eq. 1.12 into Eq. 1.1, gives:

$$e_i = g_i u_i > 0_{|d \geq 4a\beta^2} \tag{1.13}$$

where $g_i \equiv \frac{a}{2d}\left[d \mp \sqrt{d\left(d - 4a\beta^2\right)}\right] = \frac{a}{h_i}$, $i = 1, 2$, are the marginal levels of reduced current emissions from production respectively associated with u_i.

We refer to the solution that corresponds to u_1 (u_2) as the industrialist (the ecologist) strategy.

Proposition 1. The industrialist strategy entails a greater production rate and emissions reduction effort than the ecologist strategy, but generates lower current emissions per unit produced over time.

Proof Since $g_1 \leq g_2$, we should have $e_1 \leq e_2$ for any fixed production rate u_i. □

Checking the concavity property of the Hamiltonian for the two strategies to determine whether they correspond to globally or locally optimal solutions, it can be shown that neither strategy satisfies any of the standard sufficiency conditions for optimality, that is, the Mangasarian or Arrow conditions. Therefore, the existence of a globally optimal solution cannot be guaranteed, but only assumed.

To determine which strategy satisfies the concavity requirements, we check the Legendre-Clebsch necessary conditions of concavity of the Hamiltonian with respect to the control vector (u, v, w). To do so, the necessary conditions 1.6–1.7–1.8 as well as Eq. 1.11 are used to compute the Hessian matrix, that is:

$$H_H^i = \begin{bmatrix} -\dfrac{a}{u_i^2} & \dfrac{\alpha\beta h_i}{u_i} & 0 \\ \dfrac{\alpha\beta h_i}{u_i} & -d & 0 \\ 0 & 0 & -f \end{bmatrix} \tag{1.14}$$

$i = 1, 2$, from which the principal minors are:

$$\left|H_H^i\right|_1 = \left\{-\frac{a}{u_i^2}, -d, -f\right\}, \left|H_H^i\right|_2 = \left\{\frac{ad - (\alpha\beta h_i)^2}{u_i^2}, \frac{af}{u_i^2}, df\right\}, \left|H_H^i\right|_3 = \frac{f\left[(\alpha\beta h_i)^2 - ad\right]}{u_i^2}$$

$i = 1, 2$. Given $u_i(t) > 0$, $\forall\, t$, the principal minors of dimension 1 are all strictly negative. The second and third principal minors in $\left|H_H^i\right|_2$ are strictly positive. Moreover, we see that:

$$\left|H_H^i\right|_3 = -f\left[\frac{ad - (\alpha\beta h_i)^2}{u_i^2}\right],$$

$i = 1, 2$, which implies that a necessary and sufficient condition for the negative semi-definiteness of the Hessian matrix is that the first principal minor of $\left|H_H^i\right|_2$ is nonnegative. To check this requirement, we plug the expression of h_i into the first principal minor of $\left|H_H^i\right|_2$, $i = 1, 2$, respectively, which yields:

$$-\frac{d\left[d-4a\beta^2+\sqrt{d\left(d-4a\beta^2\right)}\right]}{2\beta^2}\leq 0|h_1$$

$$\frac{d\left[d-4a\beta^2-\sqrt{d\left(d-4a\beta^2\right)}\right]}{2\beta^2}\geq 0|h_2$$

Provided $d \geq 4a\beta^2$, the first principal minor of dimension 2 is nonpositive for h_1, and nonnegative for h_2. Hence, the Hessian matrix is negative semi-definite only for h_2. This implies that the Hamiltonian is concave with respect to the control vector (u, v, w) in the neighborhood of (u_2, v_2, w), but not in the neighborhood of (u_1, v_1, w). Only the path of controls (u, v, w) that converges to (u_2, v_2, w) is a (global) maximizer of the Hamiltonian with respect to (u, v, w).

The preceding results are summarized in the subsequent proposition.

Proposition 2. The ecologist strategy is optimal while the industrialist strategy is not.

Given that the technological effort for the reduction of current emissions lies outside the control region, the larger emissions reduction effort cost involved by the industrialist strategy exceeds the additional benefits drawn from the larger production rate. As a result, the industrialist strategy for mitigating emissions is clearly not an advisable solution for the preservation of the environment.

Steady State Analysis

The main issue now is to determine whether the ecological strategy for reducing emissions leads to a stable steady state located in the sustainability region, as depicted in Fig. 1.1.

To do so, we first form the canonical system related to the controllable absorption capacity configuration in the state-costate variable space. Without loss of generality, we set $f = 1$. Substituting the optimal value of e_2 from Eq. 1.13 in Eq. 1.3 and plugging the optimal expression of w in Eq. 1.4 results in Eq. 1.15 and Eq. 1.16. Finally, recalling Eq. 1.9 and Eq. 1.10 gives Eq. 1.17 and Eq. 1.18.

$$\dot{P} = -\frac{a}{\lambda_1} - \delta AP \tag{1.15}$$

$$\dot{A} = \lambda_2 - \gamma P \tag{1.16}$$

$$\dot{\lambda}_1 = cP + (r + \delta A)\lambda_1 + \gamma\lambda_2 \tag{1.17}$$

$$\dot{\lambda}_2 = r\lambda_2 + \lambda_1\delta P \tag{1.18}$$

We first compute the steady state of the system. From Eq. 1.16, we get:

$$\lambda_2^{\infty} = \gamma P^{\infty} \tag{1.19}$$

Considering Eq. 1.18, we use Eq. 1.19 to get:

$$\lambda_1^{\infty} = -\frac{r\gamma}{\delta} \tag{1.20}$$

which is strictly negative.

Considering Eq. 1.17, we use Eq. 1.19 and Eq. 1.20 to obtain:

$$P^{\infty} = \frac{r\gamma\left(r + \delta A^{\infty}\right)}{\delta(c + \gamma^2)} \tag{1.21}$$

From Eq. 1.15, we use Eqs. 1.20 and 1.21 to obtain:

$$r^2\delta\gamma^2 {A^{\infty}}^2 + r^3\gamma^2 A^{\infty} - \delta a\left(c + \gamma^2\right) = 0 \tag{1.22}$$

Solving Eq. 1.22 and imposing $P^{\infty} \geq 0$, we find:

$$A^{\infty} = \frac{k - r^2\gamma}{2r\delta\gamma} \tag{1.23}$$

where $k \equiv \sqrt{r^4\gamma^2 + 4\delta^2 a(c + \gamma^2)} > 0$, as the only valid root.

Substituting for λ_1^{∞} from Eq. 1.20 in Eq. 1.11, for A^{∞} from Eq. 1.23 in Eq. 1.21, for the resulting expression of P^{∞} in Eq. 1.19, and for the final expression of λ_2^{∞} in Eq. 1.8, respectively, leads to characterizing the system's steady state, which is formalized in the subsequent proposition.

Proposition 3. The controllable absorption capacity configuration admits a unique steady state, given by:

$$\begin{bmatrix} u^{\infty} & v^{\infty} & w^{\infty} & P^{\infty} & A^{\infty} \end{bmatrix}^T = \begin{bmatrix} \frac{\delta h_2}{r\gamma} & \frac{\alpha\beta h_2}{d} & \frac{\gamma(k + r^2\gamma)}{2\delta(c + \gamma^2)} & \frac{k + r^2\gamma}{2\delta(c + \gamma^2)} & \frac{k - r^2\gamma}{2r\delta\gamma} \end{bmatrix}^T \tag{1.24}$$

Corollary 1. The ecologist's strategy for reducing emissions reduction allows for a unique nontrivial steady-state production rate compatible with the preservation of a strictly positive environmental absorption capacity.

As the steady-state values are all interior, the variables should remain interior by continuity for initial values close enough to the steady state.[8] In contrast with the existing literature which models the environmental absorption capacity as an instantaneous function of the pollution stock (e.g., Tahvonen & Withagen, 1996), our model has a single steady state, which then implies no trade-off between "reversible" and "irreversible" state. Since it should be located either in the *sustainability region* or the *emergency region*, as illustrated in Fig. 1.1, that is, A^{∞}, $P^{\infty} > 0$, the *Soylent Green* scenario of a self-sustained pollution path can be avoided if the steady state of the current emissions rate can be reached and sustained.

Compared to the controllable absorption capacity configuration, the constant absorption capacity configuration (Table 1.2) also admits one steady state, i.e., $u^{\infty}_{|w(t)=\gamma=0} = h_2\sqrt{\rho(r+\rho)/ac}$ and $P^{\infty}_{|w(t)=\gamma=0} = \sqrt{a(r+\rho)/\rho c}$, which is also a saddle point, the determinant of the corresponding Jacobian matrix being strictly negative $(-2\rho(r + \rho) < 0)$. However, given the corresponding production rate at the steady state, the constant absorption capacity is lower than the steady state absorption capacity of the controllable absorption capacity configuration $(\rho < \delta A^{\infty}_{|w(t)>0,\gamma>0})$, despite a lower steady-state pollution stock, i.e., $P^{\infty}_{|w(t)=\gamma=0} < P^{\infty}_{|w(t)>0,\gamma>0}$.

In contrast with the controllable absorption capacity configuration, the developing absorption capacity configuration has a single, diverging steady state, which is characterized by an infinite output level and a zero-absorption capacity development effort. Accordingly, the steady state pollution stock is positive, $P^{\infty}_{|w(t)>0,\gamma=0} = \sqrt{a/c}$, while the environmental absorption capacity is infinite. That is, if the pollution stock is not destructive to the absorption capacity (i.e., the environmental absorption capacity is strongly resistant), a positive, decreasing absorption capacity development effort over time results in an infinite absorption capacity, which in turn allows for an infinite production rate. Indeed, the steady-state pollution stock in this configuration is lower than that in the controllable absorption capacity configuration.

In comparison with the controllable absorption capacity configuration, the exhaustible absorption capacity configuration, which reflects the Stern review's advocated configuration (Stern, 2006), admits a single steady state characterized by both a zero-pollution stock and production rate. The reason is that the process here is similar to that of the exploitation of an exhaustible natural resource, i.e., the production rate should be initially large and decreasing over time for a strictly positive initial environmental absorption capacity. As a result, the steady state absorption capacity is negative, and corresponds to a source fixed point. In the long run, the exhaustible absorption capacity configuration leads to an unmanageable zero production terminal state, while the *Soylent Green* scenario

[8] The subsequent numerical analysis shows that interior solutions are also granted for initial values far enough from the steady state.

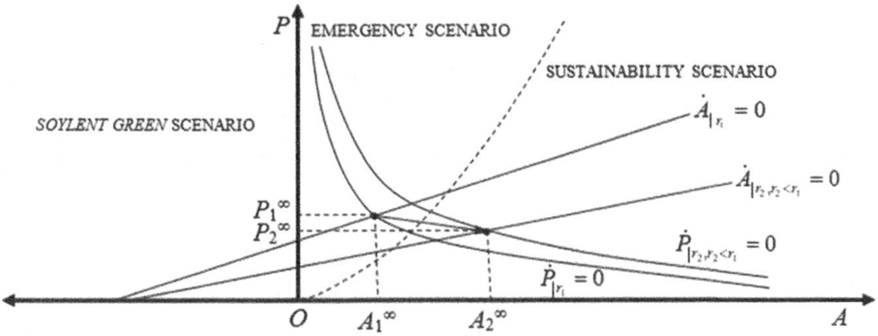

Fig. 1.2 Discounting rate and the steady state

cannot be avoided. This result clearly shows that stopping deforestation is not sufficient to prevent from irreversible evolution to a self-sustained pollution path.

Corollary 2. All things being equal, at the steady state, a smaller discounting rate should result in a higher production rate and environmental absorption capacity, a lower pollution stock and absorption development effort, and a similar emissions reduction effort.

Choosing a near-zero *pure* time preference rate, that is, a near-zero rate at which future generations are discriminated against, is basically justified by intergenerational fairness principles. *All things being equal*, sensitivity analysis confirms that it is indeed the case both for economic and environmental reasons.

In the figure above (Fig. 1.2), it is shown that a substantial decrease in the discounting rate can eventually allow a shift of the steady state from the emergency scenario $\left(A_1^\infty, P_1^\infty\right)$ to the sustainability scenario $\left(A_2^\infty, P_2^\infty\right)$. In other words, a lower discounting rate can contribute to strengthening environmental resilience. This result confirms the contentious suggestion of the Stern review (Stern, 2006) about the virtue of a small discounting rate.

In the case of an infinite discounting rate, $r \mapsto +\infty$, the unique steady state is characterized by an infinite pollution stock, zero emissions rate, and environmental absorption capacity, which corresponds to the "irreversible policy" of Tahvonen and Withagen (1996). However, this "irreversible" steady state resulting from a totally myopic social planner would require here an infinite effort in developing the environmental absorption capacity.

We now analyze the stability of the steady state. Given that the stability property available in two-dimensional state models is at best the saddle-point property (e.g., Dockner (1985)), we check the conditional stability of the steady state of the controllable absorption capacity configuration. The Jacobian matrix associated with the canonical system Eqs. 1.15–1.18 is:

$$J = \begin{bmatrix} -\delta A & -\delta P & \dfrac{a}{\lambda_1{}^2} & 0 \\ -\gamma & 0 & 0 & 1 \\ c & \delta\lambda_1 & r+\delta A & \gamma \\ \delta\lambda_1 & 0 & \delta P & r \end{bmatrix} \tag{1.26}$$

from which the determinant is:

$$|J| = \frac{k(k+r^2\gamma)}{2(c+\gamma^2)} \tag{1.27}$$

and, using Dockner's formula (Dockner, 1985), the sum of the principal minors of J of order 2 minus the squared discount rate, denoted by K, is:

$$K = -\frac{\delta^2 a(2c+\gamma^2)}{r^2\gamma^2} - \frac{\gamma(k+r^2\gamma)}{c+\gamma^2} \tag{1.28}$$

where λ_1, P, and A are evaluated at their steady-state value.

According to Dockner and Feichtinger (1991), a necessary condition for the occurrence of a Hopf bifurcation is $K > 0$. Given that $K < 0$, the existence of limit cycles is then ruled out. The necessary and sufficient condition for the steady state to be a saddle point is fulfilled, since $K < 0$ and $|J| > 0$, which implies that two eigenvalues must have negative real parts.

To determine whether the optimal path is monotonic or follows transient oscillations, we put:

$$\Omega = K^2 - 4|J| = \left[\frac{\delta^2 a(2c+\gamma^2)}{r^2\gamma^2} + \frac{\gamma(k+r^2\gamma)}{c+\gamma^2}\right]^2 - \frac{2k(k+r^2\gamma)}{c+\gamma^2} \tag{1.29}$$

The sign of Ω can be either negative or positive. In the first case, the steady state is a saddle-focus with transient oscillations. Otherwise, it is a saddle node and the optimal solution monotonically converges to the steady state. We highlight below the role played by the parameters r, δ, γ, a, and c in the type of convergence to the steady state.

From Eq. 1.29, it can be shown that: $\lim_{\gamma \mapsto 0^+} \Omega = +\infty$. On the other hand, we have:

$$\lim_{\gamma \mapsto +\infty} \Omega = \frac{\delta^2 a}{r^2}\left[\frac{\delta^2 a}{r^2} + 2\left(\sqrt{r^4 + 4\delta^2 a} - r^2\right)\right] > 0 \tag{1.30}$$

A similar analysis is made for r, a, c, δ. The final results are given in the table below (Table 1.3).

These results are summarized in the subsequent propositions.

Table 1.3 Limit value analysis

	$L_1 = 0^+$	$L_2 = +\infty$
$\lim_{r \mapsto L_j, j=1,2} \Omega$	$+\infty$	$-\infty$
$\lim_{a \mapsto L_j, j=1,2} \Omega$	$-\frac{4r^4\gamma^2 c}{(c+\gamma^2)^2}$	$+\infty$
$\lim_{c \mapsto L_j, j=1,2} \Omega$	$\frac{\delta^2 a}{r^4}\left[\delta^2 a + 2r^2\left(\sqrt{r^4 + 4\delta^2 a} - r^2\right)\right]$	$+\infty$
$\lim_{\delta \mapsto L_j, j=1,2} \Omega$	$-\frac{4r^4\gamma^2 c}{(c+\gamma^2)^2}$	$+\infty$
$\lim_{\gamma \mapsto L_j, j=1,2} \Omega$	$+\infty$	$\frac{\delta^2 a}{r^2}\left[\frac{\delta^2 a}{r^2} + 2\left(\sqrt{r^4 + 4\delta^2 a} - r^2\right)\right]$

Proposition 4. For any given r, a, c, and γ, there exists a threshold $\widetilde{\delta} > 0$ such that for any $\delta < \widetilde{\delta}$ we have $\Omega < 0$ and thus an oscillatory convergence to the steady state.

Remark As we also have $\lim_{a \mapsto 0} \Omega < 0$, the behavior of the system around $a = 0$ is similar to its behavior around $\delta = 0$.

Proposition 5. For any given a, c, δ, and γ, there exists $\widetilde{r} > 0$ such that for any $r > \widetilde{r}$ we have $\Omega < 0$ and thus an oscillatory convergence to the steady state.

This result emphasizes another feature of a small discounting rate that was not initially envisioned by the Stern review (Stern, 2006; Nordhaus, 2007), which is the fact that the monotonicity property of the optimal path to a sustainable steady state is ensured by a small discounting rate. In the converse case (i.e., large discounting rate), an oscillatory behavior results even in the case of a strongly resistant environmental absorption capacity.

Transition Behavior

An important issue suggested in the previous section is to know if a transient convergence to the steady state falls in the *Soylent Green* area for a transient period, thus imposing temporarily potentially damaging consequences. To check this, we examine the transition of the economy from the initial conditions to the steady state.

To generate numerically the state variables' optimal paths, we use the baseline parameters reported in Table 1.4. For a sensitivity analysis, the parameters r, a, δ and γ are varied in the bounds $r \in [0.05, 0.1]$, $a \in [0.001, 0.01]$, $d \in [0.0005, 0.01]$, $\gamma \in [0.002, 0.05]$. For each of the parameter values, the solution paths are calculated for the lower, upper, and intermediate parameter value. Solution paths are calculated for four initial state values, that is, $(A_0, P_0) = \{(0,0), (40,0), (0,80), (40,80)\}$.

The figure below (Fig. 1.3) confirms that, whatever the initial state, a low discounting rate results in a monotonic path to the steady state, which results in a low pollution stock and a strictly positive environmental absorption capacity. This

Table 1.4 Baseline parameters

Parameter	r	a	c	d	α	β	δ	γ
Value	0.1	0.01	0.01	0.01	1	0.5	0.001	0.01

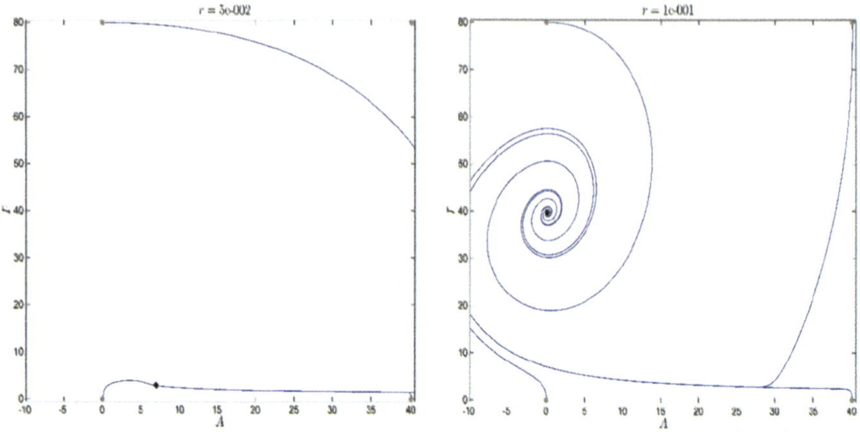

Fig. 1.3 Steady state path and the discounting rate

suggests that the steady state could fall into the *sustainability region*. Conversely, in the case of a large enough discounting rate, transient paths are observed which overflow the *emergency region*. As a result, the *Soylent Green* scenario is unavoidable during a certain period. Afterward, the steady state is characterized by a large pollution stock and a zero environmental absorption capacity, which corresponds to the emergency scenario.

According to the figure below (Fig. 1.4), small values of the production rate utility function coefficient result in paths oscillating between the *emergency region* and the *Soylent Green* scenario. An increase in the production rate utility function coefficient does not affect the steady state pollution stock, but slightly increases the environmental absorption capacity.

In the figure below (Fig. 1.5), it is shown that, whatever the initial state, a small enough value of the marginal efficiency of the absorptive capacity results in transient paths that encroach upon the *Soylent Green* region. Here also, the steady state lies in the *emergency region*. In the converse case of a large enough marginal efficiency of the absorptive capacity, a monotonic path to the steady state is obtained which ends up in the *sustainability region*.

According to the figure below (Fig. 1.6), the path to the steady state can also be affected by the marginal impact of pollution on the absorption capacity. In the case of a strongly resistant environmental absorption capacity, a monotonic path to the steady state is observed which ends up in the *sustainability region*. On the other hand, though we showed that $\lim_{\gamma \mapsto 0^+} \Omega = +\infty$ and $\lim_{\gamma \mapsto +\infty} \Omega > 0$ (cf. Table 1.3), it

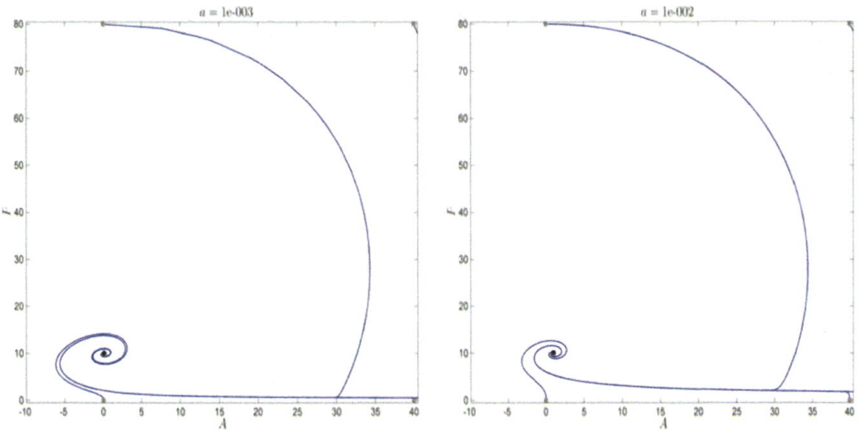

Fig. 1.4 Steady state path and the production rate utility function coefficient

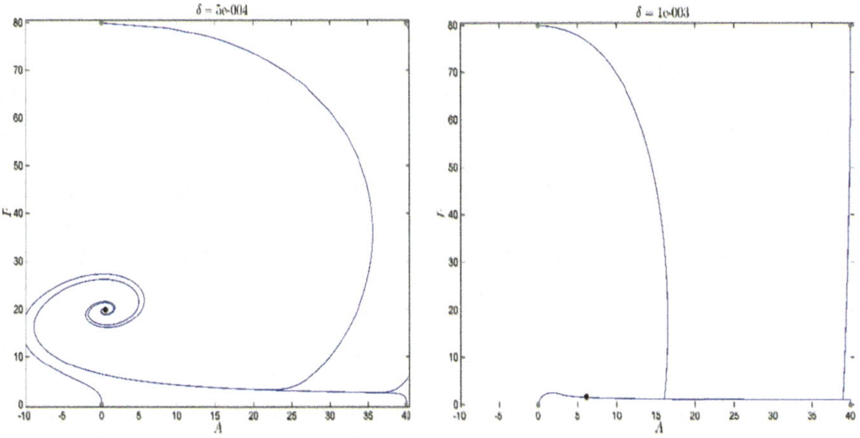

Fig. 1.5 Steady state path and the marginal efficiency of the absorptive capacity

appears that for some intervals of γ, Ω can be negative, which is the case in our example for γ. As a result, a spiraling behavior is observed. That is, a vulnerable environmental absorption capacity gives rise to widely transient paths that largely overflow the *emergency region*, notably from an initial state characterized by a low enough pollution stock. Here also, the *Soylent Green* scenario cannot be avoided during a certain period, and the steady state finally lies in the *emergency region*.

As pollution can be controlled by two policies, production reduction, and absorption capacity development, the fact that an oscillatory path can be optimal implies that they may well evolve differently over time. Indeed, the sign of Ω allows to determine the nature of the convergence to the steady state (monotone *versus* oscillatory) in the neighborhood of the steady state. For instance, when $\Omega > 0$, the

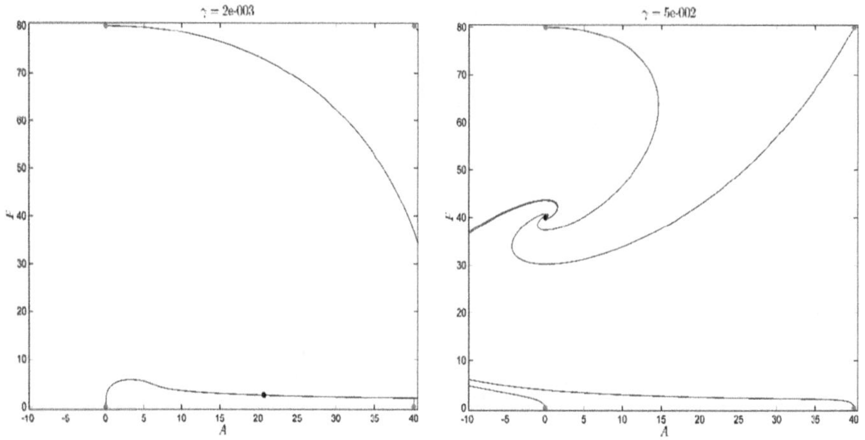

Fig. 1.6 Steady state path and the marginal impact of pollution on the absorption capacity

convergence to the steady state from initial conditions that are "close enough" to the steady state is monotonic, but a non-monotonic approach (over-shooting) to the steady state from initial conditions that are "far enough" from the steady state is not precluded. To investigate this, we compute the control time paths of the control variables according to the baseline parameters reported in Table 1.4.

According to the computations, the signs of the co-state variables are such that $\lambda_1(t) < 0$, and $\lambda_2(t) > 0$, $\forall t$, as expected, which ensures that the control variables always have interior values. The typical patterns of an optimal environmental policy depending on the initial conditions are summarized in the table above (Table 1.5).

Whatever the pattern of convergence to the steady state, when the initial stock of the environmental absorption capacity is small and the initial stock of pollution is large, which reflects the real-world situation, the optimal policy consists of initially choosing a small and increasing then decreasing production rate and a large and decreasing absorption capacity development effort. As a result, the environmental absorption capacity is increased and the pollution stock is substantially reduced in the case of monotonic convergence to the steady state. In the case of oscillatory convergence to the steady state, the environmental absorption capacity remains saturated, but with a significantly lowered pollution stock.

The graphs for the cases where we do not have an oscillatory convergence to the steady state show that, for any initial level (A_0, P_0), we have monotonic convergence to the steady state with either for A and P both decreasing or both increasing and converging to their steady values. It is interesting to note an interesting situation in these cases where the optimal path implies overshooting. That is, in the left figure of Fig. 1.6, for instance, starting from initial values of A and P that are both below (above) their steady-state values, the stock of pollution is built up (brought down) to levels above (below) its steady-state value.

Overall, the numerical results suggest that the occurrence of transient paths might be interpreted as a prediction that the steady state will end up in the *emergency*

Table 1.5 Initial conditions and optimal environmental policy

		Initial pollution stock level	
		Low	Large
Initial environmental absorption capacity level	Low	*Initially high and decreasing production rate* *Initially low and increasing then decreasing absorption capacity development effort*	*Initially low and increasing then decreasing production rate* *Initially high and decreasing absorption capacity development effort*
	Large	*Initially high and decreasing production rate* *Initially low and increasing then decreasing absorption capacity development effort*	*Initially low and increasing then decreasing production rate* *Initially high and decreasing then increasing absorption capacity development effort*

Table 1.6 State and costate steady state values

	r		a		δ		γ	
	0.05	0.1	0.001	0.01	0.0005	0.01	0.002	0.05
λ_1^∞	−0.500	−2.000	−1.000	−1.000	−2.000	−0.100	−0.200	−5.000
λ_2^∞	0.028	0.396	0.099	0.100	0.199	0.016	0.005	2.001
A^∞	7.078	0.126	0.101	1.000	0.504	6.225	20.718	0.050
P^∞	2.826	39.629	9.911	10.000	19.852	1.606	2.413	40.020

region. In the table below (Table 1.6), the steady state values of the state and costate variables are reported.

It clearly appears that the monotonic path to the steady state is more favorable than the oscillatory path as it results in both lower pollution stock and larger environmental absorption capacity, and it also allows for a larger production rate and requires less effort to build up the environmental absorption capacity. Another important result lies in the fact that the transient occurrence of the *Soylent Green* scenario, which gives rise to a self-sustained evolution of the pollution path, is not irreversible. That is, not only the development of the environmental absorption capacity can contribute to better control the stock of pollution, but it also allows the environmental absorption capacity to recover from a negative transient state.

Conclusion

To date, the political debate about pollution has two main opposing approaches: the industrialist approach and the ecologist approach. The opposition between the two approaches is explained in the simplest terms by van der Ploeg and Withagen (1991), that is: "the optimists [i.e., industrialist approach] say production must go up in order to be able to afford to clean up the environment, whilst the pessimists [i.e., ecologist approach] argue that production must go down, *even* if this leaves less scope for cleaning-up activities, as pollution as a by-product of production dominates all else."

According to our model, which accounts for a trade-off between revenues from production, and the cost of emissions reduction efforts plus the net cost of negative externalities from pollution, the technological effort cost needed to reduce the emissions resulting from a higher production rate exceeds the corresponding net revenues. In other words, the industrialist strategy is not optimal. This conclusion is confirmed by a number of studies that pointed out the oversized nature of the technological challenge associated with carbon emissions reduction (e.g., Hoffert et al., 1998; Caldeira et al., 2003; Pielke et al., 2008).

Our results clearly give support to the ecologist approach. Nevertheless, taking into account the natural limitations of the absorption capacity, a positive level of effort to build up the absorption capacity of the environment is necessary to prevent the switching of the environment from a pollution sink to a source. Merely stopping deforestation is not enough, a more active effort to regenerate the absorption capacity is warranted, not because it is economically beneficial, but basically to avoid the *Soylent Green* scenario. The possibility for a win-win situation combining both a higher production rate and technological effort of emissions reduction as described in Table 1.1 is then dominated by a strategy that complements the environmental absorption capacity-building effort with a lower production rate.

In contrast with models representing the environmental absorption capacity as an instantaneous nonlinear function of the pollution stock (e.g., Tahvonen and Withagen (1996)), our model has a single, conditionally stable steady state, compatible with both positive environmental absorption capacity and emissions rate. As a result, the possibility of a concomitant optimal "irreversible policy" along with this sustainable steady state, is ruled out.

The optimal path to the steady state can be either monotonic or oscillatory. It is noteworthy that the possibility of oscillating convergence to the steady state that we have within our framework contrasts with the monotonic convergence to the reversible steady state that is obtained within a framework such as the one of Tahvonen and Withagen (1996), where the limitation of the environmental absorption capacity is captured using a natural rate of decay of pollution.

The conditions under which the optimal solution involves a monotonic or an oscillatory convergence of the pollution stock and the environmental absorption capacity to their steady state values were identified. It turns out that the monotonic optimal path to the steady state results mainly from a farsighted policy, reflected in a low discounting rate, and/or favorable environmental conditions, induced by large marginal efficiency of the absorptive capacity or strongly resistant environmental absorption capacity. Conversely, an oscillatory optimal path to the steady state is essentially due to an impatient policymaker, and/or unfavorable environmental conditions. Given such prerequisites, the monotonic optimal path predicts a relatively clean environment with both positive environmental absorption capacity and production activity level, while the oscillatory optimal path should lead to expect a relatively polluted environment with exhausted environmental absorption capacity and low production rate.

The important question of whether pollution should be controlled by simultaneously building up absorption capacity and curbing emissions by reducing production, or by building absorption capacity and affording an increasing rate of production was also considered. Under realistic initial conditions characterized by a large initial pollution stock and a low initial environmental absorption capacity (relative to their steady-state levels), our model prescribes (1) starting from a production rate smaller than the steady state, increasing the production rate and approaching the state level from above, and (2) starting from a higher absorption capacity development effort than its steady-state level and approaching the steady state from above. These prescriptions, regrettably, depart strongly from the observed environmental policies reflected in the global trends of large emissions and deforestation levels (Canadell et al., 2007). As the combination of initial conditions considered mirrors the emergency scenario depicted in Fig. 1.1, which calls for urgent, appropriate efforts to allow for a shift to the sustainability region, one can expect a pessimistic evolution.

References

Allen, M., Frame, D., Frieler, K., Hare, W., Huntingford, C., Jones, C., Knutti, R., Lowe, J., Meinshausen, M., Meinshausen, N., & Raper, S. (2009). The exit strategy. *Nature Climate Change, 3*, 56–58.

Board on Atmospheric Sciences and Climate. (2011). *Climate stabilization targets: Emissions, Concentrations, and impacts over decades to millennia*. The National Academies Press.

Caldeira, K., Jain, A. K., & Hoffert, M. I. (2003). Climate sensitivity uncertainty and the need for energy without CO_2 emission. *Science, 299*(5615), 2052–2054.

Canadell, J. G., Le Quéré, C., Raupach, M. R., Field, C. B., Buitenhuis, E. T., Ciais, P., Conway, T. J., Gillett, N. P., Houghton, R. A., & Marland, G. (2007). Contributions to accelerating atmospheric CO_2 growth from economic activity, carbon intensity, and efficiency of natural sinks. *Proceedings of the National Academy of Sciences, 104*(47), 18866–18870.

Carpenter, S. R., & Cottingham, K. L. (1997). Resilience and restoration of lakes. *Conservation Ecology, 1*(1), 2. http://www.consecol.org/vol1/iss1/art2/

Caputo, M., & Wilen, J. (1995). Optimal cleanup of hazardous wastes. *International Economic Review, 36*(1), 217–243.

Chevé, M. (2000). Irreversibility of pollution accumulation. *Environmental and Resource Economics, 16*(1), 93–104.

Common, M., & Perrings, C. (1992). Towards an ecological economics of sustainability. *Ecological Economics, 6*(1), 7–34.

Costanza, R., & Daly, H. E. (1992). Natural capital and sustainable development. *Conservation Biology, 6*(1), 37–46.

Costanza, R., d'Arge, R., de Groot, R. S., Farber, S., Grasso, M., Hannon, B., Limburg, K., Naeem, S., O'Neill, R. V., Paruelo, J., Raskin, R. G., Sutton, P., & van den Belt, M. (1997). The value of the world's ecosystem services and natural capital. *Nature, 387*, 253–260.

Costanza, R. (1998). The value of ecosystem services. *Ecological Economics, 25*(1), 1–2.

Cox, P. M., Betts, R. A., Jones, C., Spall, S. A., & Totterdell, I. (2000). Acceleration of global warming due to carbon-cycle feedbacks in a coupled climate model. *Nature, 408*, 184–187.

Cox, P. M., Betts, R. A., Collins, M., Harris, P. P., Huntingford, C., & Jones, C. D. (2004). Amazonian forest dieback under climate-carbon cycle projections for the 21st century. *Theoretical and Applied Climatology, 78*(1/3), 137–156.

Daly, H. (1997). Georgescu-Roegen *vs.* Solow/Stiglitz. *Ecological Economics, 22*(3), 261–267.
Dasgupta, P., & Mäler, K. G. (2003). The economics of non-convex ecosystems. *Environmental and Resource Economics, 26*(4), 499–685.
Dasgupta, P., & Mäler, K. (2004). *Environmental and resource economics: Some recent developments*. SANDEE Working Papers, WP 7, ISSN 1813-1891.
Dockner, E. (1985). Local stability in optimal control problems with two state variables. In G. Feichtinger (Ed.), *Optimal control theory and economic analysis* (Vol. 2). North Holland.
Dockner, E., & Feichtinger, G. (1991). On the optimality of limit cycles in dynamic economic systems. *Journal of Economics, 53*(1), 31–50.
Farzin, Y. H. (1996). Optimal pricing of environmental and natural resource use with stock externalities. *Journal of Public Economics, 62*(2), 31–57.
Forster, B. (1975). Optimal pollution control with a nonconstant exponential rate of decay. *Journal of Environmental Economics and Management, 2*(1), 1–6.
Georgescu-Roegen, N. (1971). *The entropy law and the economic process*. Harvard University Press.
Grass, D., Caulkins, J. P., Feichtinger, G., Tragler, G., & Behrens, D. A. (2008). *Optimal control of nonlinear processes with applications in drugs, corruption, and terror*. Springer.
Grossman, G. M., & Krueger, A. B. (1995). Economic growth and the environment. *Quarterly Journal of Economics, 110*(2), 353–377.
Harris, J. (2002). *Environmental and resource economics: A contemporary approach*. Houghton Mifflin.
Hoffert, M. I., Caldeira, K., Jain, A. K., Haites, E. F., Harvey, L. D. D., Potter, S. D., Schlesinger, M. E., Schneider, S. H., Watts, R. G., Wigley, T. M. L., & Wuebbles, D. J. (1998). Energy implications of future stabilization of atmospheric CO_2 content. *Nature, 395*, 881–884.
Holling, C. (1973). Resilience and stability of ecological systems. *Annual Review of Ecology and Systematics, 4*, 1–23.
Keeler, E., Spence, M., & Zeckhauser, R. (1972). The optimal control of pollution. *Journal of Economic Theory, 4*(1), 19–34.
Krautkraemer, J. A. (1985). Optimal growth resource amenities and preservation of natural environments. *Review of Economic Studies, 52*(1), 153–170.
Leandri, M. (2009). The shadow price of assimilative capacity in optimal flow pollution control. *Ecological Economics, 68*(4), 1220–1231.
Lewis, S. L., Brando, P. M., Phillips, O. L., van der Heijden, G. M. F., & Nepstad, D. (2011). The 2010 Amazon drought. *Science, 331*(6017), 554.
Mäler, K. G. (1991). National accounts and environmental resources. *Environmental and Resource Economics, 1*(1), 1–15.
Nordhaus, W. D. (1991). To slow or not to slow: The economics of the greenhouse effect. *The Economic Journal, 101*(407), 920–937.
Nordhaus, W. D. (2007). A review of the Stern Review on the economics of climate. *Journal of Economic Literature, 45*(3), 686–702.
Pezzey, J. (1992). Sustainability: An interdisciplinary guide. *Environmental Values, 1*(4), 321–362.
Pielke, R., Jr., Wigley, T., & Green, C. (2008). Dangerous assumptions. *Nature, 452*, 531–532.
Raupach, M. R., Marland, G., Ciais, P., Le Quéré, C., Canadell, J. G., Klepper, G., & Field, C. B. (2007). Global and regional drivers of accelerating CO_2 emissions. *Proceedings of the National Academy of Sciences, 104*(24), 10288–10293.
Schaphoff, S., Lucht, W., Gerten, D., Sitch, S., Cramer, W., & Prentice, I. C. (2006). Terrestrial biosphere carbon storage under alternative climate projections. *Climatic Change, 74*(1/3), 97–122.
Schaeffer, M. (1997). *The ecology of shallow lakes*. Chapman Hall.
Scheffer, M., Bascompte, J., Brock, W. A., Brovkin, V., Carpenter, S. R., Dakos, V., Held, H., Van Nes, E. H., Rietkerk, M., & Sugihara, J. (2009). Early warning signals for critical transitions. *Nature, 461*(3), 53–59.
Stern, N. (2006). *Stern review report on the economics of climate change*. HM Treasury.

Stokey, N. (1998). Are there limits to growth? *International Economic Review, 39*(1), 1–31.
Tahvonen, O., & Salo, S. (1996). Nonconvexities in optimal pollution accumulation. *Journal of Environmental Economics and Management, 31*(2), 160–177.
Tahvonen, O., & Withagen, C. (1996). Optimality of irreversible pollution accumulation. *Journal of Economic Dynamics and Control, 20*(9), 1775–1795.
Tsur, Y., & Zemel, A. (1996). Accounting for global warming risks: Resource management under uncertainty. *Journal of Economic Dynamics and Control, 20*(6), 1289–1305.
Ulph, A., & Ulph, D. (1994). The optimal time path of a carbon tax. *Oxford Economic Papers, 46*(5), 857–868.
van der Ploeg, F., & Withagen, C. (1991). Pollution control and the Ramsey problem. *Environmental and Resource Economics, 1*(2), 215–236.
Wirl, F. (1999). Complex, dynamic environmental policies. *Resource and Energy Economics, 21*(1), 19–41.
Wirl, F. (2000). Optimal accumulation of pollution: Existence of limit cycles for the social optimum and the competitive equilibrium. *Journal of Economic Dynamics and Control, 24*(2), 297–306.
World Bank. (1992). *World Development Report: Development and the Environment*. Oxford University Press.

Chapter 2
Self-Regenerating Environmental Absorption Efficiency and the *Soylent Green* Scenario

Fouad El Ouardighi, Hassan Benchekroun, and Dieter Grass

Abstract We consider a stock pollution problem where the biosphere can transform from a sink to a source of pollution in the presence of self-regenerating environmental absorption efficiency. We examine the problem of controlling pollution and the capacity of the biosphere to absorb pollution: the regulator can mitigate emissions and can invest to build up the absorption capacity of pollution sinks. We examine conditions under which both measures, mitigation and absorption capacity investments, are substitute (complement) to each other, and the relative extent to which environmental self-regenerating capabilities affect these conditions. We also exhibit the possibility of an oscillatory approach to the steady state. Particular attention is paid to the situation where the social planner is impatient.

Keywords Pollution · Environmental absorption efficiency · Self-regeneration capabilities · Discounting rate

Introduction

Because natural CO_2 sinks absorb 55% of all anthropogenic carbon emissions, they represent "a huge subsidy to the global economy worth half a trillion US$ annually if an equivalent sink had to be created using other climate mitigation options (based on the cost of carbon in the EU-ETS)" (Canadell et al., 2007). Despite this, the efficiency of natural sinks has decreased by 5% over the last 50 years and should

F. El Ouardighi (✉)
ESSEC Business School, Cergy Pontoise Cedex, France
e-mail: elouardighi@essec.fr

H. Benchekroun
McGill University, Montréal, QC, Canada

D. Grass
Vienna University of Technology, Vienna, Austria

F. El Ouardighi (ed.), *Essays on Pollution Control in Economics and Management Science*, Lecture Notes in Operations Research,
https://doi.org/10.1007/978-3-031-78227-5_2

continue to decrease in the future (Canadell et al., 2007).[1] Notably, the terrestrial biosphere is expected to serve as a carbon sink only until 2050, and will either become carbon-saturated or turn into a carbon source thereafter (Cox et al., 2000; Cramer et al., 2001; Joos et al., 2001; Lenton et al., 2006; Schaphoff et al., 2006; Piao et al., 2008).

The main causes of the decline in the efficiency of natural sinks involve increasing polluting emissions, as well as the destruction of an essential determinant of the environmental absorption efficiency, namely, forests. Deforestation increases the release of stored pollution due to rapidly increased oxidation of soil organic matter[2] (IPCC, 2007), and reduces the carbon storage efficiency in the biosphere (Canadell & Raupach, 2008).

Given that deforestation is a major threat to the earth's future potential of CO_2 biosequestration that may both accelerate and intensify global warming, it has been claimed that stopping deforestation and reforestation could represent an efficient climate change mitigation option (UNFCC, 2008). However, due to the destructive impact of the stock of pollution on the environmental absorption efficiency, only anthropogenic efforts to restore the environmental absorption efficiency can prevent the biosphere from ultimately becoming a source of pollution in the long run (El Ouardighi et al., 2014).

Recently, some studies have emphasized that despite the "lack of data about regrowth rates of tropical forests worldwide," natural forest regrowth seems to play an important role regarding the carbon terrestrial uptake (Pan et al., 2011). This is confirmed by a recent study by FAO (FAO, 2011) according to which the net loss of forest surface between 1990 and 2005 was lower than what was previously expected due to natural regrowth, that is, a total net loss of 72.9 instead of 107.4 million hectares in 15 years. These findings suggest that the self-regeneration capabilities of forests may particularly contribute to increasing environmental absorption efficiency and therefore affect the tradeoff between the level of polluting emissions and the anthropogenic efforts to restore the environmental absorption efficiency.

Indeed, the tradeoff between polluting emissions and environmental absorption efficiency restoration efforts can also be affected by other influential exogenous parameters, among which the destructive impact of pollution on the environmental absorption efficiency, the marginal revenue generated from the polluting emissions, and the marginal pollution cost. An unexamined question is the relative extent to which environmental self-regenerating capabilities affect this tradeoff.

The main feature of this paper is to allow the environment to transform from a natural sink to a natural source contributing to a stock of pollution. We refer to such a scenario as the *Soylent Green* scenario. We examine the problem of controlling pollution and the capacity of the environment to absorb pollution: the regulator can mitigate emissions and can invest to build up the absorption efficiency of pollution

---

[1] See also Zhao and Running (2010).

[2] Deforestation already accounts for around 17% of global anthropogenic CO_2 emissions (Canadell et al., 2007).

sinks. We examine conditions under which both measures, mitigation and absorption efficiency investments, are a substitute (complement) to each other. We also exhibit the possibility of an oscillatory approach to the steady state.

We investigate the relative impact of the environmental self-regenerating capabilities on the trade-off between polluting emissions and anthropogenic efforts to restore the environmental absorption efficiency in terms of complementarity or substitutability. By complementarity (substitutability), we mean that reduction efforts of polluting emissions should be done along with (in substitution to) efforts to develop environmental absorption efficiency. We determine whether, in comparison with other influential exogenous parameters, environmental self-regenerating capabilities help prevent the natural sinks from potentially becoming a source of pollution, both in the transient path and at the steady state. Given the importance of a social planner's impatience with the nature of convergence to the steady state, we assess the sensitivity of our results to the discounting rate value.

To account for the limitations of the environmental absorption efficiency, the economic literature has extended the traditional one-dimensional pollution model (Keeler et al., 1972) by assuming nonlinear (i.e., logistic) absorption efficiency. Accordingly, once a certain level of the stock of pollution is reached, the pollution absorption efficiency vanishes [e.g., Tahvonen and Withagen (1996); Chevé (2000); and more recently Prieur (2009); Boucekkine et al. (2013a, 2013b); Prieur et al. (2013); Ayong Le Kama et al. (2014)]. This extension implies that when the initial stock of pollution is close to the saturation level of the absorption efficiency, there is a trade-off between "reversible" and "irreversible" steady states, that is, one possible optimal trajectory leads to a heavily polluted environment with exhausted absorption efficiency, while the other allows for positive absorption efficiency.[3] For example, within a growth model, Boucekkine et al. (2013b) apply the general framework developed in Boucekkine et al. (2013a) to analyze general optimal control problems with two types of regime switches to study the capital-pollution relationship and optimal technological switch problem under ecological irreversibility.

Despite nonlinear absorption efficiency, the one-dimensional representation of pollution control limits the possibility of considering additional factors that improve the absorption efficiency, as this would add enormous complexity to current difficulties concerning non-concave representation.

An alternative to a nonlinear absorption function consists of describing the absorption efficiency as a separate state variable (El Ouardighi et al., 2014). In this paper, to prevent the SGS scenario, we allow for anthropogenic restoration efforts of the absorption capacity. While throughout the paper we refer to such efforts as reforestation in the case of CO_2, the present analysis applies to general environmental problems where the environment can turn from a sink to a source of pollution.

The paper is organized as follows: The next section presents a continuous-time model where a social planner controls the pollution accumulation process directly,

[3] See also Moser et al. (2013) where a multi-stage model is used to determine the optimal switch between different technologies.

with the current emissions rate, and indirectly by investing in environmental absorption efficiency. The social planner is to maximize, over an infinite time horizon, a discounted social utility resulting from the difference between the current emissions revenues, the cost of the stock of pollution, and the cost of investing in environmental absorption efficiency, if any. The third section characterizes the optimal policy for the emissions rate and restoration effort and the related stability conditions. In the fourth section, a numerical analysis compares the impact of the environmental absorption efficiency self-regeneration capabilities with other exogenous parameters of the problem, on the steady state and the transient path under relative patience and extreme impatience. Fifth section concludes the paper.

Model

We consider an economy where the production activity generates polluting emissions. The emissions rate, $e(t) \geq 0$, depends on the production rate, $q(t) \geq 0$, such that:

$$q(t) = e(t) \tag{2.1}$$

The evolution of the stock of pollution, denoted by $P(t) \geq 0$, is described as:

$$\dot{P}(t) = e(t) - A(t)P(t) \quad P(0) = P_0 \geq 0 \tag{2.2}$$

In Eq. 2.2, the rate of change of pollution increases with the emissions rate as defined in Eq. 2.1, and decreases linearly at an environmental absorption efficiency rate, $A(t)$, which obeys the transition equation:

$$\dot{A}(t) = \alpha + w(t) - \gamma P(t) \quad A(0) = A_0 \geq 0 \tag{2.3}$$

where $\gamma > 0$. According to Eq. 2.3, the environmental absorption efficiency is decreasing with the stock of pollution. This assumption is consistent with the empirical evidence that high pollution levels may destroy environmental absorption efficiency (Holling, 1973; Scheffer, 1997). The parameter γ inversely reflects the internal capacity of the environmental absorption efficiency to resist the destructive impact of pollution. To account for this possibility that the natural regrowth of forests influences carbon terrestrial uptake (Pan et al., 2011), we assume that the baseline process of self-regeneration capabilities of the environmental absorption efficiency is a linear function of time that is reflected in the positive constant, α. On the other hand, we assume that the environmental absorption efficiency is increasing with a restoration effort, denoted by $w(t) \geq 0$. This assumption is illustrated by the suggestion of Canadell and Raupach (2008), according to which reforestation policies can help increase the potential of biosequestration. Finally, deforestation is omitted.

The properties of the model Eqs. 2.2–2.3 are now explored. We first characterize a situation with no human interference (no emissions and no restoration effort) over time, i.e., $e(t) = w(t) = 0, \forall\, t \geq 0$. In this situation, which reflects the pre-industrial state, the dynamics of the system writes:

$$\dot{P}(t) = -A(t)P(t)$$

$$\dot{A}(t) = \alpha - \gamma P(t)$$

the Jacobian of which is given by:

$$J = \begin{bmatrix} -A(t) & -P(t) \\ -\gamma & 0 \end{bmatrix}$$

for which the eigenvalues are: $\chi_{1,2} = \mp\sqrt{\alpha}$. Therefore, the steady state, given by $D: \left(A^{\infty}_{|e=w=0}, P^{\infty}_{|e=w=0}\right) = (0, \alpha/\gamma)$, has the saddle-point property.

To better understand the respective influences of emissions and restoration efforts in our model, we introduce an upper bound, $\overline{w}$, on restoration efforts, $w(t)$, such that $0 \leq w(t) \leq \overline{w}$. Given constant self-regeneration capabilities, we note that without this upper bound, it is optimal to use large values of $w(t)$ for large values of $P(t)$ to decrease pollution. This is obvious since increasing $A(t)$ is the only possibility to decrease $P(t)$. For too large values of $P(t)$, the upper bound $\overline{w}$, along with self-regeneration capabilities α, is not sufficient to enable $A(t)$ to increase. Therefore, given the value of α, for too large values of $P(t)$, no control can be chosen that enables $A(t)$ to increase. We analyze this effect for the specific case where $A(t) = 0$ to characterize the curve that determines the limit between controllable and uncontrollable situations in the state space. In the case where $A(t) = 0$, a further increase in the pollution stock $P(t)$, can be prevented by setting the emissions $e(t)$ to zero and the restoration effort $w(t)$ on its upper bound $\overline{w}$. For these control values, the system Eqs. 2.2–2.3 becomes:

$$\dot{P}(t) = -A(t)P(t)$$

$$\dot{A}(t) = \alpha + \overline{w} - \gamma P(t)$$

which has a steady state $E: \left(A^{\infty}_{|e=0,w=\overline{w}}, P^{\infty}_{|e=0,w=\overline{w}}\right) = \left(0, \frac{\alpha+\overline{w}}{\gamma}\right)$. To determine its stability, we consider the associated Jacobian matrix, which is similar and thus has equal eigenvalues to the case with no human interference, that is, $\chi_{1,2} = \mp\sqrt{\alpha}$. This implies that the steady state E also has the saddle-point property. Therefore, the corresponding stable saddle path is the linearized limiting curve between the regions where the situation is controllable and uncontrollable respectively.

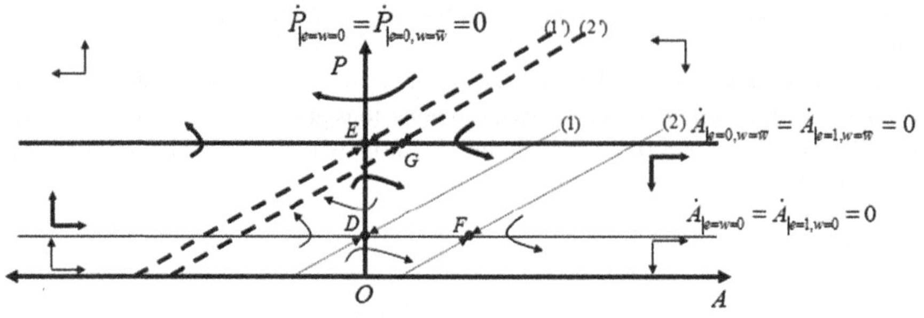

Fig. 2.1 Controllable and uncontrollable pollution-absorption dynamics (bold lines/arrows refer to $w(t) = \overline{w}$)

In the phase diagram in Fig. 2.1, the dotted arrows, denoted by Eq. 2.1, depict the line for which the linearized lower bound constraint, $w(t) = 0$, is active, provided that $e(t) = 0$, $\forall\, t \geq 0$. The area below these arrows corresponds to a *sustainability area*, because the initial absorption efficiency acts as a sink for any initial pollution stock on the right side of the arrows, while the self-regenerating capabilities of the absorption efficiency allow for its subsequent recovery. Conversely, all the paths above these arrows are diverging.

On the other hand, the dashed bold arrows, denoted by (1'), are the linearized limiting curve between controllable and uncontrollable situations, for which the upper bound constraint, $w(t) = \overline{w}$, is active, provided that $e(t) = 0$, $\forall\, t \geq 0$. Above this limiting curve, a high initial stock of pollution coupled with low initial absorption efficiency makes it possible for the biosphere to turn from a pollution sink to a source. In this region (the so-called *Soylent Green* scenario), the absorption efficiency becomes irreversibly negative, which leads to an increasing release of pollution by the biosphere. This description is consistent with a growing body of ecological literature (Lenton et al., 2006; Piao et al., 2008; Barnosky et al., 2012), as well as economic literature (Tahvonen & Withagen, 1996; Chevé, 2000; Prieur, 2009; Boucekkine et al., 2013a, 2013b; Moser et al., 2013).

The region between (1) and (1') corresponds to the area of interior solutions $(0 \leq w(t) \leq \overline{w})$, given $e(t) = 0$. This region, called the *emergency area*, is the one where the pull toward the *Soylent Green* area is imminent if no restoration efforts are made.

Now, introducing an upper bound, $\overline{e}$, on emissions, $e(t)$, such that $0 < e(t) \leq \overline{e}$, we consider the case where $A(t) > 0$ is sufficiently large so that the emissions can be set on their upper bound over the whole planning horizon, that is, $e(t) = \overline{e}, \forall t \geq 0$. Then, two (locally stable) steady states can be derived, one related to the case of upper bound constraint on restoration efforts, $w(t) = 0$, $\forall\, t \geq 0$, that is, $F : \left(A^{\infty}_{|e=\overline{e},w=0}, P^{\infty}_{|e=\overline{e},w=0}\right) = (\gamma\overline{e}/\alpha, \alpha/\gamma)$, and the other related to the case of lower bound constraint on restoration efforts, $w(t) = 0$, $\forall\, t \geq 0$, that is, $G : \left(A^{\infty}_{|e=\overline{e},w=\overline{w}}, P^{\infty}_{|e=\overline{e},w=\overline{w}}\right) = \left(\frac{\gamma\overline{e}}{\alpha+\overline{w}}, \frac{\alpha+\overline{w}}{\gamma}\right)$.

In the case of maximum emissions rate, i.e., $e(t) \equiv \overline{e}, \forall t \geq 0$, the implementation of a restoration policy moves up the steady state from F to G and therefore establishes an emergency area at the expense of the *Soylent Green* area. This emergency area prevents a switching of the biosphere from a pollution sink to a source as long as an appropriate restoration policy is employed. The extent of this area increases (decreases) with larger (smaller) values of self-regeneration capabilities α, and smaller (larger) values of the marginal destructive impact of pollution γ. Note that the optimal path to the steady state in the emergency area can be composed of segments successively involving active and inactive constraints on $w(t)$. On the other hand, in the absence of a restoration policy, a drastic emissions reduction policy leads to moving the (locally stable) steady state from F to D and therefore to extend the sustainability area at the expense of the *Soylent Green* area. A similar pattern is observed in the presence of a restoration policy, where the (locally stable) steady state moves from G to F.

The formulation suggested in Eqs. 2.2–2.3 differs from that in El Ouardighi et al. (2014) in that it also includes inherent self-regeneration capabilities of the environmental absorption efficiency. This makes the current configuration more optimistic because the *Soylent Green* scenario can be avoided even with slightly negative initial environmental absorption efficiency if the stock of pollution is sufficiently low. This configuration strongly suggests that self-regenerating environmental absorption efficiency can also prevent the *Soylent Green* scenario from occurring in a context involving human interference, possibly providing additional conveniences, such as a higher emissions rate and lower restoration efforts in the long run.

According to Fig. 2.1, restoration efforts can, in some cases, be more efficient than emissions reduction to move away the *Soylent Green* area to the left-hand side. This justifies the study of an optimal trade-off between emissions reduction and restoration efforts. In this regard, we now define an objective criterion for the social planner.

We assume that the social planner's revenue function is characterized by a constant elasticity of marginal utility of emissions equal to 1 (e.g., Chevé, 2000), that is, $a \ln e(t)$, where $a > 0$ is a revenue function coefficient. The choice of a logarithmic revenue function owes to the assumption that, in practice, a near-zero emissions level is economically infinitely costly and therefore socially not desirable. On the other hand, the negative externalities of the stock of pollution are valued as an increasing convex function of pollution (e.g., Michel and Rotillon (1995)), that is, $cP(t)^2/2$, $c > 0$. Finally, the restoration effort, which aims to develop environmental absorption efficiency, generates an increasing quadratic cost, denoted by $fw(t)^2/2, f > 0$ (El Ouardighi et al., 2014). Without loss of generality, we assume $f = 1$.

Denoting the discounting rate by $r > 0$, and assuming an infinite planning horizon, the social planner's optimal control problem is written as:

$$\underset{e(.),w(.)}{Max}\ U = \int_0^{+\infty} e^{-rt}\left(a\ln e(t) - cP(t)^2/2 - w(t)^2/2\right)dt \tag{2.4}$$

under Eqs. 2.2–2.3, and $w(t) \geq 0, \forall t$. As it will appear in our subsequent numerical study, the control constraint is never binding and we therefore do not explicitly derive the related optimality conditions.

Optimal Policy and Steady State Stability

The main issue here is to determine the nature of the steady state associated with the existence of self-regeneration capabilities of the environmental absorption efficiency.

Skipping the time index for convenience, the current-value Hamiltonian is:

$$\mathcal{H} = a\ln\ e - cP^2/2 - w^2/2 + \eta_1(e - AP) + \eta_2(a + w - \gamma P) \tag{2.5}$$

where $\eta_j \equiv \eta_j(t)$ are costate variables, $j = 1, 2$, that evolve according to:

$$\dot{\eta}_1 = cP + (r + A)\eta_1 + \gamma\eta_2 \tag{2.6}$$

$$\dot{\eta}_2 = r\eta_2 + P\eta_1 \tag{2.7}$$

The necessary conditions for optimality are:

$$\mathcal{H}_e = ae^{-1} + \eta_1 = 0 \tag{2.8}$$

$$\mathcal{H}_w = -w + \eta_2 = 0 \tag{2.9}$$

As the stock of pollution has a negative marginal influence on the social planner's objective function, its implicit price should be negative in Eq. 2.8, i.e., $\eta_1 < 0$. On the other hand, a strictly positive restoration effort in Eq. 2.9 requires $\eta_2 > 0$. A condition for $w = 0$ to be admissible is that the corresponding costate variable is nonnegative, which would require that $\eta_2 \leq 0$.

The Legendre-Clebsch condition of concavity of the Hamiltonian is satisfied, as the Hessian:

$$\begin{bmatrix} \mathcal{H}_{ee} & \mathcal{H}_{ew} \\ \mathcal{H}_{we} & \mathcal{H}_{ww} \end{bmatrix} = \begin{bmatrix} -ae^{-2} & 0 \\ 0 & -1 \end{bmatrix}$$

is negative definite. This guarantees a maximum of the Hamiltonian. Note however that the sufficiency conditions for optimality may not hold, which thus does not guarantee the existence of a globally optimal solution.

Plugging the value of e and w from Eqs. 2.8 and 2.9 in Eqs. 2.2 and 2.3, respectively, the equations:

$$\dot{P} = -a\eta_1{}^{-1} - AP \tag{2.10}$$

$$\dot{A} = \alpha + \eta_2 - \gamma P \tag{2.11}$$

along with Eqs. 2.6 and 2.7, form the canonical system in the state-costate space of the case with both self-regeneration capabilities and restoration efforts.

Proposition 1. The steady state is unique, given by:

$$e^{\infty} = \frac{[\gamma(r^2+\alpha)+k]a}{r[r^2\gamma^2-\alpha(2c+\gamma^2)+\gamma k]} \tag{2.12a}$$

$$w^{\infty} = \frac{r^2\gamma^2-\alpha(2c+\gamma^2)+\gamma k}{2(c+\gamma^2)} \tag{2.12b}$$

$$P^{\infty} = \frac{\gamma(r^2+\alpha)+k}{2(c+\gamma^2)} \tag{2.12c}$$

$$A^{\infty} = \frac{2(c+\gamma^2)a}{r[r^2\gamma^2-\alpha(2c+\gamma^2)+\gamma k]} \tag{2.12d}$$

where $k = \sqrt{\gamma^2(r^2+\alpha)^2 + 4(a-r^2\alpha)(c+\gamma^2)} \in \mathfrak{R}|\alpha \leq a/r^2$.

Proof Equating the RHS of Eqs. 2.6–2.7–2.10–2.11 to 0, using Eqs. 2.8 and 2.9, and solving by identification and substitution, we get Eqs. 2.12a–2.12d. □

From Eqs. 2.12a–2.12d, we observe that e^{∞} and w^{∞} have nonnegative values if (and only if) both $\alpha \leq a/r^2$ and $\alpha < \gamma\sqrt{a/c}$, which imply that $r^2\gamma > \sqrt{ac}$. Under these conditions, the unique steady state is characterized by positive values of the control and state variables, which suggests that self-regenerating capabilities combined with efforts to restore the environmental absorption efficiency both prevent the *Soylent Green* scenario from occurring in the long run. Note that $w^{\infty} = 0$ can never arise as a possible outcome because this would result in diverging steady-state values for both e^{∞} and A^{∞}.

The results in Eqs. 2.12a–2.12d differ significantly from those without self-regeneration capabilities, i.e., $\alpha = 0$, $\forall\, t \geq 0$, which are developed below.

Corollary 1. The steady-state without self-regeneration capabilities is given by:

$$\left[e^{\infty}_{|\alpha=0}\; w^{\infty}_{|\alpha=0}\; P^{\infty}_{|\alpha=0}\; A^{\infty}_{|\alpha=0}\right]^T = \left[\frac{a}{r\gamma}\quad \frac{\gamma(r^2\gamma+k')}{2(c+\gamma^2)}\quad \frac{r^2\gamma+k'}{2(c+\gamma^2)}\quad \frac{2(c+\gamma^2)a}{r\gamma(r^2\gamma+k')}\right]^T \tag{2.13}$$

where $k' = \sqrt{r^4\gamma^2 + 4a(c+\gamma^2)}$.

Note also that the steady state involved by the corner solution $w(t) = 0, \forall\, t \geq 0$, is given by:

$$\left[e^{\infty}_{|w=0}\; w^{\infty}_{|w=0}\; P^{\infty}_{|w=0}\; A^{\infty}_{|w=0}\right]^T = \left[\frac{\alpha\gamma a(\alpha-r^2)}{r(\gamma^2 a-\alpha^2 c)}\quad 0\quad \frac{\alpha}{\gamma}\quad \frac{\gamma^2 a(\alpha-r^2)}{r(\gamma^2 a-\alpha^2 c)}\right]^T \tag{2.14}$$

In this case, a feasible, non-trivial steady state requires both $\alpha > (<) r^2$ and $\alpha < (>) \gamma\sqrt{a/c}$, which implies that $r < (>) \gamma\sqrt{a/c}$. In other words, the steady state is viable if a strictly positive self-regeneration capabilities coefficient of the environmental absorption capacity is compatible with both a sufficiently low (high) discounting rate, and a sufficiently high (low) revenue function coefficient.

Using Eqs. 2.12a–2.12d, the limiting transversality conditions are satisfied for saddle-paths (solutions converging to the steady state) since:

$$\lim_{t\mapsto+\infty} e^{-rt}\eta_1(t)P(t) = \lim_{t\mapsto+\infty}\left\{-\frac{r[r^2\gamma^2-\alpha(2c+\gamma^2)+\gamma k]e^{-rt}}{2(c+\gamma^2)}\right\} = 0 \tag{2.15}$$

$$\lim_{t\mapsto+\infty} e^{-rt}\eta_2(t)A(t) = \lim_{t\mapsto+\infty} ae^{-rt}/r = 0 \tag{2.16}$$

Corollary 2. At the steady state, the social planner's utility is given by:

$$u^{\infty} = \frac{1}{r}\left\{a\ln\frac{[\gamma(r^2+\alpha)+k]a}{r[r^2\gamma^2-\alpha(2c+\gamma^2)+\gamma k]} - \frac{c[\gamma(r^2+\alpha)+k]^2+[r^2\gamma^2-\alpha(2c+\gamma^2)+\gamma k]^2}{8(c+\gamma^2)^2}\right\} \tag{2.17}$$

Equation 2.17 represents the future generations' utility when the self-regeneration capabilities are positive.

Corollary 3. At the steady state, the social planner's utility without self-regeneration capabilities is given by:

$$u^{\infty}_{|\alpha=0} = \frac{1}{r}\left[a \ln\frac{a}{r\gamma} - \frac{(r^2\gamma + k')^2}{8(c+\gamma^2)}\right] \tag{2.18}$$

Proposition 2. The steady-state exhibits a (local) two-dimensional stable manifold.

Proof To analyze the stability of the steady state, we compute the Jacobian matrix of the canonical system Eqs. 2.6–2.7–2.10–2.11, that is:

$$J = \begin{bmatrix} -A & -P & \frac{a}{\eta_1^2} & 0 \\ -\gamma & 0 & 0 & 1 \\ c & \eta_1 & r+A & \gamma \\ \eta_1 & 0 & P & r \end{bmatrix}$$

Given that η_1, P, and A are evaluated at their steady-state value, we get:

$$|J| = -(2\eta_1^{\infty} + r\gamma)A^{\infty} P^{\infty} - (\eta_1^{\infty} + r\gamma)\left(rP^{\infty} + \frac{a}{\eta_1^{\infty}}\right) + (\gamma^2 + c)P^{\infty\,2}$$

$$= 2a - r^2\alpha + \frac{\gamma[\gamma(r^4+\alpha^2) + (r^2+\alpha)k]}{2(c+\gamma^2)}$$

from which a sufficient condition for a positive value of $|J|$ is $\alpha \leq 2a/r^2$.

Using Dockner's formula (Dockner, 1985), the sum of the principal minors of J of order 2 minus the squared discounting rate, that is:

$$K = -A^{\infty}(r + A^{\infty}) - 2\gamma P^{\infty} - \frac{ac}{\eta_1^{\infty\,2}}$$

$$= -\frac{a}{r[r^2\gamma^2 - \alpha(2c+\gamma^2) + \gamma k]} \times \left\{2r(c+\gamma^2) + \frac{4(c+\gamma^2)^2 a + c[\gamma(r^2+\alpha)+k]^2}{r[r^2\gamma^2 - \alpha(2c+\gamma^2)+\gamma k]}\right\} - \frac{\gamma[\gamma(r^2+\alpha)+k]}{(c+\gamma^2)}$$

is negative. Thus, two eigenvalues have negative real parts and two have positive real parts. □

Corollary 4. The steady-state without self-regeneration capabilities also exhibits a (local) two-dimensional stable manifold.

Proof In this case, we get:

$$|J|_{|\alpha=0} = 2a + \frac{r^2\gamma(r^2\gamma + k)}{2(c+\gamma^2)} > 0$$

$$K_{|a=0} = -\frac{a}{r\gamma(r^2\gamma+k)}\left[2r\left(c+\gamma^2\right)+\frac{4(c+\gamma^2)^2 a + c(r^2\gamma+k)^2}{r\gamma(r^2\gamma+k)}\right]-\frac{\gamma(r^2\gamma+k)}{(c+\gamma^2)}<0$$

which proves that two eigenvalues have negative real parts and two have positive real parts. □

To determine whether the optimal path is monotonic or follows transient oscillations, we compute (Dockner & Feichtinger, 1991):

$$\begin{aligned}\Omega &= K^2 - 4|J| \\ &= A^{\infty\,2}\left(r+A^{\infty}\right)^2 + 4\gamma\left(2r+A^{\infty}\right)A^{\infty}\,P^{\infty} + \frac{a^2c^2}{{\eta_1^{\infty}}^{4}} \\ &\quad + \frac{2a}{\eta_1^{\infty}}\left[\frac{c\left(r+A^{\infty}\right)A^{\infty}}{\eta_1^{\infty}} + 2r\gamma - 2\eta_1^{\infty}\right] + 4\left(r^2\gamma + \frac{\gamma ac}{{\eta_1^{\infty}}^{2}} + r\eta_1^{\infty} - cP^{\infty}\right)P^{\infty}\end{aligned} \quad (2.19)$$

A positive (negative) sign of Ω indicates that the saddle-path, i.e., the path converging to the saddle-point, is monotonic (spiraling) near the steady state. The sign of Ω in Eq. 2.19 is ambiguous. Note that an important feature of the model at hand, with no self-regenerating capabilities of the absorption efficiency, is that the value of the discounting rate strongly affects both the nature of the convergence to the steady state and the steady state value (El Ouardighi et al., 2014). That is, for a given constellation of parameters, a heavily discounted policy is more capable than a lightly discounted policy of producing a spiraling transient path to the steady state. As a spiraling saddle path encroaches upon the *Soylent Green* scenario during a certain time interval, thus leading to the biosphere turning temporarily into a source of pollution, this has the effect of deteriorating the steady state, which finally falls into the upper *emergency area* depicted in Fig. 2.1. Therefore, a monotonic convergence is less detrimental to the future generations than a spiraling convergence.

Proposition 3. At the steady state, it holds that:

$$e^{\infty}_{|a>0} > e^{\infty}_{|a=0} \quad (2.20)$$

$$w^{\infty}_{|a>0} < w^{\infty}_{|a=0} \quad (2.21)$$

$$P^{\infty}_{|a>0} \lesseqgtr P^{\infty}_{|a=0} | \gamma \lesseqgtr r^2\sqrt{c/a} \quad (2.22)$$

$$A^{\infty}_{|a>0} > A^{\infty}_{|a=0} \quad (2.23)$$

Proof Using Eq. 2.12a to compare $e^{\infty}_{|\alpha>0}$ and $e^{\infty}_{|\alpha=0}$, we get:

$$\frac{\gamma^2(r^2+\alpha)+\gamma\sqrt{\gamma^2(r^2+\alpha)^2+4(a-r^2\alpha)(c+\gamma^2)}}{r^2\gamma^2-\alpha(2c+\gamma^2)+\gamma\sqrt{\gamma^2(r^2+\alpha)^2+4(a-r^2\alpha)(c+\gamma^2)}}>1$$

which proves Eq. 2.20. Similarly, using Eq. 2.13 to compare $w^{\infty}_{|\alpha>0}$ with $w^{\infty}_{|\alpha=0}$ yields:

$$2\alpha c(c+\gamma^2)+\gamma(2c+\gamma^2)\left[r^2\gamma+\sqrt{r^4\gamma^2+4a(c+\gamma^2)}\right]>0$$

which proves Eq. 2.21. Using Eq. 2.14, to compare $P^{\infty}_{|\alpha>0}$ with $P^{\infty}_{|\alpha=0}$ gives:

$$\gamma^2 a<r^4 c$$

which proves Eq. 2.22. Finally, using Eq. 2.15, to compare $A^{\infty}_{|\alpha>0}$ with $A^{\infty}_{|\alpha=0}$ gives:

$$2\alpha c(c+\gamma^2)+\gamma(2c+\gamma^2)\left[r^2\gamma+\sqrt{r^4\gamma^2+4a(c+\gamma^2)}\right]>0$$

which proves Eq. 2.23. □

At the steady state, the elasticity of the emissions rate with respect to environmental self-regenerating capabilities is positive, equal to:

$$\begin{aligned}\psi_{e^{\infty}\ /\alpha}&=\frac{\partial e^{\infty}}{\partial\alpha}\frac{\alpha}{e^{\infty}}\\&=\frac{\alpha}{k}\left\{\frac{[\gamma(\alpha\gamma+k)-r^2(2c+\gamma^2)][\gamma(r^2\gamma+k)-\alpha(2c+\gamma^2)]}{\gamma(r^2+\alpha)+k}\right.\\&\qquad\left.+(r^2\gamma+k)(2c+\gamma^2)-\gamma^3\alpha\right\}\end{aligned}\tag{2.24}$$

Similarly, the steady-state elasticity of the restoration effort with respect to the self-regenerating capabilities of the environment is negative, equal to:

$$\psi_{w^{\infty}\ /\alpha}=\frac{\partial w^{\infty}}{\partial\alpha}\frac{\alpha}{w^{\infty}}=-\frac{\alpha[(2c+\gamma^2)(r^2\gamma+k)-\alpha\gamma^3]}{k[\gamma(r^2\gamma+k)-\alpha(2c+\gamma^2)]}<0\tag{2.25}$$

On the other hand, the elasticity of the pollution stock to environmental self-regenerating capabilities is:

$$\psi_{P^\infty}\ /\alpha = \frac{\partial P^\infty}{\partial \alpha}\frac{\alpha}{P^\infty} = \frac{\alpha[\gamma(k+\gamma\alpha)-r^2(2c+\gamma^2)]}{k[\gamma(r^2+\alpha)+k]} \tag{2.26}$$

which is either positive or negative depending on whether the marginal destructive impact of pollution on the absorption efficiency, γ, is greater or lower than $r^2\sqrt{c/a}$. This condition is compatible with the inequality implied by the feasibility conditions, i.e., $r > \left(\sqrt{ac}/\gamma\right)^{1/2}$ if $r > a^{1/4}$.

Finally, the elasticity of the restoration effort with respect to environmental self-regenerating capabilities is equal to:

$$\psi_{A^\infty}\ /\alpha = \frac{\partial A^\infty}{\partial \alpha}\frac{\alpha}{A^\infty} = \frac{\alpha\{k(2c+\gamma^2)+\gamma[r^2(2c+\gamma^2)-\gamma^2\alpha]\}}{k[r^2\gamma^2-\alpha(2c+\gamma^2)+\gamma k]} \tag{2.27}$$

which is always positive.

An important question now is to determine the relative magnitude of the influence of environmental self-regenerating capabilities on the control and state variables and the social utility. This issue is tackled in the following section.

Numerical Analysis

To evaluate the extent to which self-regenerating capabilities of the absorption efficiency can, in comparison with other exogenous parameters, affect the transient path and the steady state, we conduct a numerical analysis and select a broad range of values for the parameters (Table 2.1).

The set of lower bound values reflects the base case, which is subscripted by *bc*. The range of parameter values selected aims to contrast some important configurations, that is:

- Relative patience *vs.* extreme impatience of the decision-maker (i.e., low *vs.* very high discounting rate, r_l *vs.* r_u).[4] Relative patience (or simply patience) corresponds to the market interest in normal time (that is, 5%), and extreme impatience (or simply impatience) is an approximation of peak levels of the market interest in times of liquidity crisis.

Table 2.1 Parameter values

Parameter	r	a	c	α	γ
Lower bound value	0.05	0.001	0.01	0.0001	0.05
Upper bound value	1	0.01	0.1	0.001	0.5

[4]The idyllic but unlikely case of *extremely patient behavior* reflected in an interest rate of 0.1% advocated by Stern (2006) is disregarded.

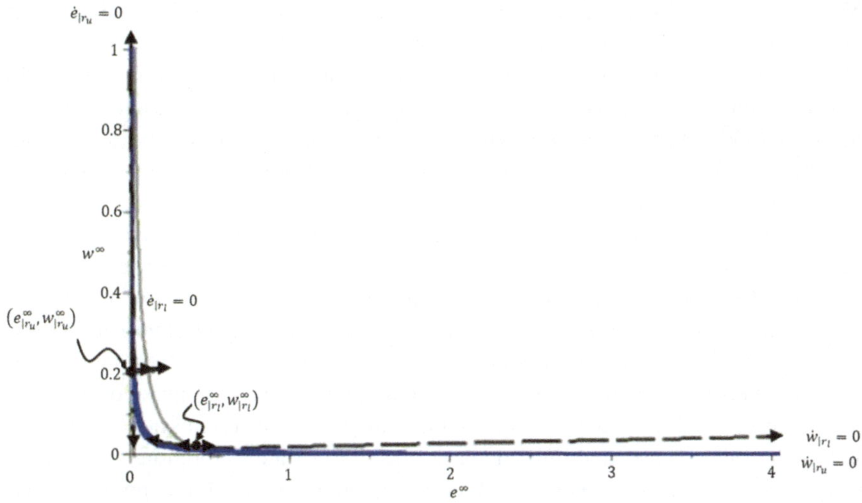

Fig. 2.2 Sensitivity of emissions rate and restoration effort steady states under patient and impatient behavior

- Low *vs.* high self-regeneration capabilities of the environmental absorption efficiency, under the constraint that $\alpha \leq a/r^2$ (i.e., low *vs.* high self-regeneration capabilities, α_l *vs.* α_u).
- Strong *vs.* weak environmental resilience (i.e., low *vs.* high destructive impact of pollution on the absorption efficiency, γ_l *vs.* γ_u),
- Low *vs.* high marginal revenue (i.e., low *vs.* high revenue function coefficient, a_l *vs.* a_u), and
- Low *vs.* high marginal pollution cost (i.e., low *vs.* high-cost function coefficient, c_l *vs.* c_u).

For the numerical resolution, a boundary value approach is used (Grass et al., 2008; Grass, 2012). The results are computed with Maple 18.0 and OCMat (MATLAB package for the computation of optimal control problems). Neither the feasibility nor the stability of the solutions generated is affected by the variations from the baseline parameter values.

For the baseline parameters, the saddle-point property is fulfilled with two positive ($\chi_{2lbc} = 1.1345$, $\chi_{4lbc} = 0.0697$) and two negative eigenvalues ($\chi_{1lbc} = -1.0845$, $\chi_{3lbc} = -0.0197$), which ensures a locally stable steady state.

Using Eqs. 2.6–2.9 and 2.14–2.15, Fig. 2.2 shows the sensitivity of the control variables' steady state to the parameter values respectively for low and high values of the discounting rate. Each arrow starting from the lightly and heavily discounted steady state, $\left[e^{\infty}_{|bc}, w^{\infty}_{|bc}\right]$ and $\left[e^{\infty}_{|r_u}, w^{\infty}_{|r_u}\right]$, respectively, indicates the direction toward which the steady state is moved and the magnitude of the impact of the corresponding parameter on the steady state as reflected in the length of the arrow.

Note that the results in the following figures are consistent with those that would result from an elasticity analysis based on Eqs. 2.24–2.27.

An impatient behavior consists of a higher valuation of emissions during the transient path than in the steady state. Accordingly, the steady-state emissions rate decreases with an increasing discounting rate. Correlatively, when the steady-state restoration effort is increased, this partly offsets the destructive impact of pollution on the steady-state absorption efficiency. In both patient and impatient cases, an increase in the self-regeneration capabilities of the absorption efficiency, α, moves the steady state to the right-hand side very slightly (shortest arrows). Hence, whatever the level of the discounting rate and the constraint that $\alpha \leq a/r^2$ is satisfied, self-regeneration capabilities of the absorption efficiency neither significantly reinforce nor undermine the substitutability between the steady state emissions rate and restoration effort. In comparison, a greater marginal destructive impact of pollution on the absorption efficiency, γ, only reduces the steady-state emissions rate in the lightly discounted case, and only increases the steady-state restoration efforts in the heavily discounted case. On the other hand, an increase in the pollution cost coefficient, c, in the lightly discounted case leads to moving the steady state to the left-hand side very slightly (shortest arrows). That is, compared with the base case, a heavier valuation of the negative impact of the stock of pollution under patient behavior does not entail any significant economic sacrifice at the steady state, neither in terms of reducing the emissions rate nor in terms of increasing the restoration effort cost. In contrast, under impatience, it lowers the restoration effort cost substantially but offers no latitude for enhancing the emissions rate. Finally, an increase in the revenue function coefficient, a, under patient behavior allows for the highest steady-state emissions, as well as a slight increase in the steady-state restoration effort.

These results are confirmed in Fig. 2.3 which is obtained from Eqs. 2.6–2.7, 2.10–2.11, and 2.12a–2.13. The two steady states depicted, i.e., $\left[A^{\infty}_{|bc}, P^{\infty}_{|bc}\right]$ and $\left[A^{\infty}_{|r_u}, P^{\infty}_{|r_u}\right]$ are indeed mutually exclusive. For a given parameter constellation, the initial state will always be in the basin of attraction of the same steady state. A change of parameter would therefore be needed to move from the initial state to the basin of the other steady state. That is, due to path dependence, each steady state can be reached only along its corresponding transient path.

Under patient behavior, the self-regeneration capabilities of the absorption efficiency, α, also move the steady state very slightly to the right-hand side (shortest arrow). Actually, to get a similar impact to the pollution cost coefficient, c, or the revenue function coefficient, a, on the steady state, the self-regenerating capabilities of the environmental absorption efficiency should be enhanced by as much as 100, rather than only 10 times its base case value, as the other parameters. Similar results are observed for the heavily discounted steady state, yet with a greater increase in the stock of pollution. As for the previous figure, the nature of the influence of the pollution cost coefficient and the revenue function coefficient on the steady state depends on the decision-maker's relative patience. While a greater

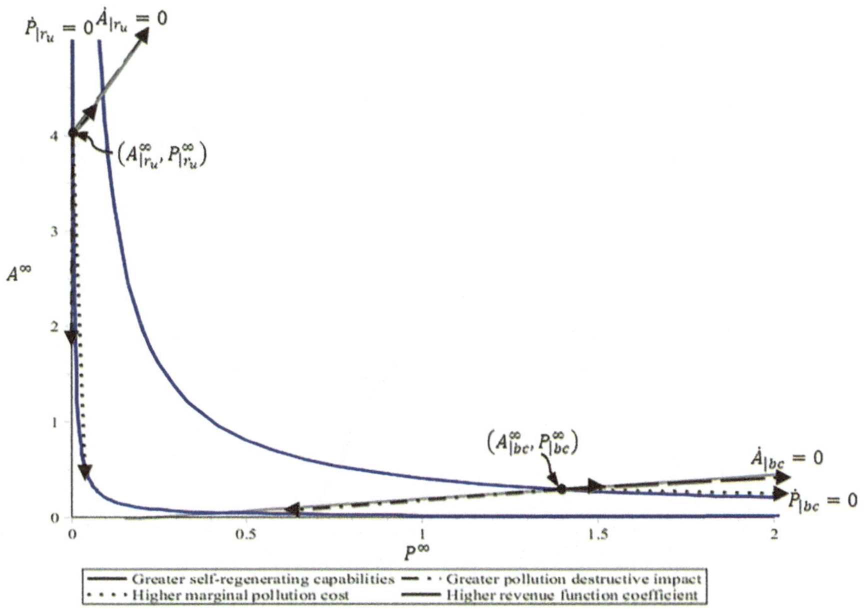

Fig. 2.3 Sensitivity of absorption efficiency and pollution steady states under the patient and impatient behavior

pollution cost coefficient increases the absorption efficiency with negligible impact on the stock of pollution in the case of the patient decision-maker, it decreases the stock of pollution with insignificant impact on the absorption efficiency in the case of the impatient decision-maker. Similarly, a greater revenue function coefficient increases the absorption efficiency and keeps the stock of pollution almost unchanged under a patient decision-maker, but increases the stock of pollution with a limited increase in the absorption efficiency under an impatient social planner. In other words, the leveraging of the pollution cost and revenue function coefficients improves the sustainability of the lightly discounted steady state, while it fails to move the heavily discounted steady state out of the emergency area. Finally, an increase in the marginal destructive impact of pollution on the absorption efficiency has the effect of reducing both the steady state absorption efficiency and stock of pollution in the lightly discounted case. This pattern is related to a decrease in the steady-state emissions rate. In the impatient case, despite a considerable increase in the steady state restoration effort, the only result is a decrease in the steady state stock of pollution.

Using Eq. 2.19, Fig. 2.4 shows the sensitivity of the convergence of the saddle-path to the parameters of the model across the discounting rate. Self-regenerating capabilities of the absorption efficiency, α, do not significantly influence the transient behavior that leads to the steady state of both under-patient and impatient behaviors. Similarly, a lower environmental resilience, i.e., a greater value of γ, does

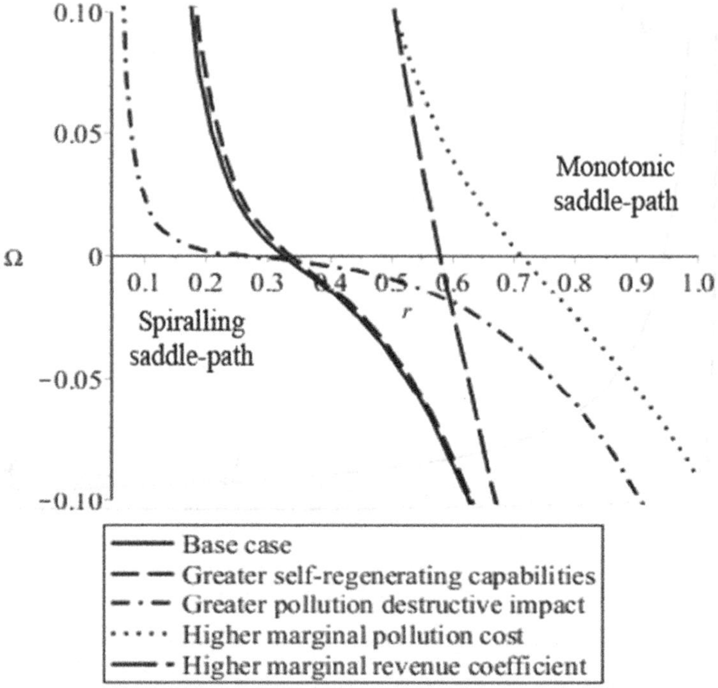

Fig. 2.4 Sensitivity of the convergence to the steady state under the patient and impatient behavior

not appear to significantly change the threshold of impatience above which a spiraling behavior is present. In contrast, an increased marginal revenue coefficient, a, raises the threshold of impatience above which the saddle path begins to spiral. An even stronger influence is observed for an increased pollution cost coefficient, c. This emphasizes the benefit of a proper quantification of the stock of pollution's destructive impact in terms of the monotonicity of the convergence to the steady state.

We now generate time paths for the control and state variables using the range of values in Table 2.1. To do so, we assume adverse initial conditions reflected by a low initial value for the absorption efficiency (i.e., lower than the base case steady state) and a large initial value for the stock of pollution (i.e., larger than the base case steady state), that is, $A_0 = 0.5$ and $P_0 = 1$. For a compact representation, the control variables are paired within one figure, with the emissions (expressed as a logarithmic function) along the X-axis and the restoration efforts along the Y-axis, both evolving over time.

The first important observation that emerges from Fig. 2.5 is that emissions and restoration efforts are substitutes rather than complements throughout the time path to the steady state, under both patience and impatience. The related policy implications are that efforts to reduce polluting emissions should be made along with efforts to replant trees.

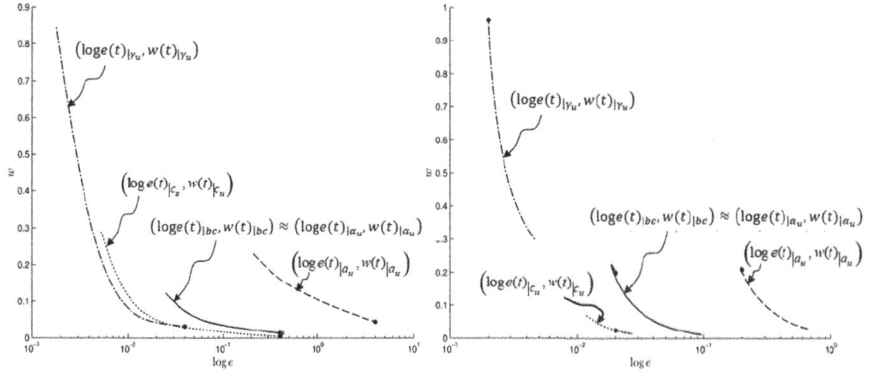

a. Transient paths under patient behaviour

b. Transient paths under impatient behaviour

Fig. 2.5 Transient paths of emissions rate and restoration effort under a patient (**a**) and impatient (**b**) behavior

Under patient behavior (5a), the optimal policy consists of initially low and increasing emissions and initially large and decreasing restoration efforts over time. This pattern is particularly pronounced for a greater destructive pollution impact, i.e., a greater value of γ. In contrast, a heavily discounted policy (5b) leads, in most cases—except the one with greater marginal pollution cost, c—to initially large and decreasing emissions and initially low and increasing restoration efforts over time. That is, except for a greater pollution cost coefficient impatient behavior implies an *inversion of the transient path* of the control policy that prevails under patient behavior. This result is driven by the fact that impatient behavior involves a higher valuation of both emissions and restoration costs in the short run than in the long run. A greater marginal pollution cost modifies this pattern, as it first imposes lowering the stock of pollution to allow for a partially inversed, patient policy. Finally, an increase in the self-regenerating capabilities of the environmental absorption efficiency, α, does not affect the time path of the control policy, neither under patient nor under impatient behavior.

In Fig. 2.6, we see that the patient behavior leads first to a reduction in the stock of pollution and then to an increase in the environmental absorption efficiency (6a). This leads to strengthening the biosphere as pollution sinks considerably, notably in the cases of greater pollution cost and revenue function coefficients. In contrast, the impatient behavior most often leads to a reduction in the environmental absorption efficiency and an increase in the stock of pollution either simultaneously or sequentially along a spiraling path that crosses the *Soylent Green* area over a finite but significant time interval (6b). During this interval, the biosphere acts as a source of pollution, which leads to a steady state much worse than the yet adverse initial state because the absorption efficiency is exhausted in a heavily polluted environment, including in the case of greater self-regenerating capabilities of the environmental absorption efficiency, α. An exception to this pattern is the case with a greater pollution cost coefficient, c, for which the spiraling saddle path finally results in a

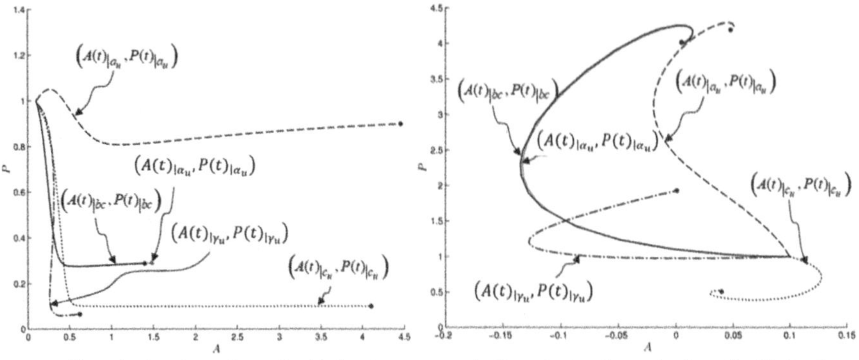

a. Transient paths under patient behaviour b. Transient paths under impatient behaviour

Fig. 2.6 Transient path of absorption efficiency and stock of pollution under a patient (**a**) and impatient (**b**) behavior

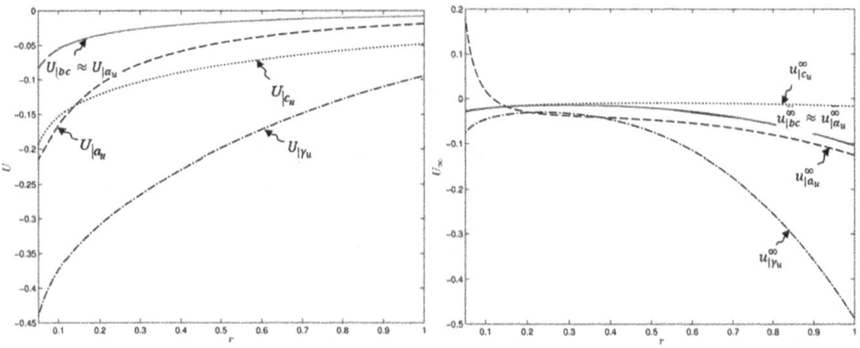

a. b.

Fig. 2.7 Sensitivity analysis of (**a**) overall and (**b**) steady-state utilities under patient and impatient behavior

reduction of both the environmental absorption efficiency and the stock of pollution. In contrast with the influence of the other parameters, the transient path in this case never enters the *Soylent Green* area, which makes a heavier valuation of the negative impact of the stock of pollution as the best antidote against the effects of impatience, because it avoids a transformation from a pollution sink to a source of pollution in the short run and ensures a positive absorption efficiency in a lightly polluted environment in the long run.

Finally, Fig. 2.7 describes the sensitivity of the overall utility (7a) and the steady state utility (7b) to the parameters' values in terms of discounting rate.

In Fig. 2.7a, the overall utility is lower under patient than under impatient behavior in all cases, mainly because of a difference in the valuation of revenues and costs between the short and long run. This difference in valuation comes from the fact that the overall utility under patient (impatient) behavior results from lower (greater) transient revenues from emissions and higher (lower) transient pollution and restoration costs than at the steady state where the corresponding values are smoothed. The difference in valuation between the short and the long run justifies a specific evaluation of the steady state utility, based on Eq. 2.27, along with the discounting rate (7b), for which an opposite pattern to that of the overall utility is obtained. That is, the steady-state utility is greater under patient than under impatient behavior in all cases. Note that, while a higher revenue function coefficient, a, is most beneficial in terms of steady-state utility under patient behavior, it is most detrimental under impatient behavior.

On the other hand, unless its magnitude is huge (e.g., 100 times the base case value), an increase in the self-regenerating capabilities of the absorption efficiency, α, does not mitigate the impact of impatience on the steady-state utility. This is consistent with our previous results. Actually, only an increase in the pollution cost coefficient, c, attenuates the effect of impatience on the steady-state utility. This result is driven by the fact that, though a greater discounting rate decreases the steady-state absorption efficiency, an increase in the pollution cost coefficient also reduces the stock of pollution. As a result, the pollution cost increases less than with any other parameter increase under impatient behavior, and the deterioration of the steady state utility is therefore limited. The conclusion is clear: impatient decision makers should evaluate more heavily the negative impact of pollution, rather than targeting an increase in revenue drawn from current emissions (e.g., through greater added value from emissions), and vice versa for the patient social planners.

Conclusions

In this paper, we take into account the case of self-regenerating capabilities of the environmental absorption efficiency. The proposed configuration, which accounts for the possibility that the natural regrowth of forests has an influence on the carbon terrestrial uptake (Pan et al., 2011), is optimistic as it may prevent the biosphere from transitioning from a pollution sink to a source including slightly negative initial environmental absorption efficiency if the stock of pollution is sufficiently low.

The comparison between the impact of self-regenerating capabilities of the environmental absorption efficiency and other exogenous influential parameters of the model suggests clear-cut results: we have shown that, unless their magnitude is huge, self-regenerating capabilities of the absorption efficiency are unlikely neither to improve the outcome and dynamics associated with patient decision-making, nor to prevent the negative effects associated with impatient decision-making, i.e., the biosphere temporarily turning into a source of pollution and a deteriorated steady state.

In contrast, an increase in the valuation of the negative impact of the stock of pollution seems to be more effective in avoiding a transition from a pollution sink to a source in the short run and in maintaining positive absorption efficiency in a lightly polluted environment in the long run. This result suggests that the method of imputation of the negative externalities of pollution should be rethought to reach a more exhaustive assessment of the cost of pollution, as exemplified by the Obama administration's current effort to quantify—though approximately—the social cost of pollution to include damages like rising sea levels, the increased number of Katrina-level hurricanes or sustained droughts (Wald, 2013). Such a change in the marginal pollution cost would imply greater robustness of the model outcomes and dynamics to variations of the discounting rate and impatience. The main implication is that the optimal policy consisting of initially low and increasing emissions and initially large and decreasing restoration efforts over time remains broadly valid including for extreme values of the discounting rate, with a relatively limited sacrifice in terms of overall and steady state utilities.

This paper contributed to improving consistency between the economic analysis of pollution and the scientific analyses on the evolution of the efficiency of natural sinks (Cramer et al., 2001; Joos et al., 2001; Lenton et al., 2006; Piao et al., 2008). We have examined the case of deterministic dynamics of the stock of pollution and the absorption capacity. It would be interesting to enrich the model by allowing a stochastic evolution of the stock of pollution (Benchekroun & Long, 2014) and/or the absorption capacity (Ayong Le Kama et al., 2014) for the case of a stochastic irreversible threshold of an ecosystem. The scope of our model could also be extended to the case where the impact of restoration efforts on the evolution of absorption efficiency depends on the level of absorption efficiency or pollution stock. This assumption would require multiplicative separability between control and state variables, which would allow the emergence of a Skiba threshold (Grass et al., 2008) with multiple steady states and lead to the determination of history-dependent policies. Another interesting research direction could involve adding deforestation activities as an additional policy instrument generating both emissions and revenue. Such extension would impose a more complex nonlinear structure that would essentially involve numerical methods of resolution.

Acknowledgements This research was supported by the Centre for Research of ESSEC Business School, the Canadian Social Sciences and Humanities Research Council (SSHRC), the Fonds Québécois de la Recherche sur la Société et la Culture [Quebec Fund of Cultural and Societal Research] (FQRSC), and the Austrian Science Fund (FWF) under grant No. P23084-N13. The paper was written while the first author was visiting the Department of Economics of McGill University in Canada, whose hospitality is gratefully acknowledged. The authors are grateful to one anonymous reviewer for constructive comments and suggestions on a previous draft. The usual disclaimer applies.

References

Ayong Le Kama, A., Pommeret, A., & Prieur, F. (2014). Optimal emission policy under the risk of irreversible pollution. *Journal of Public Economic Theory, 16*(6), 959–980.

Barnosky, A. D., Hadly, E. A., Bascompte, J., Berlow, E. L., Brown, J. H., Fortelius, M., Getz, W. M., Harte, J., Hastings, A., Marquet, P. A., Martinez, N. D., Mooers, A., Roopnarine, P., Vermeij, G., Williams, J. W., Gillespie, R., Kitzes, J., Marshall, C., Matzke, N., Mindell, D. P.,

Revilla, E., & Smith, A. B. (2012). Approaching a state shift in Earth's biosphere. *Nature, 486*, 52–58.

Benchekroun, H., & Long, N. V. (2014). Do increases in risk mitigate the tragedy of the commons? *Strategic Behavior and the Environment, 4*(1), 1–14.

Boucekkine, R., Pommeret, A., & Prieur, F. (2013a). Optimal regime switching and threshold effects. *Journal of Economic Dynamics and Control, 37*(12), 2979–2997.

Boucekkine, R., Pommeret, A., & Prieur, F. (2013b). Technological vs ecological switch and the environmental Kuznets curve. *American Journal of Agricultural Economics, 95*(2), 252–260.

Canadell, J. G., & Raupach, M. R. (2008). Managing forests for climate change. *Science, 320*(5882), 1456–1457.

Canadell, J. G., Le Quéré, C., Raupach, M. R., Field, C. B., Buitenhuis, E. T., Ciais, P., Conway, T. J., Gillett, N. P., Houghton, R. A., & Marland, G. (2007). Contributions to accelerating atmospheric CO_2 growth from economic activity, carbon intensity, and efficiency of natural sinks. *Proceedings of the National Academy of Sciences, 104*(47), 18866–18870.

Chevé, M. (2000). Irreversibility of pollution accumulation. *Environmental and Resource Economics, 16*(1), 93–104.

Cox, P. M., Betts, R. A., Jones, C., Spall, S. A., & Totterdell, I. (2000). Acceleration of global warming due to carbon-cycle feedbacks in a coupled climate model. *Nature, 408*, 184–187.

Cramer, W., Bondeau, A., Woodward, F. I., Prentice, I. C., Betts, R. A., Brovkin, V., Cox, P. M., Fisher, V., Foley, J. A., Friend, A. D., Kucharik, C., Lomas, M. R., Ramankutty, N., Sitch, S., Smith, B., White, A., & Young-Molling, C. (2001). Global response of terrestrial ecosystem structure and function to CO_2 and climate change: Results from six dynamic global vegetation models. *Global Change Biology, 7*(4), 357–373.

Dockner, E. (1985). Local stability in optimal control problems with two state variables. In G. Feichtinger (Ed.), *Optimal control theory and economic analysis* (Vol. 2). North Holland.

Dockner, E., & Feichtinger, G. (1991). On the optimality of limit cycles in dynamic economic systems. *Journal of Economics, 53*(1), 31–50.

El Ouardighi, F., Benchekroun, H., & Grass, D. (2014). Controlling pollution and environmental absorption capacity. *Annals of Operations Research, 220*(1), 111–134.

FAO (Food and Agriculture Organization). (2011). *Global forest land-use change from 1990 to 2005*. http://foris.fao.org/static/data/fra2010/RSS_Summary_Report_lowres.pdf

Grass, D. (2012). Numerical computation of the optimal vector field: Exemplified by a fishery model. *Journal of Economic Dynamics and Control, 36*(10), 1626–1658.

Grass, D., Caulkins, J. P., Feichtinger, G., Tragler, G., & Behrens, D. A. (2008). *Optimal control of nonlinear processes with applications in drugs, corruption, and terror*. Springer.

Holling, C. (1973). Resilience and stability of ecological systems. *Annual Review of Ecology and Systematics, 4*, 1–23.

IPCC (Intergovernmental Panel on Climate Change). (2007). *Fourth IPCC assessment report: Climate change 2007*. Cambridge University Press.

Joos, F., Prentice, I. C., Sitch, S., Meyer, R., Hooss, G., Plattner, G.-K., Gerber, S., & Hasselmann, K. (2001). Global warming feedbacks on terrestrial carbon uptake under the Intergovernmental Panel on Climate Change (IPCC) emission scenarios. *Global Biogeochemical Cycles, 15*(4), 891–907.

Keeler, E., Spence, M., & Zeckhauser, R. (1972). The optimal control of pollution. *Journal of Economic Theory, 4*(1), 19–34.

Lenton, T. M., Williamson, M. S., Edwards, N. R., Marsh, R., Price, A. R., Ridgwell, A. J., Shepherd, J. G., Cox, S. J., & The GENIE team. (2006). Millennial timescale carbon cycle and climate change in an efficient Earth system model. *Climate Dynamics, 26*(7/8), 687–711.

Michel, P., & Rotillon, G. (1995). Disutility of pollution and endogenous growth. *Environmental and Resource Economics, 6*, 279–300.

Moser, E., Seidl, A., & Feichtinger, G. (2013). History-dependence in production-pollution-trade-off models: a multi-stage approach. *Annals of Operations Research*, 1–23. https://doi.org/10.1007/s10479-013-1349-9

Pan, Y., Birdsey, R. A., Fang, J., Houghton, R., Kauppi, P. E., Kurz, W. A., Phillips, O. L., Shvidenko, A., Lewis, S. L., Canadell, J. G., Ciais, P., Jackson, R. B., Pacala, S., McGuire, A. D., Piao, S., Rautiainen, A., Sitch, S., & Hayes, D. (2011). A large and persistent carbon sink in the world's forests. *Science*, 1–10.

Piao, S. L., Ciais, P., Friedlingstein, P., Peylin, P., Reichstein, M., Luyssaert, S., Margolis, H., Fang, J. Y., Barr, A., Chen, A. P., Grelle, A., Hollinger, D. Y., Laurila, T., Lindroth, A., Richardson, A. D., & Vesala, T. (2008). Net carbon dioxide losses of northern ecosystems in response to autumn warming. *Nature, 451*, 49–52.

Prieur, F. (2009). The environmental Kuznets curve in a world of irreversibility. *Economic Theory, 40*(1), 57–90.

Prieur, F., Tidball, M., & Withagen, C. (2013). Optimal emission-extraction policy in a world of scarcity and irreversibility. *Resource and Energy Economics, 35*(4), 637–658.

Schaphoff, S., Lucht, W., Gerten, D., Sitch, S., Cramer, W., & Prentice, I. C. (2006). Terrestrial biosphere carbon storage under alternative climate projections. *Climatic Change, 74*(1/3), 97–122.

Scheffer, M. (1997). *The ecology of shallow lakes*. Kluwer Academic Publishers.

Stern, N. (2006). *Stern review report on the economics of climate change*. HM Treasury.

Tahvonen, O., & Withagen, C. (1996). Optimality of irreversible pollution accumulation. *Journal of Economic Dynamics and Control, 20*(9), 1775–1795.

UNFCCC (United Nations Framework Convention on Climate Change). (2008, December 10). *Reducing emissions from deforestation in developing countries: Approaches to stimulate action*. (UNFCCC/SBSTA/2008/L.23, Paragraphs 8 and 9).

Wald, M. L. (2013, June 18). New efforts to quantify 'social cost' of pollution. *New York Times*.

Zhao, M., & Running, S. W. (2010). Drought-induced reduction in global terrestrial net primary production from 2000 through 2009. *Science, 329*, 940–943.

Chapter 3
Transboundary Pollution Control and Environmental Absorption Efficiency Management

Fouad El Ouardighi, Konstantin Kogan, Giorgio Gnecco, and Marcello Sanguineti

Abstract In this paper, we suggest a two-player differential game model of transboundary pollution that accounts for time-dependent environmental absorption efficiency, which allows the biosphere to switch from a carbon sink to a source. We investigate the impact of negative externalities resulting from a transboundary pollution noncooperative game wherein countries are dynamically involved. Based on a linear-quadratic specification for the instantaneous revenue function, we assess transient path and steady-state differences between cooperative solution, open-loop, and Markov perfect Nash equilibria (MPNE). Regarding the methodological contribution of the paper, we suggest a particular structure of the conjectured value function to solve MPNE problems with multiplicative interaction between state variables in one state equation, so that third-order terms that arise in the Hamilton-Jacobi-Bellman equation are made negligible. Using a collocation procedure, we confirm the validity of the particular structure of the conjectured value function. The results suggest unexpected contrasts in terms of pollution control and environmental absorption efficiency management: (1) in the long run, an open-loop Nash equilibrium (OLNE) allows equivalent emissions to the social optimum but requires greater restoration efforts; (2) while an MPNE is likely to end up with lower emissions and greater restoration efforts than an OLNE, it has a much greater chance of falling in the emergency area; (3) the absence of cooperation and or precommitment becomes more costly as the initial absorption efficiency decreases; (4) more heavily

F. El Ouardighi (✉)
ESSEC Business School, Cergy Pontoise Cedex, France
e-mail: elouardighi@essec.edu

K. Kogan
Bar-Ilan University, Ramat-Gan, Israel

G. Gnecco
IMT – School for Advanced Studies, Lucca, Italy

M. Sanguineti
University of Genova, Genova, Italy

F. El Ouardighi (ed.), *Essays on Pollution Control in Economics and Management Science*, Lecture Notes in Operations Research,
https://doi.org/10.1007/978-3-031-78227-5_3

discounted MPNE strategies are less robust than OLNE to prevent irreversible switching of the biosphere from a carbon sink to a source.

Keywords Transboundary pollution · Environmental absorption efficiency · Hamilton-Jacobi-Bellman approximation

Introduction

Because natural CO_2 sinks absorb 55% of all anthropogenic carbon emissions, they represent "a huge subsidy to the global economy worth half a trillion US$ annually" (Canadell et al., 2007). However, the efficiency of natural sinks in absorbing CO_2 has decreased by 5% in the last 50 years, and this pattern is expected to continue (Canadell et al., 2007). One illustration of this trend is provided by the oceanic carbon sink, the decreased efficiency of which is believed to have caused a decline of up to 30% in the efficiency of the Southern Ocean sink over the last 20 years (Le Quéré et al., 2007). As consequences of this trend, greenhouse warming and the associated release of soil carbon favor a transition of the biosphere from a carbon sink to a source of pollution in the long run (Cox et al., 2000; Cramer et al., 2001; Joos et al., 2001; Lenton et al., 2006; Piao et al., 2008). According to Joos et al. (2001), "in the most extreme cases, the terrestrial biosphere becomes a source of carbon during the second half of the century."

In the absence of a transnational institution that can impose an environmental policy, and due to the lack of enforceable global coordination, it is likely that decentralized decision-making processes at the national level related to polluting emissions will aggravate the decline in the efficiency of carbon sinks.

With rare exceptions, the problem of declining environmental absorption efficiency has been disregarded in dynamic game models of transboundary pollution, which are typically based on the assumption of a linear environmental absorption function (e.g., Li, 2014; Benchekroun & Martin-Herrán, 2016; Fünfgelt & Schulze, 2016; Huang et al., 2016). More recently, a nonlinear absorption function has been proposed to account for declining environmental absorption efficiency (Mäler et al., 2003; Kossioris et al., 2008). This formulation, which combines the linear negative influence of the pollution stock with nonlinear positive feedback related to the release of past accumulated pollution, involves multiple steady-state equilibria with different stability properties. Although this change in perspective allows more accurate handling of potentially irreversible environmental degradation, it relies on the assumption of an instantaneous change in the absorption efficiency that precludes investing in a restoration effort whose impact is, indeed, not instantaneous.

Economic study of these issues is of prime importance because, just like emissions reduction, restoration of carbon sinks' efficiency is also a public good. With two superimposed free-rider problems, the failure of the reduction policy might also impede the restoration policy.

In this paper, we suggest a two-player differential game model of transboundary pollution that accounts for time-dependent environmental absorption efficiency, which allows switching of the biosphere from a carbon sink to a source, and the ability to invest in the restoration of environmental absorption efficiency. In this setup, we investigate the impact of negative externalities resulting from the transboundary pollution control noncooperative game wherein countries are dynamically involved. To do so, we assess differences related to both transient and steady states between cooperative and noncooperative pollution control and environmental absorption efficiency management.

To date, the theoretical debate on transboundary pollution control has led to the emergence of two main streams (cf. Jørgensen et al., 2010; Long, 2010; Benchekroun & Long, 2011). The first stream recommends that polluters commit to a predetermined plan of action over time whenever environmental absorption efficiency is linear (Van der Ploeg & de Zeeuw, 1991, 1992) or nonlinear (Mäler et al., 2003; Kossioris et al., 2008). The second stream of literature prescribes that, in the context of linear environmental absorption efficiency, polluters use softly discounted, nonlinear strategies to make decisions contingent on the current pollution level over time (Dockner & Long, 1993).[1]

In practice, only voluntary "contributions" to global emissions reduction, rather than commitments, were agreed to by the countries attending the successive International Conferences on the Climate since the implementation of the Kyoto Protocol in 2005. A recent conference, which took place in Warsaw in November 2013, illustrated how difficult it is to reach a global, enforceable agreement on emissions reduction. It is therefore important to analyze whether the absence of commitment at an international level on pollution control is likely to aggravate the decline in the efficiency of carbon sinks.

In this perspective, we compare the outcomes related to an open-loop Nash equilibrium (OLNE), which reflects the situation where polluters commit to a predetermined plan of action, and a Markov perfect Nash equilibrium (MPNE), which corresponds to the case where polluters make decisions contingent on the current state of the biosphere (Dockner et al., 2000; Long, 2010). To characterize an MPNE solution near its steady state, we use both a carefully designed local quadratic approximation, and, to get a comparison, a numerical approximation based on the collocation method for PDEs (Judd, 1998; Doraszelski, 2003). Setting the cooperative solution as a benchmark, we identify which decision rule, OLNE or MPNE, better prevents the durable or transient switching of the biosphere from a pollution sink to a source, if any. The results show that the players' transient behavior has a critical impact on the resulting steady state in terms of pollution stock and environmental absorption efficiency. More importantly, unexpected contrasts are found between the cooperative and noncooperative steady states.

The paper proceeds as follows. Second section develops the differential game model and its properties. In third and fourth sections, we derive the cooperative and open-loop Nash equilibria, respectively. In fifth section, we study the Markov

[1]This prescription, which results in multiple equilibria, is confirmed by Rowat (2006), moderated by Rubio and Casino (2002), and opposed by Wirl (2007).

perfect Nash equilibrium for a locally quadratic value function and evaluate its accuracy. In sixth section, we compare the results. Seventh section concludes the paper.

Model Formulation

We extend the nonlinear quadratic model in El Ouardighi et al. (2014) to a differential game of transboundary pollution. In the setup of a two-player game where players are, e.g., nations, we interpret the biosphere as the place of interaction between pollution and environmental absorption efficiency. The pollution stock is supposed to increase with anthropogenic emissions and depreciate at an environmental absorption efficiency rate. The pollution stock has a destructive impact on absorption efficiency, which in turn increases the pollution stock. In contrast with the existing literature, we model environmental absorption efficiency as a state variable, which allows for negative absorption efficiency that would stimulate pollution even in the absence of anthropogenic emissions. This is in line with scientific evidence that the biosphere may switch from a carbon sink to a source (e.g., Cox et al., 2000; Canadell et al., 2007). To counteract such a switch, we consider two policy instruments: anthropogenic emissions and efforts to restore environmental absorption efficiency.

The rate of change of the pollution stock over time is written as:

$$\dot{P}(t) = \sum_i e_i(t) - A(t)P(t) \qquad P(0) = P_0 > 0 \tag{3.1}$$

where $P(t) \geq 0$ denotes the pollution stock at time t, $e_i(t) \geq 0$, $i = 1, 2$, player i's current emissions, and $A(t)$ environmental absorption efficiency.[2]

Further, the rate of change of environmental absorption efficiency over time is:

$$\dot{A}(t) = \sum_i w_i(t) - \gamma P(t) \qquad A(0) = A_0 \geq 0 \tag{3.2}$$

$w_i(t) \geq 0$ being player i's effort to restore environmental absorption efficiency, $i = 1$, 2, and $\gamma > 0$ the marginal destructive impact of pollution on the evolution of absorption efficiency. The efforts to restore environmental absorption efficiency may include reforestation (UNFCC, 2008), and geo-engineering technologies (such as iron fertilization, biochar), which aim at accelerating natural processes of carbon removal (Vaughan & Lenton, 2011).

The formulation in Eqs. 3.1–3.2 includes the pollution accumulation model with linear absorption efficiency as a special case if environmental absorption efficiency

[2]By an application of Gronwall's lemma to the differential inequality $\dot{P}(t) \geq -A(t)P(t)$ which is implied by Eq. 3.1, one obtains, indeed, $P(t) \geq 0$, $\forall t \geq 0.9$.

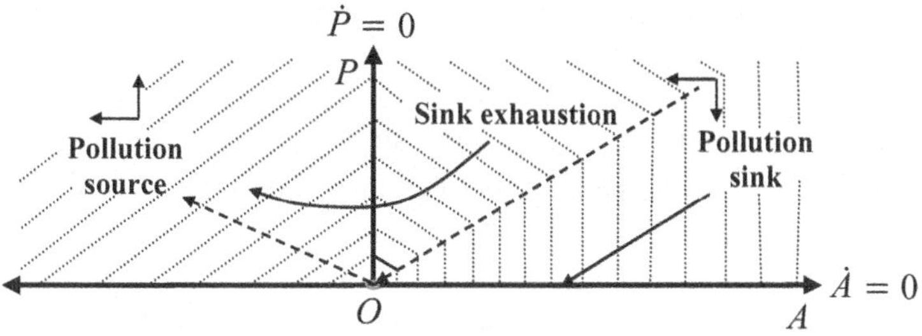

Fig. 3.1 Pollution-absorption dynamics without human interference

is totally insensitive to the destructive impact of the pollution stock, $\gamma = 0$, and there is no restoration effort of environmental absorption efficiency, $w_i(t) = 0, \forall t \geq 0$.[3] The formulation in Eqs. 3.1–3.2 also extends the pollution accumulation model with zero absorption efficiency if environmental absorption efficiency is zero over the whole planning horizon, $A(t) = 0, \forall t \geq 0$ (El Ouardighi et al., 2016).

Assuming that there is no anthropogenic intervention (which corresponds to the pre-industrial era), over an infinite horizon, i.e., $e_i(t) = w_i(t) = 0, \forall t \geq 0, i = 1, 2$, the properties of the system Eqs. 3.1–3.2 can be described in the phase diagram below. The system admits infinite steady states on the isocline $\dot{A} = 0$. The eigenvalues associated with such a system are $-A_\infty$ and 0, which means that the positive horizontal axis is a sink, and the negative horizontal axis is a source.

Based on the steady state $(A_\infty, P_\infty) = (0, 0)$, the phase diagram can be divided into three regions (Fig. 3.1):

- A sustainability area, where the initial environmental absorption capacity acts as a sink for any pollution stock starting on the right side of the (dashed) saddle path.
- An emergency area, where any path starting on the left side of the (dashed) saddle path and the right side of the $\dot{P} = 0$ isocline, though declining, tends to exhaust the pollution sink.
- An irreversible area (the so-called *Soylent Green* scenario), on the left side of the $\dot{P} = 0$ isocline, where the environment becomes a source of formerly stored pollution.

The phase diagram shows that a large initial pollution stock combined with sufficiently low initial environmental absorption efficiency is a sufficient condition for switching from a pollution sink to a source.

Regarding the objective functions, we assume that each player seeks to maximize its intertemporal utility, U_i. In this regard, we make the following assumptions. First, each player's instantaneous revenue is a concave function of its own current rate of emissions, $e_i(t)\left(a_i - \frac{e_i(t)}{2}\right)$, where $2a_i > 0$ is an upper bound for emissions, $i = 1$, 2. This choice comes from the fact that the quadratic revenue function has been often

[3] For more details, see El Ouardighi et al. (2014).

used in dynamic game models with a linear absorption function (Rubio & Casino, 2002; Wirl, 2007).[4] The instantaneous social cost incurred by each player due to the pollution stock is a quadratic function, $\frac{c_i P(t)^2}{2}$, where $c_i > 0$ is a pollution cost coefficient, $i = 1, 2$ (Rubio & Casino, 2002; Wirl, 2007; Boucekkine et al., 2013). Finally, each player incurs an instantaneous increasing convex cost of effort to restore environmental absorption efficiency, $\frac{f_i w_i(t)^2}{2}$, where $f_i > 0$ is a restoration cost coefficient, $i = 1, 2$ (El Ouardighi et al., 2014, 2016).

Letting $r_i > 0$ denote player i's discounting rate, and assuming an infinite planning horizon, player i's objective criterion is written as:

$$\underset{e_i(t),\, w_i(t) \geq 0}{\text{Max}} U_i = \int_0^{+\infty} e^{-r_i t} \left[e_i(t) \left(a_i - \frac{e_i(t)}{2} \right) - \frac{c_i P(t)^2}{2} - \frac{f_i w_i(t)^2}{2} \right] \mathrm{d}t \tag{3.3}$$

subject to Eqs. 3.1–3.2, $e_i(t) \geq 0$ and $w_i(t) \geq 0$, $\forall t \geq 0$, $i = 1, 2$. In the remainder of the paper, we set $f_i = 1$, $i = 1, 2$, without loss of generality. Further, invoking the symmetry assumption, which is standard in the literature on pollution control (e.g., Jørgensen et al., 2010; Long, 2010), we use: $a_i \equiv a$, $c_i \equiv c$, and, $r_i \equiv r$, $i = 1, 2$.

Cooperative Solution

Cooperative solution strategies require that the players align their respective interests and maximize the overall utility over the planning horizon, that is:

$$U = \int_0^{+\infty} e^{-rt} \left[\sum_i e_i(t) \left(a - \frac{e_i(t)}{2} \right) - cP(t)^2 - \sum_i \frac{w_i(t)^2}{2} \right] \mathrm{d}t \tag{3.4}$$

Accordingly, the current-value Hamiltonian is written as:

$$H = \sum_i e_i \left(a - \frac{e_i}{2} \right) - cP^2 - \sum_i \frac{w_i^2}{2} + \eta_1 \left(\sum_i e_i - AP \right) + \eta_2 \left(\sum_i w_i - \gamma P \right) \tag{3.5}$$

where $\eta_j \equiv \eta_j(t)$ are (current value) costate variables, $j = 1, 2$.

[4]Unlike most game studies with nonlinear absorption functions, our study does not use a logarithmic revenue function (Mäler et al., 2003; Kossioris et al., 2008; Dockner & Wagener, 2014). Note that a logarithmic function provides quite similar results to those obtained with our quadratic function.

Assuming an interior solution and applying Pontryagin's principle in the discounted case, the necessary conditions for a cooperative solution are:

$$H_{e_i} = a - e_i + \eta_1 = 0 \Rightarrow e_i = a + \eta_1 \tag{3.6}$$

$$H_{w_i} = -w_i + \eta_2 = 0 \Rightarrow w_i = \eta_2 \tag{3.7}$$

$i = 1, 2$, where the evolution of the costate variables is given by:

$$\dot{\eta}_1 = 2cP + (r + A)\eta_1 + \gamma\eta_2 \tag{3.8}$$

$$\dot{\eta}_2 = r\eta_2 + \eta_1 P \tag{3.9}$$

Equations 3.8 and 3.9 confirm that the players' equilibrium strategies are symmetric.

Using Eqs. 3.6–3.7 in Eqs. 3.1–3.2, we obtain:

$$\dot{P} = 2(a + \eta_1) - AP \tag{3.10}$$

$$\dot{A} = 2\eta_2 - \gamma P \tag{3.11}$$

Let us assume a steady state to study optimal control at this state. Assuming that $\dot{P} = \dot{A} = 0$ gives $\sum_i e_i = AP$ and $\sum_i w_i = \gamma P$. Differentiating these equations with respect to time and accounting for $\dot{P} = \dot{A} = 0$ yields steady efforts $\dot{e}_i = \dot{w}_i = 0$, $i = 1$, 2. Differentiating Eqs. 3.6–3.7 over any time interval and accounting for $\dot{e}_i = \dot{w}_i = 0$, $i = 1, 2$, the costate variables are time-invariant. This implies that if the pollution stock and the absorption efficiency are both steady, then the related costate variables are also steady, and each player performs time-invariant efforts.

Proposition 1. There exists a nontrivial cooperative steady state that is unique, given by:

$$[P_\infty \; A_\infty \; e_{i\infty} \; w_{i\infty}]^T = \left[\frac{r^2\gamma + h}{2(4c + \gamma^2)} \quad \frac{2(2a - r\gamma)(4c + \gamma^2)}{r^2\gamma + h} \quad \frac{2a - r\gamma}{2} \quad \frac{\gamma(r^2\gamma + h)}{4(4c + \gamma^2)}\right]^T \tag{3.12}$$

where $h \equiv \sqrt{r^4\gamma^2 + 4r\gamma(2a - r\gamma)(4c + \gamma^2)} > 0$, $i = 1, 2$.

Proof See Appendix A1.

The cooperative solution allows for a feasible steady-state emissions rate compatible with the preservation of a strictly positive environmental absorption rate if $a > r\gamma/2$. In Eq. 3.12, we observe that a lower discounting rate should result in a higher steady-state emissions rate $e_{i\infty}$ and environmental absorption efficiency A_∞ as well as lower pollution stock P_∞ and restoration efforts $w_{i\infty}$. Note that there exists a trivial steady-state solution characterized by zero emissions, pollution stock and restoration efforts and negative absorption efficiency, i.e.,

$\begin{bmatrix} P'_\infty & A'_\infty & e'_{i\infty} & w'_{i\infty} \end{bmatrix}^T = \begin{bmatrix} 0 & -r & 0 & 0 \end{bmatrix}^T$. This solution is different from the irreversible solution found in Tahvonen and Withagen (1996), which implies a "high" pollution concentration that can be optimal. In our model, this second steady state is unstable and therefore cannot be optimal.

Corollary 1. At the steady state, the instantaneous cooperative utility is:

$$u_\infty = \frac{4a^2 - r^2\gamma^2}{8} - \frac{(r^2\gamma + h)^2}{16(4c + \gamma^2)} \tag{3.13}$$

Proof See Appendix A2.

The expression in Eq. 3.13 characterizes the utility of future generations in the cooperative case. It can be shown that $\frac{\partial u_\infty}{\partial r} < 0$, which implies that a larger discounting rate results in a lower utility of future generations under cooperation. Similarly, as $\frac{\partial u_\infty}{\partial \gamma} < 0$, more vulnerable environmental absorption efficiency negatively affects the utility of future generations.

Proposition 2. The cooperative steady state is a saddle-point and the optimal path to the steady state is either monotonic or oscillatory.

Proof See Appendix A3.

That is, the saddle path can either converge monotonically or oscillate toward the steady state. This result is consistent with previous findings where a clockwise spiraling path can go through the *Soylent Green* area, with negative environmental absorption efficiency during a finite time interval (El Ouardighi et al., 2014, 2016).

Lemma 1. The saddle paths of the control and state variables near the cooperative steady state are:

$$e_i(t) = a + \eta_{1\infty} + B_1 e^{\chi_1 t} + B_2 e^{\chi_3 t} \tag{3.14}$$

$$w_i(t) = w_{i\infty} + B_3 e^{\chi_1 t} + B_4 e^{\chi_3 t} \tag{3.15}$$

$$P(t) = P_\infty + B_5 e^{\chi_1 t} + B_6 e^{\chi_3 t} \tag{3.16}$$

$$A(t) = A_\infty + B_7 e^{\chi_1 t} + B_8 e^{\chi_3 t} \tag{3.17}$$

where $B_1 \ldots B_8$ are constants of integration and χ_1, χ_3 are eigenvalues with negative real part.

Proof See Appendix A4.

Equations 3.14–3.17 are useful to characterize the transient paths for the control and state variables in the cooperative solution.

Open-Loop Nash Equilibrium

An analysis of the OLNE strategy, which can be justified whenever the players are capable of precommitting themselves to announced time paths, offers valuable insights into the impact of negative externalities resulting from the transboundary noncooperative game of pollution control and environmental absorption efficiency management. In addition, open-loop strategies provide a useful benchmark for subsequent assessment of the strategic effects related to Markovian strategies (Fudenberg & Levine, 1988).

Using the objective function in Eq. 3.3, the current-value Hamiltonian of player i is written as:

$$H^i = e_i\left(a - \frac{e_i}{2}\right) - \frac{cP^2}{2} - \frac{w_i^2}{2} + \lambda_{i1}(e_1 + e_2 - AP) + \lambda_{i2}(w_1 + w_2 - \gamma P) \tag{3.18}$$

where $\lambda_{ij} \equiv \lambda_{ij}(t)$ are player i's (current value) costate variables, $i = 1, 2, j = 1, 2$.

Assuming an interior solution and applying Pontryagin's principle in the discounted case, player i's necessary conditions for OLNE are:

$$H^i_{e_i} = a - e_i + \lambda_{i1} = 0 \Rightarrow e_i^{on} = a + \lambda_{i1} \tag{3.19}$$

$$H^i_{w_i} = -w_i + \lambda_{i2} = 0 \Rightarrow w_i^{on} = \lambda_{i2} \tag{3.20}$$

$i = 1, 2$, where the superscript "*on*" stands for open-loop Nash equilibrium.

The evolution of the costate variables is given by:

$$\dot{\lambda}_{i1} = cP + (r + A)\lambda_{i1} + \gamma\lambda_{i2} \tag{3.21}$$

$$\dot{\lambda}_{i2} = r\lambda_{i2} + \lambda_{i1}P \tag{3.22}$$

In the present case, the costate variables of both players coincide; that is, $\lambda_{i1} \equiv \lambda_1$ and $\lambda_{i2} \equiv \lambda_2$, $i = 1, 2$, and Eqs. 3.21 and 3.22 confirm that the players' equilibrium strategies are symmetric. Therefore, we are left with only two costate equations.

Using Eqs. 3.21–3.22, 3.1–3.2, we obtain:

$$\dot{P} = 2(a + \lambda_1) - AP \tag{3.23}$$

$$\dot{A} = 2\lambda_2 - \gamma P \tag{3.24}$$

Note that here also, if the pollution stock and the absorption efficiency are both steady, then each player performs time-invariant control efforts.

Proposition 3. There exists a nontrivial OLNE steady state that is unique, given by:

$$\begin{bmatrix} P_{\infty}^{on} & A_{\infty}^{on} & e_{i\infty}^{on} & w_{i\infty}^{on} \end{bmatrix}^T = \begin{bmatrix} \frac{r^2\gamma + k}{2(2c+\gamma^2)} & \frac{2(2a-r\gamma)(2c+\gamma^2)}{r^2\gamma+k} & \frac{2a-r\gamma}{2} & \frac{\gamma(r^2\gamma+k)}{4(2c+\gamma^2)} \end{bmatrix}^T \tag{3.25}$$

where $k \equiv \sqrt{r^4\gamma^2 + 4r\gamma(2a - r\gamma)(2c + \gamma^2)} > 0$, $i = 1, 2$.

Proof See Appendix A5.

Therefore, precommitment strategies also allow a feasible steady-state emissions rate compatible with the preservation of a strictly positive environmental absorption rate under the feasibility condition $a > r\gamma/2$. In Eq. 3.25, a lower discounting rate also increases the steady-state emissions rate $w_{i\infty}^{on}$ and environmental absorption efficiency A_{∞}^{on}, and lowers the pollution stock P_{∞}^{on} and restoration efforts $w_{i\infty}^{on}$. Here also, there exists a trivial steady-state solution such that $\begin{bmatrix} P_{\infty}^{on\prime} & A_{\infty}^{on\prime} & e_{i\infty}^{on\prime} & w_{i\infty}^{on\prime} \end{bmatrix}^T = \begin{bmatrix} 0 & -r & 0 & 0 \end{bmatrix}^T$, which is diverging and therefore not optimal.

Corollary 2. At the steady state, player i's OLNE instantaneous utility is:

$$u_{i\infty}^{on} = \frac{4a^2 - r^2\gamma^2}{8} - \frac{(4c + \gamma^2)(r^2\gamma + k)^2}{32(2c + \gamma^2)^2} \tag{3.26}$$

Proof See Appendix A6.

The expression in Eq. 3.26 represents the utility of future generations from the perspective of player i in the OLNE case. It has the same properties as in the cooperative case regarding the influence of the discounting rate, r, and the marginal impact of pollution on the evolution of absorption efficiency, γ.

Proposition 4. The OLNE steady state has the saddle-point property and the optimal path to the steady state is either monotonic or oscillatory.

Proof See Appendix A7.

As in the cooperative setting, the saddle path can either converge monotonically or follow an oscillatory trajectory toward the steady state.

Lemma 2. The saddle paths of the control and state variables near the OLNE steady state are:

$$e_i^{on}(t) = a + \lambda_{1\infty} + D_1 e^{\xi_1 t} + D_2 e^{\xi_3 t} \tag{3.27}$$

$$w_i^{on}(t) = w_{i\infty}^{on} + D_3 e^{\xi_1 t} + D_4 e^{\xi_3 t} \tag{3.28}$$

$$P(t) = P_{\infty}^{on} + D_5 e^{\xi_1 t} + D_6 e^{\xi_3 t} \tag{3.29}$$

$$A(t) = A_{\infty}^{on} + D_7 e^{\xi_1 t} + D_8 e^{\xi_3 t} \tag{3.30}$$

where $D_1 .. D_8$ are constants of integration and ξ_1, ξ_3 are eigenvalues with negative real part.

Proof See Appendix A8.

Equations 3.27–3.30 enable the computation of transient paths for the control and state variables in an OLNE.

Markov Perfect Nash Equilibrium

The MPNE has the desirable property of subgame perfectness because current actions depend on the current state vector and time, and not only on time, as OLNE does. However, the nonlinear quadratic structure of the game makes the searches for an analytical solution of MPNE more difficult than for usual linear quadratic models. The main reason for this is that the two-state variables interact multiplicatively in the one-state equation, which involves intricate third-order terms in the resolution process. Thus, the design of an approximation for the MPNE requires particular attention.

Using the objective function in Eq. 3.3, the Hamilton-Jacobi-Bellman (HJB) equations are:

$$rV^i = \underset{e_i, w_i}{\text{Max}} \left[e_i\left(a - \frac{e_i}{2}\right) - \frac{cP^2}{2} - \frac{w_i^2}{2} + V_P^i(e_1 + e_2 - AP) + V_A^i(w_1 + w_2 - \gamma P) \right] \tag{3.31}$$

where $V^i(A, P)$ is player i's value function, $i = 1, 2$.

Player i's necessary conditions for Markov perfect Nash equilibrium in the case of an interior solution are written as:

$$e_i^{mp} = a + V_P^i(A, P) \tag{3.32}$$

$$w_i^{mp} = V_A^i(A, P) \tag{3.33}$$

After substituting the necessary conditions of optimality into the HJB equations, we obtain:

$$rV^i = \left(a + V_P^i\right)^2/2 - cP^2/2 + \left(V_A^i\right)^2/2 + V_P^i\left(V_P^j - AP\right) + V_A^i\left(V_A^j - \gamma P\right) + aV_P^i \tag{3.34}$$

$i, j = 1, 2, i \neq j$. Invoking the symmetry assumption, we write: $V^i = V$, $V_P^i = V_P$ and $V_A^i = V_A$, $i = 1, 2$, which leads us to rewrite Eq. 3.34 as follows:

$$rV^i = a^2/2 + (3V_P/2 + 2a)V_P + 3(V_A)^2/2 - cP^2/2 - \gamma V_A P - V_P AP \tag{3.35}$$

In linear-quadratic transboundary pollution models, a quadratic approximation of the value function for the evaluation of the MPNE is usually considered (see, e.g., Long (2010), Sect. 1.2.4). In the context of our nonlinear quadratic model, assuming a local quadratic approximation of the form, with real parameters $\alpha_1..\alpha_6$, that is:

$$V = \alpha_1 + \alpha_2 A + \alpha_3 P + \alpha_4 A^2 + \alpha_5 P^2 + \alpha_6 AP$$

may however not be relevant because it would lead to the inclusion of two third-order terms, respectively proportional to A^2P and AP^2, in the approximation of the HJB Eq. 3.35, that are likely non-negligible near the steady state. So, some care is needed to design the local quadratic approximation, to make such third-order terms locally negligible.

Assumption 1. We consider for the value function a local quadratic approximation with real parameters $\alpha_1..\alpha_6$, that is:

$$\begin{aligned} V = {} & \alpha_1 + \alpha_2\left(A - A_\infty^{mp}\right) + \alpha_3\left(P - P_\infty^{mp}\right) + \alpha_4\left(A - A_\infty^{mp}\right)^2 \\ & + \alpha_5\left(P - P_\infty^{mp}\right)^2 + \alpha_6\left(A - A_\infty^{mp}\right)\left(P - P_\infty^{mp}\right) \end{aligned} \tag{3.36}$$

where A_∞^{mp} and P_∞^{mp} are the steady state values (to be determined) of A and P for the MPNE. We assume that the parameters of the problem are such that the steady-state values of all the state and control variables are positive. Note that the steady state utility, or social welfare of future generations, is approximated by $r\alpha_1$.

By substituting the expression in Eq. 3.36 and its respective partial derivatives in Eq. 3.35, two third-order terms, respectively proportional to $\left(A - A_\infty^{mp}\right)^2\left(P - P_\infty^{mp}\right)$ and $\left(A - A_\infty^{mp}\right)\left(P - P_\infty^{mp}\right)^2$, appear in the approximation of the HJB equation, which are clearly negligible near the steady state and can therefore be neglected. Though this is sufficient to ensure the accuracy of our local quadratic approximation in Eq. 3.36, additional justifications based on a numerical procedure are provided hereafter; they compare this local approximation with the one obtained by a different procedure.

Proposition 5. Under Assumption 1, the MPNE strategies near the steady state are given by:

$$e^{mp} = a + \alpha_3 + 2\alpha_5\left(P - P_\infty^{mp}\right) + \alpha_6\left(A - A_\infty^{mp}\right) \tag{3.37}$$

$$w^{mp} = \alpha_2 + 2\alpha_4\left(A - A_\infty^{mp}\right) + \alpha_6\left(P - P_\infty^{mp}\right) \tag{3.38}$$

and the MPNE steady state is given by:

$$\begin{bmatrix} P_\infty^{mp} & A_\infty^{mp} & e_\infty^{mp} & w_\infty^{mp} \end{bmatrix}^T = \begin{bmatrix} 2\alpha_2/\gamma & \gamma(a+\alpha_3)/\alpha_2 & a+\alpha_3 & \alpha_2 \end{bmatrix}^T \tag{3.39}$$

Proof See Appendix A9.

From Eqs. 3.38 and 3.39, we note that both emissions and restoration effort MPNE strategies are locally linear. However, $\alpha_2 > 0$ and $a < \alpha_3 < 0$ are required for a feasible steady-state solution. The steady-state solution involved by Eq. 3.39, if any, should be globally asymptotically stable for the linearized dynamical system presented in the proof of Proposition 5 (see Appendix A9). The Jacobian matrix associated with the MPNE-controlled system is:

$$J^{mp} = \begin{bmatrix} 4\alpha_5 - A_\infty^{mp} & 2\alpha_6 - P_\infty^{mp} \\ 2\alpha_6 - \gamma & 4\alpha_4 \end{bmatrix} \tag{3.40}$$

Using Eq. 3.39, we get the trace and determinant of the Jacobian matrix, that is, respectively:

$$TrJ^{mp} = 4(\alpha_4 + \alpha_5) - \frac{\gamma(a+\alpha_3)}{\alpha_2}, |J^{mp}| = 2\left\{2\alpha_4\left[4\alpha_5 - \frac{\gamma(a+\alpha_3)}{\alpha_2}\right] - (2\alpha_6 - \gamma)\left(\alpha_6 - \frac{\alpha_2}{\gamma}\right)\right\}$$

The global asymptotic stability property imposes that $Tr\, J^{mp} < 0$ and $|J^{mp}| > 0$, and that both the eigenvalues $\psi_{1,\,2}$ of the Jacobian matrix, that is:

$$\psi_{1,2} = \frac{\gamma[4\alpha_2(\alpha_4 + \alpha_5) - \gamma(a+\alpha_3)] \pm \sqrt{\gamma^2[4\alpha_2(\alpha_4 - \alpha_5) + \gamma(a+\alpha_3)]^2 + 8\gamma\alpha_2^2(2\alpha_6 - \gamma)(\gamma\alpha_6 - \alpha_2)}}{2\gamma\alpha_2}$$

have negative real parts.

Because our feedback solution is approximated around a small neighborhood of the steady state, convergence to the actual solution, if it exists, should be quite satisfactory. To confirm the accuracy of the quadratic model adopted in Assumption 1, we also use an alternative approach based on the collocation method for PDEs (e.g., Judd, 1998; Doraszelski, 2003; Dawid et al., 2015; Jaakkola, 2015) to compute the MPNE strategies numerically (see Appendix A10). For the following constellation of parameters, that is:

$$r = 0.075, \gamma = 0.05, a = 0.006, c = 0.035,$$

we used the approach described in Appendix A9 and identified two MPNE strategies, each of them resulting in a globally asymptotically stable steady state for the associated linearized system. We selected the most profitable Nash equilibrium strategy according to the objective criterion in Eq. 3.3. The selected MPNE is a more profitable (higher) equilibrium because it is characterized by greater environmental absorption efficiency and lower pollution stock at the steady state than the less profitable (lower) MPNE. Note that, in practice, convergence to the higher equilibrium is not granted, as the lower equilibrium may well arise as a poverty trap (e.g., Azariadis & Stachurski, 2005) if the players are unable to coordinate. Therefore, a coordination device on the higher equilibrium is needed to escape from the poverty trap and therefore allow for an increase in welfare to both players. The coefficient values of the local quadratic approximation of the value function for the higher equilibrium are given in Table 3.1.

The corresponding steady state is given by:

$$\left(P_\infty^{mp}, A_\infty^{mp}, e_\infty^{mp}, w_\infty^{mp}\right)$$
$$= \left(2.880045 \cdot 10^{-2}, 1.978042 \cdot 10^{-1}, 2.848426 \cdot 10^{-3}, 7.200114 \cdot 10^{-4}\right)$$

along with the corresponding negative eigenvalues:

$$(\psi_1, \psi_2) = \left(-3.103781 \cdot 10^{-1},\ -5.767552 \cdot 10^{-3}\right)$$

By numerically comparing the local quadratic approximation of the value function on a neighborhood of the steady state solution with its local approximation obtained using the collocation method on the same neighborhood (see Appendix A10) with *Matlab*, a very small difference between the two approximations is obtained, as shown in Fig. 3.2.

Figure 3.3 compares the two approximations in terms of the relative error of the quadratic approximation with respect to the one obtained by the collocation method. Such a relative error is defined as the absolute value of their difference, divided by the average on the neighborhood of the absolute value of the approximation obtained by the collocation method (the average of the absolute value is introduced to avoid an infinite relative error in case of zero crossings of the approximation). As one can see from the figure, a negligible relative error is obtained, which suggests that the initial conjectured quadratic approximation was so good that only a negligible update was performed by the collocation method to reach the default tolerance.[5]

For the local quadratic approximation, Fig. 3.4 shows the Bellman residual error, defined as:

$$rV - \left[a^2/2 + (3V_P/2 + 2a)V_P + 3(V_A)^2/2 - cP^2/2 - \gamma V_A P - V_P AP\right]$$

[5]The result depends partly on the neighborhood size. For larger sizes, a larger relative error is usually obtained.

Table 3.1 Coefficient values of the quadratic approximation of the value function for the higher equilibrium

α_1	α_2	α_3	α_4	α_5	α_6
$-2.321435 \cdot 10^{-5}$	$7.200114 \cdot 10^{-4}$	$-3.151574 \cdot 10^{-3}$	$-1.602299 \cdot 10^{-3}$	$-2.798305 \cdot 10^{-2}$	$1.093377 \cdot 10^{-2}$

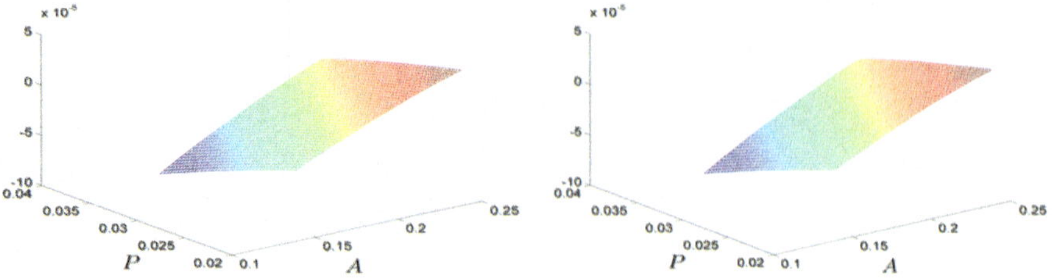

a. b.

Fig. 3.2 Compared approximations of the value function with (**a**) local quadratic conjecture and (**b**) collocation method

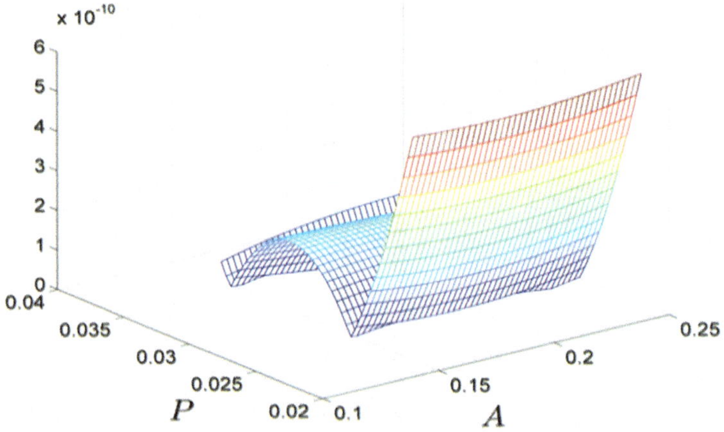

Fig. 3.3 Relative error of the local quadratic approximation of the value function with respect to the local approximation obtained with the collocation method

and provides a measure of the quality of the approximation, as it expresses the pointwise violation of the HJB Eq. 3.35. The figure shows a very small Bellman residual error. Moreover, for the local quadratic approximation, when it is computed in $\left(A_{\infty}^{mp}, P_{\infty}^{mp}\right)$, the absolute value of the ratio between the Bellman residual error and the left-hand side of the HJB Eq. 3.35 is $3.4766 \cdot 10^{-7}$. Similar results hold for the approximation by the collocation method.

Finally, for the same choice of the problem parameters, we compare the value α_1 assumed in $\left(A_{\infty}^{mp}, P_{\infty}^{mp}\right)$ by the local quadratic approximation of the value function, with the expression:

$$\left[e_{\infty}^{mp}\left(a - e_{\infty}^{mp}/2\right) - c\left(P_{\infty}^{mp}\right)^2/2 - \left(w_{\infty}^{mp}\right)^2/2\right]/r$$

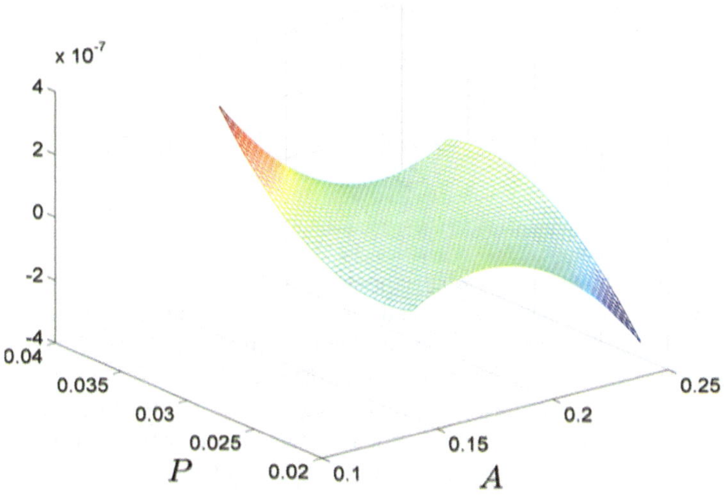

Fig. 3.4 Bellman residual error for the local quadratic approximation of the value function

that is obtained by replacing in Eq. 3.3 the resulting approximation of the steady state solution for the MPNE: the two values so obtained ($-2.321435 \cdot 10^{-5}$ and $-2.321434 \cdot 10^{-5}$, respectively) are practically coincident. The results obtained clearly support the validity of the local quadratic approximation in Assumption 1.[6]

Comparative Analysis

In this section, we determine the main differences between the cooperative, OLNE and approximated MPNE strategies in terms of steady-state and transient paths. To do so, we use both analytical and numerical results.

Based on our analytical results for the cooperative and OLNE cases, we first set some important comparisons.

Proposition 6. It holds that:

$$e_{i\infty}^{on} = e_{i\infty} \tag{3.41}$$

[6]Figures 3.2–3.4, as the final check of the validity of the local quadratic approximation, refer to a specific choice of the parameter values inside the ranges considered in Table 3.1, for which it was possible to find a solution of the nonlinear system in Proposition 7 that also satisfies the feasibility requirements provided in Appendix A9. However, the robustness of our approach was also confirmed by the results obtained for other choices of the parameter values inside such ranges, for which a feasible solution was obtained.

$$P_\infty^{on} > P_\infty \tag{3.42}$$

$$w_{i\infty}^{on} > w_{i\infty} \tag{3.43}$$

$$A_\infty^{on} < A_\infty \tag{3.44}$$

$$\sum_i u_{i\infty}^{on} = u_\infty \tag{3.45}$$

Proof See Appendix A10.

That is, the steady-state global emissions in the cooperative and the OLNE settings are equal. The same conclusion can be obtained with a logarithmic revenue function of the current rate of emissions, that is, $a_i \ln e_i$, $a_i > 0$, $i = 1, 2$, and with a linear pollution cost, that is, $cP(t)$, which makes it a robust result in that it is driven neither by the functional form of revenue function nor by that of pollution costs. Actually, this result is due to the particular transition equations assumed, which allow the switching of the biosphere from a carbon sink to a source, and the ability to invest in the restoration of environmental absorption efficiency. The main implication of Eq. 3.41 is that the OLNE steady state imposes no sacrifice on future generations in terms of emissions compared with the cooperative steady state. This result departs from other transboundary pollution game models in which environmental absorption efficiency is represented as an instantaneous function of the pollution stock in that these models suggest that the steady-state emissions in the cooperative setting should be lower than in an OLNE.

On the other hand, the result in Eq. 3.42 states that the steady state pollution is lower in the cooperative solution than in the OLNE. Intuitively, the difference between the two outcomes results from the fact that the players in the OLNE game do not take into account the *social* cost of a marginal increment in the pollution stock as in the cooperative game, but only their *private* marginal user cost. The result in Eq. 3.43 indicates that, compared with the cooperative solution, an OLNE leads each player to over-invest in environmental absorption efficiency. Note that the dominance of the OLNE investment in absorption efficiency over the cooperative investment effort refers to the steady state, which by definition is an asymptotic outcome where player i's investment is needed only to make up for the destructive impact of the pollution stock on environmental absorption efficiency. Relatedly, in Eq. 3.44, steady-state absorption efficiency is greater in the cooperative case than in the OLNE case. The intuition behind this result is that over-investment in absorption efficiency involved by the OLNE solution at the steady state compared to the cooperative solution does not suffice to negate the impact of the excess emissions during the transient path. This point will be checked on a numerical basis below. At the steady state, cooperative and OLNE emissions are equivalent, which implies equivalent revenues. However, the noncooperative steady state involves not only a greater cost of pollution than the cooperative setting but also a greater cost of investment effort in environmental absorption efficiency. Therefore, future generations are better off if the players decide to cooperate rather than to pre-commit, as shown in Eq. 3.45.

Table 3.2 Base case and range values

Parameter	r	γ	a	c
Base case	0.05	0.05	0.006	0.01
Range	[0.05, 0.1]	[0.05, 0.1]	[0.001, 0.006]	[0.01, 0.06]

Similar analytical comparisons involving the MPNE case are unavailable. To compare the cooperative, OLNE and MPNE solutions on a numerical basis, we select the base case parameter values in Table 3.2.

The base case in Table 3.2 reflects a favorable configuration which is characterized *in relative terms* by players' patience (i.e., low discounting rate, r), strong environmental resilience (i.e., low destructive impact of pollution on absorption efficiency, γ), the high marginal incentive for emissions (i.e., a high upper bound on emissions, $2a$) and low marginal pollution cost (i.e., low pollution cost function coefficient, c). The numerical solutions of the cooperative, OLNE and MPNE cases were computed with *Maple* 18.0.

To assess the sensitivity of the steady states, a broad range of values was used for all parameters (Table 3.2). In the MPNE case, the solutions were calculated for 10,000 combinations of the parameters with 10 values for each parameter r, γ, a, c. Among the 10,000 trials, 141 combinations of parameters gave rise to a single locally asymptotically stable solution and 162 combinations resulted in two locally asymptotically stable solutions. In the remaining 9697 combinations of parameters, no MPNE stable solution could be found.

We proceed in four steps. We start by identifying the stable steady states respectively associated with the cooperative, OLNE and MPNE solutions in the base case. Then we assess the sensitivity of these steady states to the various parameters in terms of comparative statics, whenever possible. We also represent the phase diagram in the state space, and the transient paths of the emissions rate and restoration effort for each case. Finally, we characterize the nature of convergence to the steady state as a function of the magnitude of the discounting rate for each case.

To compare the transient behaviors associated with the cooperative, OLNE and MPNE solutions, we assumed in all the subsequent cases a large initial pollution stock, i.e., significantly larger than the maximum corresponding stable steady state value resulting from the three solutions ($P_0 = 0.1$). In contrast, to assess the sensitivity of the transient paths to the magnitude of the initial absorption efficiency, we used two initial values, one being greater than the maximum stable steady state value among the three solutions, that is, $A_0 = 0.45$, and the other being lower than the minimum stable steady state value among the three solutions, that is, $A_0 = 0.15$.

We are primarily interested in stable interior solutions. As for the cooperative and OLNE solutions, the system of eight nonlinear equations derived for the MPNE gave rise to a single Nash equilibrium with a globally asymptotically stable steady state. In the cooperative, OLNE and MPNE cases, the respective stability conditions were all satisfied. That is, the saddle-point property was fulfilled with two positive and two negative eigenvalues with no imaginary part in the cooperative case, $(\chi_1, \chi_2) = (-0.427, -0.004)$, and in the OLNE case, $(\xi_1, \xi_2) = (-0.305, -0.008)$.

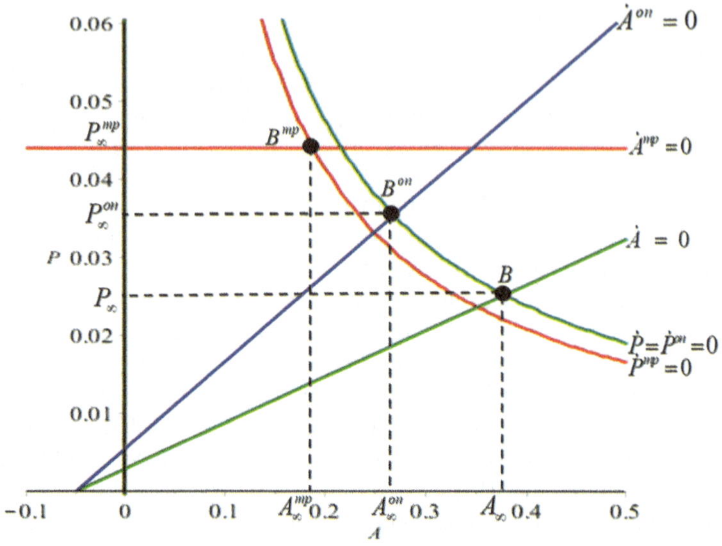

Fig. 3.5 Comparison between cooperative, OLNE and MPNE steady states

In the MPNE case, the global asymptotic stability criterion was granted with two negative eigenvalues with no imaginary part, $(\psi_1, \psi_2) = (-0.01, -0.235)$.

The main results are illustrated in Fig. 3.5. A unique (nontrivial) steady state is observed in each case, that is, B (i.e., intersection of the isoclines $\dot{P} = 0$ and $\dot{A} = 0$) for the cooperative solution with $(A_\infty, P_\infty) = (0.3776, 0.025)$, B^{on} (i.e., $\dot{P}^{on} = 0$ and $\dot{A}^{on} = 0$) for the OLNE with $\left(A_\infty^{on}, P_\infty^{on}\right) = (0.2684, 0.035)$, and B^{mp} (i.e., $\dot{P}^{mp} = 0$ and $\dot{A}^{mp} = 0$) for the MPNE with $\left(A_\infty^{mp}, P_\infty^{mp}\right) = (0.1883, 0.0439)$.

For the chosen constellation of parameters, the phase diagram for the MPNE (Fig. 3.6) confirms that all paths starting from a reasonable neighborhood around the steady state, B^{mp}, converge to the steady state. From (slightly) negative initial absorption efficiency, the convergence to the steady state is also granted with initially increasing then decreasing pollution stock. This pattern is driven by the fact that emissions reduction, though useful, is insufficient under negative absorption efficiency and only heavy restoration efforts can turn the biosphere into a pollution sink. Therefore, the pollution stock can never decrease before a strengthening of the absorption efficiency. A similar requirement is observed in a cooperative solution with clockwise oscillatory convergence to the steady state (e.g., El Ouardighi et al., 2014).

In Fig. 3.5, we obtain $P_\infty^{mp} > P_\infty^{on} > P_\infty$. This result is consistent with other differential game models that represent environmental absorption efficiency as an instantaneous function of the pollution stock because our model shows that the steady state pollution stock is lower in an OLNE than in an MPNE. Moreover, we also get $A_\infty^{mp} < A_\infty^{on} < A_\infty$, so that the absorption efficiency in an MPNE is substantially lower than in an OLNE despite lower emissions and greater restoration efforts

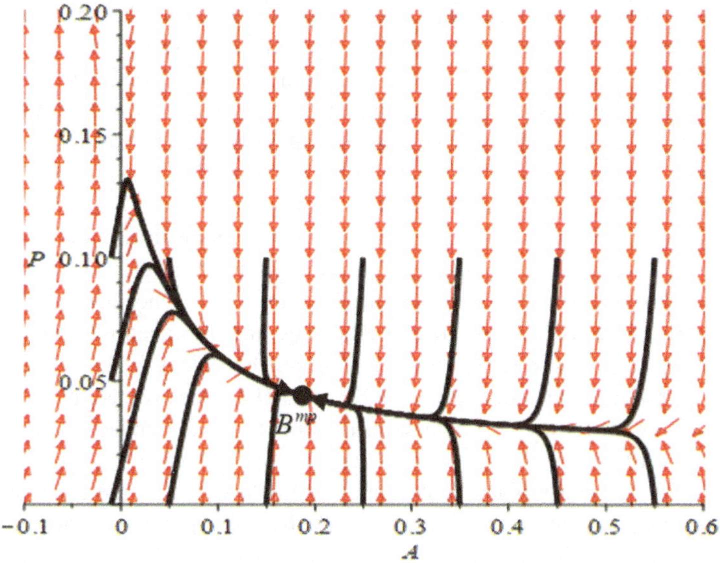

Fig. 3.6 Phase diagram in the state space (A, P) for Markov perfect Nash equilibrium

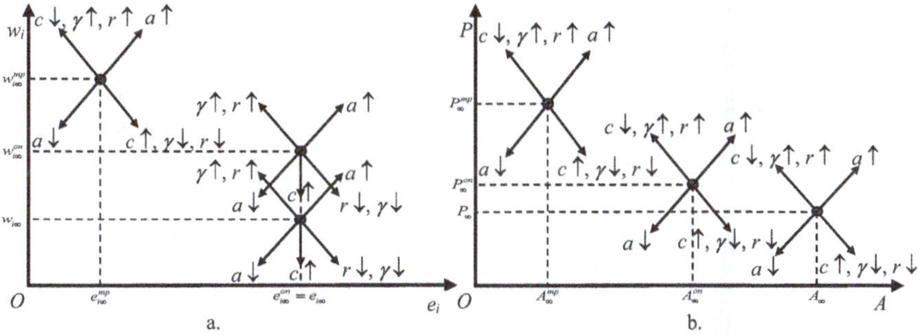

Fig. 3.7 Sensitivity analysis for cooperative, OLNE and MPNE stable steady states. (**a**) Control variables; (**b**) State variables

at the steady state. Therefore, the MPNE strategy leads to a stable steady state, which is more likely to be located in the emergency area, whereas the cooperative and OLNE strategies result in a steady state that has a better chance of falling into the sustainability area.

The figures below describe the sensitivity of the steady state to the parameters' values for the control variables (Fig. 3.7a) and the state variables (Fig. 3.7b) in the three equilibria.

In the cooperative, OLNE and MPNE cases, a lower discounting rate, r, moves the steady state further toward the sustainability area (7b). This suggests that players' patience enhances the long-run efficiency of the biosphere as a pollution sink because it leads to the substitution of lower emissions during the transient path for increased emissions at the steady state. Correlatively, the restoration effort should be reduced to a steady state (7a). A greater pollution cost coefficient, c, also moves the steady state further toward the sustainability area in the cooperative, OLNE and MPNE cases (7b). Though it reduces the restoration effort at the steady state, it does not affect the emissions in the cooperative and OLNE cases, which suggests that emissions are actually reduced during the transient path only. Regarding the influence of an increase in the destructive impact of pollution on absorption efficiency, γ, results in lower emissions and greater restoration efforts at the steady state in all equilibria (7a). A move of the steady state toward the emergency area is observed in all equilibria, with reduced absorption efficiency and an increased pollution stock (7b). Finally, an increase in the revenue function coefficient, a, leads to a complementary increase in the cooperative, OLNE and MPNE steady state pollution stock and absorption efficiency (7a), which results in both greater emissions and restoration efforts (7b).

In Fig. 3.8, the phase plane in the state space is depicted for the cooperative and noncooperative solutions. In the case of (relatively) high initial absorption efficiency, i.e., $A_0 = 0.45$, the cooperative, OLNE and MPNE strategies consist of

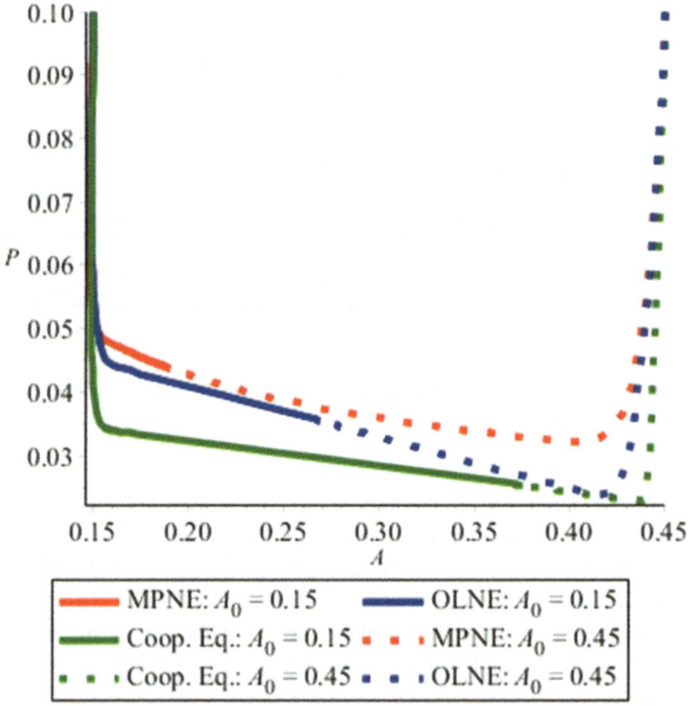

Fig. 3.8 Phase plane in the state space for the cooperative, open-loop and Markov perfect Nash cases from low and high absorption efficiency

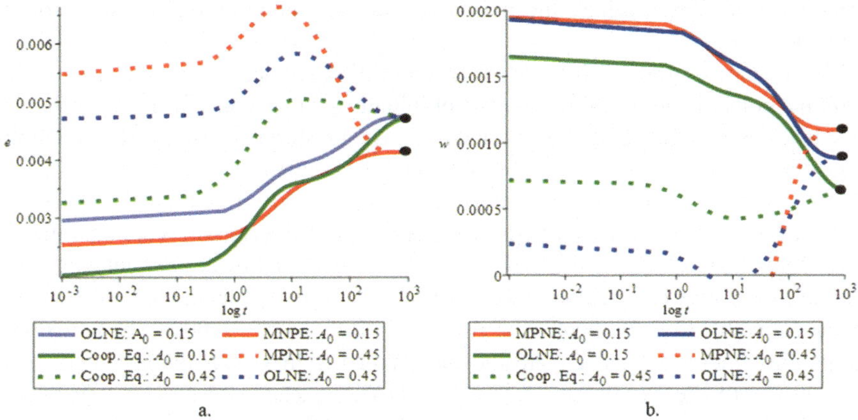

Fig. 3.9 Transient paths of (**a**) emissions rate and (**b**) restoration effort for the cooperative, OLNE and MPNE cases from low and high absorption efficiency

decreasing the absorption efficiency slightly and the pollution stock dramatically in a finite time, and then decreasing the absorption efficiency significantly and increasing the pollution stock slightly until the corresponding steady state is reached, that is, B for the cooperative solution, B^{on} for the OLNE, and B^{mp} for the MPNE. In the case of (relatively) low initial absorption efficiency, i.e., $A_0 = 0.15$, the cooperative, OLNE and MPNE strategies also consist in substantially decreasing the pollution stock first, and then in increasing the absorption efficiency while converging to the steady state. Overall, the cooperative, OLNE and MPNE transient paths are quite similar and differ only in their steady-state respective values. Whatever the initially (positive) efficiency of pollution sinks, it is optimal in all cases to reduce the pollution stock substantially before strengthening the biosphere as a pollution sink. However, although a contingent strategy preserves the long-run absorption efficiency somewhat, OLNE is more effective than an MPNE in managing pollution sinks' efficiency.

The patterns in Fig. 3.8 are explained by the time paths of emissions rate and restoration effort shown in Fig. 3.9, where time is expressed as a logarithmic function along the X-axis. We first observe that in all equilibria, the greater the initial absorption efficiency, the greater the emissions rate and the lower the restoration effort.

In the case of (relatively) high initial absorption efficiency, i.e., $A_0 = 0.45$, the OLNE strategy exhibits greater emissions and initially lower and then greater restoration efforts than the cooperative strategy. In contrast, the MPNE strategy shows initially greater then lower emissions, and initially lower then greater restoration efforts than the cooperative strategy. The noncooperative equilibria under high initial absorption efficiency are consistent with the tragedy of the commons in that they reflect both excessive emissions and underinvestment in absorption efficiency over the most economically influential part of the planning horizon, i.e., the time interval during which the discounted instantaneous profits are still strictly positive.

In the case of (relatively) low initial absorption efficiency, i.e., $A_0 = 0.15$, the optimal policy in all equilibria consists of initially small and increasing emissions in gradual stages and initially positive and decreasing restoration efforts. The first jump in emissions occurs quickly, immediately after the dramatic decrease in the pollution stock. Along the time path, cooperative emissions are lower than OLNE emissions over a wide time interval, until the steady state where the emissions are the same in the cooperative solution and OLNE ($e_{i\infty}^{on} = e_{i\infty} = 0.0047$). Lower transient emissions in the cooperative case require fewer restoration efforts than in an OLNE during the transient path and in the steady state ($w_{i\infty} = 0.0006 < w_{i\infty}^{on} = 0.0008$). In an MPNE, emissions alternate between greater and smaller values than in the cooperative solution until the steady state ($e_{\infty}^{mp} = 0.0041$) is reached. Restoration efforts are always greater than in the cooperative solution until the steady state ($w_{\infty}^{mp} = 0.001$) is attained. The emissions in an MPNE are lower and restoration efforts alternate between greater and smaller values than in the OLNE until the steady state.

Because transient emissions in an OLNE are greater than in a cooperative solution because of greater transient restoration efforts, the OLNE strategy under low initial absorption efficiency transiently promotes both excessive emissions and overinvestment in absorption efficiency over the planning horizon. By comparison, an MPNE promotes lumpier substitutability between the two policy instruments than an OLNE does. At the steady state, the magnitude of emissions in an MPNE is limited by low absorption efficiency.

Figure 3.10 reports the overall utility associated with the cooperative, OLNE and MPNE cases under (relatively) high and low initial absorption efficiency, respectively.

In the context of high initial absorption efficiency, the players' joint overall utility is positive for all cases. Note that the overall utility under cooperation and the joint overall utility under noncooperative equilibrium with commitment are almost similar. That is, the welfare cost for players with OLNE strategies that are not able to cooperate is limited. In comparison, the welfare cost for MPNE strategies regarding cooperation or commitment is greater. Under low initial absorption efficiency, the players' overall utility is positive only for cooperation, while it is zero for OLNE and negative for MPNE. This decline in joint overall utility of noncooperative strategies might be seen as a confirmation of the presence of a voracity effect. Here, the welfare cost for noncooperative players that are unable to cooperate, and the welfare cost for players with MPNE strategies that are unable to commit are both significantly greater than previously. In terms of elasticity of the overall joint utility with respect to the initial absorption efficiency, denoted by ϕ and defined as the percentage of decrease of the overall utility divided by the percentage of decrease of the initial absorption efficiency, we get $\phi_{U/A_0} = 0.876$ in the cooperative case, $\phi^{on}_{\sum_i U_i/A_0} = 1.49$ in the OLNE case and $\phi^{mp}_{\sum_i U_i/A_0} = 2.399$ in the MPNE case. These results imply that contingent strategies are more vulnerable in terms of social welfare to a decrease

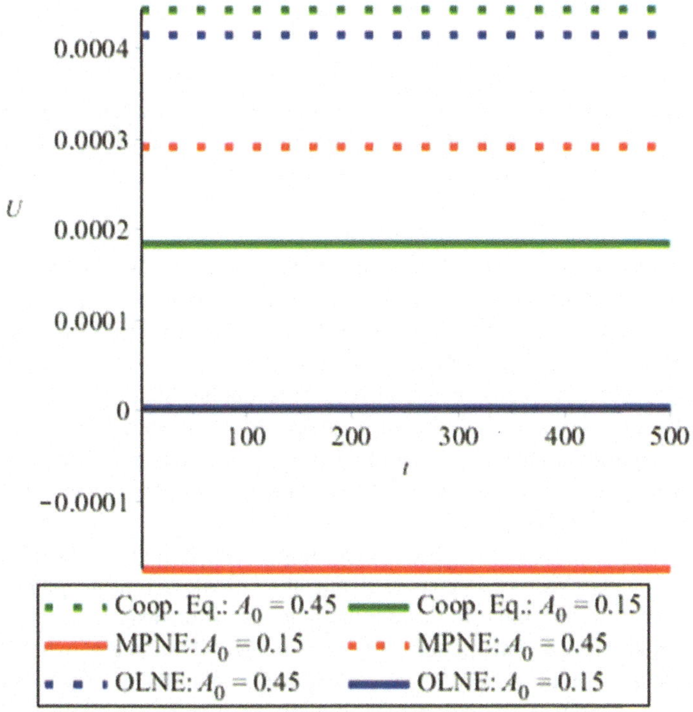

Fig. 3.10 Overall utility for the cooperative, OLNE and MPNE cases from low and high initial absorption efficiency

in initial absorption efficiency than commitment strategies are, which in turn are more affected than cooperative strategies are. That is, the absence of cooperation or commitment becomes more costly as the initial absorption efficiency of pollution sinks tends to exhaustion. This result explains why nations have been urged in recent years to commit to pollution control at an international level (Stern, 2006, 2015; IPCC, 2007).

Finally, the steady-state utility, or social welfare of future generations, is positive in the cooperative, OLNE and MPNE cases and its ranking is consistent with that reported in Fig. 3.10 from high initial absorption efficiency. The welfare cost of future generations is limited for OLNE strategies regarding cooperative strategy and significant for MPNE strategies regarding cooperation or commitment. In all cases, the decrease in initial absorption efficiency is more costly to current generations than to future ones because the future generations neither benefit from the revenues nor incur the costs related to the transient path toward the steady state.

Table 3.3 provides insights into the nature of convergence to the steady state depending on the discounting rate value for the set of lower bound values. The value of the discounting rate above, with which the path to the steady state becomes diverging in the case of the MPNE, is relatively low ($r = 0.15$). In comparison

Table 3.3 Threshold of the impact of the discounting rate on the nature of convergence to the steady state

Discounting rate	0.1	0.15	0.2	0.25
Cooperative solution	Monotonic	Monotonic	Monotonic	Diverging
OLNE	Monotonic	Monotonic	Monotonic	Diverging
MPNE	Monotonic	Diverging	Diverging	Diverging

with this threshold, the value of the discounting rate above, with which the saddle path to the steady state becomes diverging, is greater both in the cooperative solution and OLNE (i.e., $r = 0.24$). As a result, more heavily discounted MPNE strategies are less robust to a definitive switching of pollution sinks to a source, while such switching would not occur in an OLNE. This implies that relatively impatient players should not opt for contingent strategies because they are more likely to make the evolution of the biosphere irreversibly uncontrollable. It is noteworthy that the threshold value of the discounting rate from which cooperative and OLNE paths are diverging is still monotonic.

Figure 3.11 confirms the irreversible impact of a higher discounting rate ($r = 0.15$) on the MPNE trajectory. Along the MPNE diverging path, which is also monotonic, restoration efforts are initially low and convexly increasing, but less rapidly than emissions. This pattern suggests that MPNE strategies where transient emissions and restoration efforts act as complements rather than substitutes are likely to result in a definitive switching of pollution sinks to a source. In comparison, although both cooperative and OLNE paths remain monotonic, they are more likely to end up in the emergency area. This result confirms that although commitment and cooperative strategies are more robust than MPNE strategies to a definitive switching of pollution sinks to a source, they are nevertheless vulnerable to players' impatience, which directly affects future generations' social welfare.

Conclusion

In this paper, we suggest a nonlinear quadratic differential game model of pollution control with a time-dependent environmental absorption efficiency that allows the biosphere to switch from a carbon sink to a source.

Using a cooperative solution as a benchmark, we showed that the fact that the players do not properly internalize the social cost of the pollution stock in the noncooperative cases incites them to excessively substitute their transient path for steady state utility. This leads them to disproportionate use of strategic substitutability between polluting emissions and restoration of environmental absorption efficiency during the transient path, which reduces the efficiency of carbon sinks and increases the pollution stock at the steady state. In both cooperative and noncooperative steady states, the magnitude of emissions increases with absorption efficiency, while restoration efforts increase with the pollution stock, and vice versa.

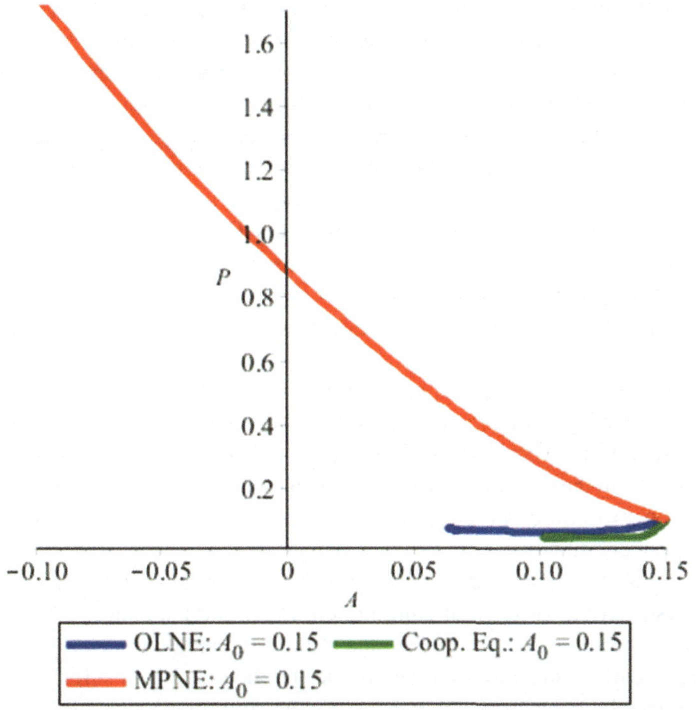

Fig. 3.11 Phase plane in the state space for the cooperative, OLNE and MPNE cases from low absorption efficiency and higher discounting rate

We were then able to identify which decision rule, open-loop or Markov perfect Nash strategy, best prevents the biosphere from switching from a pollution sink to a source. Although it fails to reach the same environmental absorption efficiency as a cooperative solution, an OLNE imposes limited economic sacrifice at the steady state notably because it allows equivalent polluting emissions to the social optimum. This result justifies the optimistic tone of the call by some authors for rapid action against climate change (Stern, 2015). However, an OLNE requires a greater restoration effort at the steady state than a cooperative solution does to partly compensate for the excessive emissions released during the transient path.

If the biosphere is still an efficient pollution sink, it is optimal in all cases to reach the steady state both by an overshooting emissions trajectory and an undershooting restoration efforts trajectory. If the efficiency of the biosphere as a pollution sink is already low, it is optimal to set initially low emissions and high restoration efforts, and then to increase emissions gradually as absorption efficiency increases, and decrease restoration efforts as the pollution stock declines. Our results suggest that the absence of cooperation and/or precommitment becomes more costly as the initial efficiency of pollution sinks tends to exhaustion. Also, a decrease in the initial efficiency of pollution sinks is more costly to current generations than to future ones in both cooperative and noncooperative strategies.

When both players are sufficiently patient, an MPNE and an OLNE do not differ much in terms of transient path and steady state. In this context, both strategies can prevent pollution sinks from turning into a source during the transient path and the steady state, including in the case where the absorption efficiency is low. However, although an MPNE may generate lower emissions and greater restoration efforts than an OLNE in the case of low absorption efficiency, it has a much greater chance of falling into the emergency area. In contrast, when both players are relatively impatient, MPNE strategies are less robust than OLNE strategies at preventing irreversible switching of the biosphere from a carbon sink to a source. Although more heavily discounted OLNE strategies can prevent pollution sinks from transiently turning into a source, these strategies are more likely to end up in the emergency area.

Regarding the methodological contribution of the paper, we suggest a particular structure of the conjectured value function to solve MPNE problems with multiplicative interaction between state variables in one state equation, so that third-order terms that arise in the Hamilton-Jacobi-Bellman equation are made negligible. Using a collocation procedure, we confirm the validity of the particular structure of the conjectured value function.

Overall, our results provide an update in the debate on pollution control in that they suggest that the absence of commitment at an international level on pollution control can preserve both current and future generations from experiencing switching of carbon sinks to a source only if the players remain sufficiently patient. More importantly, regardless of whether polluters commit to a plan of action or make contingent decisions, our study emphasizes the need for environmental policies where efforts to restore environmental absorption efficiency are an essential policy instrument.

The scope of our model could be extended to the case where the marginal destructive impact of the pollution stock on the evolution of absorption efficiency can be reduced by an effort to improve environmental resilience. Another possible extension of the model presented here could assume that the impact of restoration efforts on the evolution of absorption efficiency depends concavely on the level of absorption efficiency. These assumptions would require multiplicative separability between control and state variables, which would allow the emergence of a Skiba threshold (Grass et al., 2008) with multiple steady states and lead to the determination of history-dependent policies. Other future directions of research include an analysis of the Pareto optimality of the obtained equilibria and an extension of the model to the case of taxation rules.

Acknowledgements This research was supported by the ESSEC Business School Research Centre (France). Giorgio Gnecco and Marcello Sanguineti are members of GNAMPA-INdAM (Gruppo Nazionale per l'Analisi Matematica, la Probabilità e le loro Applicazioni - Istituto Nazionale di Alta Matematica). The authors acknowledge very helpful comments from two anonymous referees. They also thank Hassan Benchekroun and Michèle Breton for constructive suggestions on an early draft.

Appendix

A1 To determine the (nontrivial) cooperative steady state, we solve the canonical system Eqs. 3.8–3.11 in the state-costate space. Using Eqs. 3.6 and 3.7 leads to the system's steady state in Eq. 3.12, $i = 1, 2$. Note that since:

$$\lim_{t\to+\infty} e^{-rt}\eta_1(t)P(t) = \lim_{t\to+\infty}\left[-\frac{r\gamma(r^2\gamma+h)e^{-rt}}{4(4c+\gamma^2)}\right] = 0$$

$$\lim_{t\to+\infty} e^{-rt}\eta_2(t)A(t) = \lim_{t\to+\infty}\left[\frac{\gamma(2a-r\gamma)e^{-rt}}{2}\right] = 0$$

the cooperative steady-state solution satisfies the limiting transversality conditions. ■

A2 Plugging the expressions of $e_{i\infty}$, P_∞ and $w_{i\infty}$ from Eq. 3.12 in the integrand of Eq. 3.4 and simplifying gives Eq. 3.13. ■

A3 Writing the variables in the order P, A, η_1, η_2, the Jacobian matrix associated with the canonical system Eqs. 3.8–3.11 is:

$$J = \begin{bmatrix} -A & -P & 2 & 0 \\ -\gamma & 0 & 0 & 2 \\ 2c & \eta_1 & r+A & \gamma \\ \eta_1 & 0 & P & r \end{bmatrix}$$

The determinant of the Jacobian matrix is:

$$|J| = (4c+\gamma^2){P_\infty}^2 + r\gamma A_\infty P_\infty = \frac{(r^2\gamma+h)^2}{4(4c+\gamma^2)} + r\gamma(2a-r\gamma)$$

Using Dockner's formula (Dockner, 1985), the sum of the principal minors of J of order 2 minus the squared discount rate, denoted by K, is:

$$K = -A_\infty(r+A_\infty) - 4c - 2\gamma P_\infty = -\frac{[2(2a-r\gamma)(4c+\gamma^2)+r(r^2\gamma+h)]^2}{(r^2\gamma+h)^2}$$
$$-\frac{\gamma h + 4c(4c+\gamma^2-r^2)}{4c+\gamma^2}$$

Because $|J| > 0$ and $K < 0$, the cooperative steady state is a saddle point. We compute:

$$\Omega = K^2 - 4|J|$$
$$= A_\infty(r + A_\infty)[A_\infty(r + A_\infty) + 8c] + 4\gamma A_\infty{}^2 P_\infty + 16c(c + \gamma P_\infty) - 16cP_\infty{}^2$$

The sign of Ω can be either negative or positive. In the first case, the steady state is a saddle-focus with transient oscillations. Otherwise, it is a saddle-node and the optimal solution monotonically converges to the steady state. ■

A4 The linear approximation of the system Eqs. 3.8–3.11 around its steady state is:

$$\dot{\eta}_1 = 2cP + (r + A_\infty)\eta_1 + (A - A_\infty)\eta_{1\infty} + \gamma\eta_2$$
$$\dot{\eta}_2 = r\eta_2 + \eta_{1\infty}P + P_\infty(\eta_1 - \eta_{1\infty})$$
$$\dot{P} = 2(a + \eta_1) - (A - A_\infty)P_\infty - A_\infty P$$
$$\dot{A} = 2\eta_2 - \gamma P$$

The four eigenvalues associated with the Jacobian matrix of the canonical system are:

$${}_1^3\chi_2^4 = \frac{r}{2} \pm \sqrt{\frac{r^2}{4} - \frac{K}{2} \pm \frac{1}{2}\sqrt{K^2 - 4|J|}}$$

$$= \frac{r}{2} \pm \sqrt{\frac{(r + A_\infty)^2 + A_\infty{}^2}{4} + \gamma P_\infty + \frac{8ac}{r^2\gamma^2} \pm \sqrt{\left(A_\infty{}^2 + \frac{8ac}{r^2\gamma^2}\right)\left(\gamma P_\infty + \frac{8ac}{r^2\gamma^2}\right) + \left[\frac{(r + A_\infty)^2 A_\infty}{4} + \frac{8ac}{r\gamma^2}\right]A_\infty - 4cP_\infty{}^2}}$$

where A_∞ and P_∞ are given in Eq. 3.12. As expected, two eigenvalues, χ_1 and χ_3, have negative real part and two have positive real part, χ_2 and χ_4. Choosing the roots with negative real part for convergence, the time paths of the costate and state variables are written as:

$$\eta_1(t) = \eta_{1\infty} + B_1 e^{\chi_1 t} + B_2 e^{\chi_3 t}$$
$$\eta_2(t) = \eta_{2\infty} + B_3 e^{\chi_1 t} + B_4 e^{\chi_3 t}$$
$$P(t) = P_\infty + B_5 e^{\chi_1 t} + B_6 e^{\chi_3 t}$$
$$A(t) = A_\infty + B_7 e^{\chi_1 t} + B_8 e^{\chi_3 t}$$

These equations involve 10 unknowns ($B_1 \ldots B_8, \eta_1(0)$, $\eta_2(0)$) that can be solved with the 10 following equations, which are drawn from the above expressions and the linearized versions of Eqs. 3.8–3.11:

$$\eta_1(0) = \eta_{1\infty} + B_1 + B_2$$

$$\eta_2(0) = \eta_{2\infty} + B_3 + B_4$$

$$P_0 = P_\infty + B_5 + B_6$$

$$A_0 = A_\infty + B_7 + B_8$$

$$\dot{\eta}_1(0) = 2cP_0 + (r + A_\infty)\eta_1(0) + (A_0 - A_\infty)\eta_{1\infty} + \gamma\eta_2(0) = B_1\chi_1 + B_2\chi_3$$

$$\dot{\eta}_2(0) = r\eta_2(0) + \eta_{1\infty}(P_0 - P_\infty) + \eta_1(0)P_\infty = B_3\chi_1 + B_4\chi_3$$

$$\dot{P}(0) = 2(a + \eta_1(0)) - (A_0 - A_\infty)P_\infty - A_\infty P_0 = B_5\chi_1 + B_6\chi_3$$

$$\dot{A}(0) = 2\eta_2(0) - \gamma P_0 = B_7\chi_1 + B_8\chi_3$$

$$2(B_3\chi_1 + B_4\chi_3) - \gamma(B_5\chi_1 + B_6\chi_3) = B_7{\chi_1}^2 + B_8{\chi_3}^2$$

$$2(B_1\chi_1 + B_2\chi_3) - A_\infty(B_5\chi_1 + B_6\chi_3) - P_\infty(B_7\chi_1 + B_8\chi_3) = B_5{\chi_1}^2 + B_6{\chi_3}^2$$

This system of equations can be solved numerically to determine real-valued solutions. Finally, using Eqs. 3.6–3.7 yields Eqs. 3.14–3.17. ■

A5 To compute the (nontrivial) OLNE steady state, we solve the system in the state-costate space (Eqs. 3.21–3.24). Using Eqs. 3.19 and 3.20 leads us to derive the steady state in Eq. 3.25, $i = 1, 2$. Note that since:

$$\lim_{t\to+\infty} e^{-rt}\lambda_1(t)P(t) = \lim_{t\to+\infty}\left[-\frac{r\gamma(r^2\gamma + k)e^{-rt}}{4(2c + \gamma^2)}\right] = 0$$

$$\lim_{t\to+\infty} e^{-rt}\lambda_2(t)A(t) = \lim_{t\to+\infty}\left[\frac{\gamma(2a - r\gamma)e^{-rt}}{2}\right] = 0$$

the OLNE steady-state solution satisfies the limiting transversality conditions. ■

A6 Plugging the expressions of $e^{on}_{i\infty}$, P^{on}_{∞} and $w^{on}_{i\infty}$ from Eq. 3.25 in the integrand of Eq. 3.3 and simplifying gives (Eq. 3.26). ■

A7 Writing the variables in the order P, A, λ_1, λ_2, the Jacobian matrix associated with the canonical system (Eqs. 3.21–3.24) is:

$$J^{on} = \begin{bmatrix} -A & -P & 2 & 0 \\ -\gamma & 0 & 0 & 2 \\ c & \lambda_1 & r + A & \gamma \\ \lambda_1 & 0 & P & r \end{bmatrix}$$

The determinant of J^{on} is:

$$|J| = (2c+\gamma^2)P_\infty^{on2} + r\gamma A_\infty^{on} P_\infty^{on} = \frac{(r^2\gamma+k)^2}{4(2c+\gamma^2)} + r\gamma(2a - r\gamma)$$

Further, the sum of the principal minors of J^{on} of order 2 minus the squared discount rate is:

$$\begin{aligned} K^{on} &= -A_\infty^{on}\left(r + A_\infty^{on}\right) - 2c - 2\gamma P_\infty^{on} \\ &= -\frac{2(2a-r\gamma)(2c+\gamma^2)}{r^2\gamma+k}\left[r + \frac{2(2a-r\gamma)(2c+\gamma^2)}{r^2\gamma+k}\right] - \frac{\gamma(r^2\gamma+k)}{2c+\gamma^2} - 2c \end{aligned}$$

which is strictly negative. Hence, the OLNE steady state is a saddle point. We now compute:

$$\Omega^{on} = A_\infty^{on2}\left[\left(r+A_\infty^{on}\right)^2 + 4\gamma P_\infty^{on}\right] + \frac{16ac}{r^2\gamma^2}\left[A_\infty^{on}\left(r+A_\infty^{on}\right) + \frac{4ac}{r^2\gamma^2}\right] + 8c\left(\frac{4a}{r^2\gamma} - P_\infty^{on}\right)P_\infty^{on}$$

where A_∞^{on} and P_∞^{on} are given in Eq. 3.25. The sign of this expression, which is unclear, determines whether the saddle path converges monotonically or oscillates towards the steady state. ■

A8 The linear approximation of the system (Eqs. 3.21–3.24) around its steady state is:

$$\begin{aligned} \dot{\lambda}_1 &= cP + \left(r + A_\infty^{on}\right)\lambda_1 + \left(A - A_\infty^{on}\right)\lambda_{1\infty} + \gamma\lambda_2 \\ \dot{\lambda}_2 &= r\lambda_2 + \lambda_{1\infty}P + P_\infty^{on}\left(\lambda_1 - \lambda_{1\infty}\right) \\ \dot{P} &= 2(a + \lambda_1) - \left(A - A_\infty^{on}\right)P_\infty^{on} - A_\infty^{on}P \\ \dot{A} &= 2\lambda_2 - \gamma P \end{aligned}$$

The eigenvalues associated with the Jacobian of the noncooperative canonical system are:

$$\xi_{1\,2}^{3\,4} = \frac{r}{2} \pm \sqrt{\frac{\left(r+A_\infty^{on}\right)^2 + A_\infty^{on2}}{4} + \gamma P_\infty^{on} + \frac{4ac}{r^2\gamma^2} \pm \sqrt{\left[\frac{A_\infty^{on}\left(r+A_\infty^{on}\right)}{2} + \frac{2ac}{r^2\gamma^2} + \gamma P_\infty^{on}\right]^2 - (2c+\gamma^2)P_\infty^{on2} - 4a}}$$

where A_∞^{on} and P_∞^{on} are given in Eq. 3.25. Choosing roots with negative real part, ξ_1 and ξ_3, and using similar methods as in the cooperative case leads us to find $D_1 \ldots D_8$, and $\lambda_1(0)$, $\lambda_2(0)$. ■

A9 Assuming a second-order Taylor-polynomial approximation of the value function centered on $\left(A_\infty^{mp}, P_\infty^{mp}\right)$, i.e.,

$$V = \alpha_1 + \alpha_2\left(A - A_\infty^{mp}\right) + \alpha_3\left(P - P_\infty^{mp}\right) + \alpha_4\left(A - A_\infty^{mp}\right)^2 + \alpha_5\left(P - P_\infty^{mp}\right)^2 + \alpha_6\left(A - A_\infty^{mp}\right)\left(P - P_\infty^{mp}\right)$$

we obtain constant second-order derivatives for V. We assume that its unknown coefficients $\alpha_1 \ldots \alpha_6$ and the problem parameters are such that the steady state is positive. Plugging Eqs. 3.37 and 3.38 in Eqs. 3.1 and 3.2, respectively, gives the MPNE controlled system, that is:

$$\dot{P} = 2\left[a + \alpha_3 + 2\alpha_5\left(P - P_\infty^{mp}\right) + \alpha_6\left(A - A_\infty^{mp}\right)\right] - AP$$
$$\dot{A} = 2\left[\alpha_2 + 2\alpha_4\left(A - A_\infty^{mp}\right) + \alpha_6\left(P - P_\infty^{mp}\right)\right] - \gamma P$$

The resulting approximation of the HJB Eq. 3.35 is:

$$\begin{aligned} & r\Big[\alpha_1 + \alpha_2\left(A - A_\infty^{mp}\right) + \alpha_3\left(P - P_\infty^{mp}\right) + \alpha_4\left(A - A_\infty^{mp}\right)^2 + \alpha_5\left(P - P_\infty^{mp}\right)^2 \\ & + \alpha_6\left(A - A_\infty^{mp}\right)\left(P - P_\infty^{mp}\right)\Big] \\ & = \left[\alpha_3 + 2\alpha_5\left(P - P_\infty^{mp}\right) + \alpha_6\left(A - A_\infty^{mp}\right)\right]\left\{\frac{3}{2}\left[\alpha_3 + 2\alpha_5\left(P - P_\infty^{mp}\right) + \alpha_6\left(A - A_\infty^{mp}\right)\right] - AP + 2a\right\} \\ & + \left[\alpha_2 + 2\alpha_4\left(A - A_\infty^{mp}\right) + \alpha_6\left(P - P_\infty^{mp}\right)\right]\left\{\frac{3}{2}\left[\alpha_2 + 2\alpha_4\left(A - A_\infty^{mp}\right) + \alpha_6\left(P - P_\infty^{mp}\right)\right] - \gamma P\right\} \\ & - \frac{cP^2}{2} + \frac{a^2}{2} \end{aligned}$$

Rearranging its terms, this can be expressed as a third-order polynomial in $\left(A - A_\infty^{mp}\right)$ and $\left(P - P_\infty^{mp}\right)$. Then, the unknown parameters of the approximation of the value function are found by setting to 0 the coefficients, up to the second order, of such a third-order polynomial (its third-order terms are locally negligible, by construction). To do so, one solves the system:

$$r\alpha_1 - \frac{3\alpha_2{}^2}{2} - \alpha_3\left(\frac{3\alpha_3}{2} + 2a\right) - \frac{a^2}{2} + \gamma\alpha_2 P_\infty^{mp} + \alpha_3 A_\infty^{mp} P_\infty^{mp} + \frac{c P_\infty^{mp2}}{2} = 0$$
$$\alpha_2(\gamma - 3\alpha_6) + r\alpha_3 - 2\alpha_5(3\alpha_3 + 2a) + \alpha_3 A_\infty^{mp} + (c + \gamma\alpha_6)P_\infty^{mp} + 2\alpha_5 A_\infty^{mp} P_\infty^{mp} = 0$$
$$\frac{c}{2} + \alpha_5\left(r - 6\alpha_5 + 2A_\infty^{mp}\right) + \alpha_6\left(\gamma - \frac{3\alpha_6}{2}\right) = 0$$
$$\alpha_2(r - 6\alpha_4) - \alpha_6(3\alpha_3 + 2a) + (\alpha_3 + 2\gamma\alpha_4)P_\infty^{mp} + \alpha_6 A_\infty^{mp} P_\infty^{mp} = 0$$
$$\alpha_3 + 2\gamma\alpha_4 - 6\alpha_6(\alpha_4 + \alpha_5) + \alpha_6\left(r + A_\infty^{mp}\right) + 2\alpha_5 P_\infty^{mp} = 0$$

$$\alpha_4(r-6\alpha_4)-\alpha_6\left(\frac{3\alpha_6}{2}-P_\infty^{mp}\right)=0$$

We have six equations in eight unknowns $\alpha_1...\alpha_6, A_\infty^{mp}, P_\infty^{mp}$. To get two other equations, we use Eqs. 3.37–3.38, imposing a steady-state solution for the MPNE:

$$2(a+\alpha_3)-A_\infty^{mp}P_\infty^{mp}=0$$

$$2\alpha_2-\gamma P_\infty^{mp}=0$$

Finally, we obtain a nonlinear system of eight equations in eight unknowns, which in general might be solved numerically, but not analytically. When it can be solved, its solution provides:

$$\begin{bmatrix} P_\infty^{mp} & A_\infty^{mp} & e_\infty^{mp} & w_\infty^{mp} \end{bmatrix}^T=\begin{bmatrix} 2\alpha_2/\gamma & \gamma(a+\alpha_3)/\alpha_2 & a+\alpha_3 & \alpha_2 \end{bmatrix}^T$$

Clearly, the necessary conditions for the positivity of the steady state values and the positivity of each player's instantaneous revenue are $\alpha_2>0$ and $\alpha_3>-a$. In general, the nonlinear system above may admit multiple solutions (or even no solutions at all). To solve the system, we use the *Matlab* function `fsolve`, using default options, and starting with a random initialization of the eight unknowns, initially chosen as realizations of independent and uniformly distributed random variables on the interval [0, 10]. To check whether a numerically found solution of the nonlinear system above is likely also a steady-state solution for the MPNE, we perform the following additional checks:

(1) Positivity of the state and control variables, and positivity of each player's instantaneous revenue: $\alpha_2>0$ and $-a<\alpha_3<0$;
(2) Global asymptotical stability for the linearized dynamical system: the eigenvalues of the Jacobian matrix (Eq. 3.40) should have negative real parts;
(3) Comparison between the value of α_1 assumed in $(A_\infty^{mp}, P_\infty^{mp})$ by the local quadratic approximation of the value function, and the one obtained by replacing the resulting steady state values in Eq. 3.3, the latter being equal to $\left[e_\infty^{mp}(a-e_\infty^{mp}/2)-c(P_\infty^{mp})^2/2-(w_\infty^{mp})^2/2\right]/r$: in the case where $(P_\infty^{mp}, A_\infty^{mp}, e_\infty^{mp}, w_\infty^{mp})$ is really an approximation of a steady-state solution for the MPNE, and the local quadratic approximation is a good approximation of the value function, then the two values should be approximately the same. In practice, we check whether the relative error of the first expression with respect to the second one is smaller than 0.01.

In the case where multiple solutions are obtained that satisfy the requirements 1), 2), and 3) above, one can choose the one associated with the largest value of $\left[e_\infty^{mp}(a-e_\infty^{mp}/2)-c(P_\infty^{mp})^2/2-(w_\infty^{mp})^2/2\right]/r$. ■

A10 To perform an additional check for the validity of the local quadratic approximation of the value function, we also exploit an alternative method to approximate the HJB equation and then we check whether the two obtained approximations are similar. As the alternative approximation method,[7] we use the collocation method for PDEs, applied to the HJB Eq. 3.35, on a small neighborhood of the steady-state solution determined by the local quadratic approximation (in our implementation, we chose the box $\left[0.75A_{\infty}^{mp}, 1.25A_{\infty}^{mp}\right] \times \left[0.75P_{\infty}^{mp}, 1.25P_{\infty}^{mp}\right]$ as such a neighborhood). The collocation method approximates locally on that neighborhood the unknown value function as a high-order polynomial in $\left(A - A_{\infty}^{mp}\right)$ and $\left(P - P_{\infty}^{mp}\right)$, with unknown coefficients. These are determined by imposing that the HJB Eq. 3.35 holds exactly on a finite number of fixed points ("collocation points") inside the neighborhood (or equivalently, that its Bellman residual error is 0 on such points). The number of these points is equal to the number of unknown coefficients of the high-order polynomial. In this way, to find the values of the coefficients, one is reduced to solving a nonlinear system (due to the nonlinearity of the HJB equation), where each equation is associated with one collocation point. For the implementation of the collocation method, we follow the procedure proposed in Doraszelski (2003), that is:

- We construct a polynomial approximation of order $M = 12$, using a tensor-product basis of univariate Chebyshev polynomials. This choice of the approximation order is guided by the need to find a suitable trade-off between the accuracy of the approximation, and the computational effort needed to find it (as in the cited reference, the number of coefficients in the expansion is $(M + 1)^2$);
- Likewise, in Doraszelski (2003), we employ a two-dimensional expanded Chebyshev array [as defined in Judd (1998)] to define the $(M + 1)^2$collocation points.

To initialize the collocation method, at first, we fit the initial local quadratic approximation of the value function by a linear combination of the same basis functions used by the collocation method itself (in practice, we apply the collocation method itself to this function approximation problem, instead of to the PDE). Then, we apply the collocation method to the PDE, starting from the resulting vector of coefficients. Finally, to check for the validity of the resulting high-order polynomial approximation and for the validity of the original local quadratic approximation, we compute, for both approximations, the Bellman residual error on a larger uniform grid of N^2 points in the same neighborhood (with $N = 30$), which are different from the collocation points, checking whether this error is small. We also evaluate, on

[7]In an initial phase, we also tested the application of the method of successive approximations, using a discrete-time version of the HJB Eq. 3.39, and following a procedure similar to the one proposed in Cacace et al. (2013) for another dynamic game model. However, that approach demonstrated to be much slower than the collocation method, and suffered from boundary effects, due to the need to discretize and bound all the state and control variables. A similar finding about such boundary effects was reported in Cacace et al. (2013).

each grid point, the relative error of the quadratic approximation with respect to the high-order polynomial approximation obtained using the collocation method. The results of all these comparisons are reported in fifth section, for a specific choice of the problem parameters. We conclude by mentioning that, in contrast with Doraszelski (2003), it is not necessary to check a-posteriori whether the partial derivatives of the approximation of the value function obtained by the collocation method are good approximations of the corresponding partial derivatives of the value function, which are needed to express the optimal controls (see the expression Eqs. 3.37 and 3.38). Indeed, in the case where the local quadratic approximation and the one found by the collocation method are similar, we can use the partial derivatives of the local quadratic approximation itself. In any case, it is worth mentioning that the numerical results have shown good agreement between the approximations of the partial derivatives obtained by the two methods, as one can infer from Fig. 3.2 in fifth section. ■

A11 The results in Eqs. 3.41–3.45 follow from the respective comparisons of Eqs. 3.12 and 3.25. ■

References

Azariadis, C., & Stachurski, J. (2005). Poverty traps. In S. Durlauf & P. Aghion (Eds.), *Handbook of economic growth*. North-Holland.

Benchekroun, H., & Martin-Herran, G. (2016). The impact of foresight in a transboundary pollution game. *European Journal of Operational Research, 251*(1), 300–309.

Benchekroun, H., & Long, N. V. (2011). Game theory: Static and dynamic games. In A. A. Batabyal & P. Nijkamp (Eds.), *Research tools in natural resource and environmental economics*. World Scientific Publishing.

Boucekkine, R., Pommeret, A., & Prieur, F. (2013). Optimal regime switching and threshold effects. *Journal of Economic Dynamics and Control, 37*(12), 2979–2997.

Cacace, S., Cristiani, E., & Falcone, M. (2013). Numerical approximation of Nash equilibria for a class of non-cooperative differential games. In L. Petrosjan & V. Mazalov (Eds.), *Game theory and applications* (Vol. 16). Nova Publishers.

Canadell, J. G., Le Quéré, C., Raupach, M. R., Field, C. B., Buitenhuis, E. T., Ciais, P., Conway, T. J., Gillett, N. P., Houghton, R. A., & Marland, G. (2007). Contributions to accelerating atmospheric CO_2 growth from economic activity, carbon intensity, and efficiency of natural sinks. *Proceedings of the National Academy of Sciences, 104*(47), 18866–18870.

Cox, P. M., Betts, R. A., Jones, C., Spall, S. A., & Totterdell, I. (2000). Acceleration of global warming due to carbon-cycle feedbacks in a coupled climate model. *Nature, 408*, 184–187.

Cramer, W., Bondeau, A., Woodward, F. I., Prentice, I. C., Betts, R. A., Brovkin, V., Cox, P. M., Fisher, V., Foley, J. A., Friend, A. D., Kucharik, C., Lomas, M. R., Ramankutty, N., Sitch, S., Smith, B., White, A., & Young-Molling, C. (2001). Global response of terrestrial ecosystem structure and function to CO_2 and climate change: results from six dynamic global vegetation models. *Global Change Biology, 7*(4), 357–373.

Dawid, H., Keoula, M. Y., Kopel, M., & Kort, P. M. (2015). Product innovation incentives by an incumbent firm: A dynamic analysis. *Journal of Economic Behavior and Organization, 117*, 411–438.

Dockner, E. (1985). Local stability in optimal control problems with two state variables. In G. Feichtinger (Ed.), *Optimal control theory and economic analysis* (Vol. 2). North Holland.

Dockner, E., Jørgensen, S., Long, N. V., & Sorger, G. (2000). *Differential games in economics and management science*. Cambridge University Press.
Dockner, E. J., & Long, N. V. (1993). International pollution control: Cooperative versus non-cooperative strategies. *Journal of Environmental Economics and Management, 25*(1), 13–29.
Dockner, E. J., & Wagener, F. (2014). Markov perfect Nash equilibria in models with a single capital stock. *Economic Theory, 56*(3), 585–625.
Doraszelski, U. (2003). An R&D race with knowledge accumulation. *RAND Journal of Economics, 34*(1), 19–41.
El Ouardighi, F., Benchekroun, H., & Grass, D. (2014). Controlling pollution and environmental, absorption capacity. *Annals of Operations Research, 220*(1), 111–133.
El Ouardighi, F., Benchekroun, H., & Grass, D. (2016). Self-regenerating environmental absorption efficiency and the *Soylent Green* scenario. *Annals of Operations Research, 238*(1), 179–198.
Fudenberg, D., & Levine, D. K. (1988). Open-loop and closed-loop equilibria of dynamic games with many players. *Journal of Economic Theory, 44*(1), 1–18.
Fünfgelt, J., & Schulze, G. G. (2016). Endogenous environmental policy for small open economies with transboundary pollution. *Economic Modelling, 57*, 294–310.
Grass, D., Caulkins, J. P., Feichtinger, G., Tragler, G., & Behrens, D. A. (2008). *Optimal control of nonlinear processes with applications in drugs, corruption, and terror*. Springer.
Huang, X., He, P., & Zhang, W. (2016). A cooperative differential game of transboundary industrial pollution between two regions. *Journal of Cleaner Production, 120*(1), 43–52.
IPCC (Intergovernmental Panel on Climate Change). (2007). *Fourth IPCC Assessment Report: Climate Change 2007*. Cambridge University Press.
Jaakkola, N. (2015). Green technologies and the protracted end to the age of oil: A strategic analysis. *OxCarre Research Paper*, 99, 50 p.
Joos, F., Prentice, I. C., Sitch, S., Meyer, R., Hooss, G., Plattner, G.-K., Gerber, S., & Hasselmann, K. (2001). Global warming feedbacks on terrestrial carbon uptake under the Intergovernmental Panel on Climate Change (IPCC) emission scenarios. *Global Biogeochemical Cycles, 15*(4), 891–907.
Jørgensen, S., Martín-Herrán, G., & Zaccour, G. (2010). Dynamic games in the economics and management of pollution. *Environmental Modelling and Assessment, 15*(6), 433–467.
Judd, K. L. (1998). *Numerical methods in economics*. MIT Press.
Kossioris, G., Plexousakis, M., Xepapadeas, A., de Zeeuw, A. J., & Mäler, K.-G. (2008). Feedback Nash equilibria for non-linear differential games in pollution control. *Journal of Economic Dynamics and Control, 32*(4), 1312–1331.
Lenton, T. M., Williamson, M. S., Edwards, N. R., Marsh, R., Price, A. R., Ridgwell, A. J., Shepherd, J. G., Cox, S. J., & The GENIE team. (2006). Millennial timescale carbon cycle and climate change in an efficient Earth system model. *Climate Dynamics, 26*(7/8), 687–711.
Le Quéré, C., Rödenbeck, C., Buitenhuis, E. T., Conway, T. J., Langenfelds, R., Gomez, A., Labuschagne, C., Ramonet, M., Nakazawa, T., Metzl, N., Gillett, N., & Heimann, M. (2007). Saturation of the Southern ocean CO_2 sink due to recent climate change. *Science, 316*(5832), 1735–1738.
Li, S. (2014). A differential game of transboundary industrial pollution with emission permits trading. *Journal of Optimization Theory and Applications, 163*(2), 642–659.
Long, N. V. (2010). *A survey of dynamic games in economics*. World Scientific.
Mäler, K.-G., Xepapadeas, A., & de Zeeuw, A. J. (2003). The economics of shallow lakes. *Environmental and Resource Economics, 26*(4), 603–624.
Piao, S. L., Ciais, P., Friedlingstein, P., Peylin, P., Reichstein, M., Luyssaert, S., Margolis, H., Fang, J. Y., Barr, A., Chen, A. P., Grelle, A., Hollinger, D. Y., Laurila, T., Lindroth, A., Richardson, A. D., & Vesala, T. (2008). Net carbon dioxide losses of northern ecosystems in response to autumn warming. *Nature, 451*, 49–52.
Rowat, C. (2006). Nonlinear strategies in a linear quadratic differential game. *Journal of Economic Dynamics and Control, 31*(10), 3179–3202.

Rubio, S., & Casino, B. (2002). A note on cooperative versus noncooperative strategies in international pollution control. *Resource and Energy Economics, 24*(3), 251–261.
Stern, N. (2006). *Stern review report on the economics of climate change*. HM Treasury.
Stern, N. (2015). *Why are we waiting? The logic, urgency, and promise of tackling climate change*. MIT Press.
Tahvonen, O., & Withagen, C. (1996). Optimality of irreversible pollution accumulation. *Journal of Economic Dynamics and Control, 20*(9), 1775–1795.
UNFCC. (2008). Poznan Climate Change Conference. https://unfccc.int/conference/poznan-climate-changeconference-december-2008.
Van der Ploeg, F., & de Zeeuw, A. J. (1991). A differential game of international pollution control. *Systems & Control Letters, 17*(6), 409–414.
Van der Ploeg, F., & de Zeeuw, A. J. (1992). International aspects of pollution control. *Environmental and Resource Economics, 2*(2), 117–139.
Vaughan, N. E., & Lenton, T. M. (2011). A review of climate geoengineering proposals. *Climatic Change, 109*(3–4), 745–790.
Wirl, F. (2007). Do multiple Nash equilibria in Markov strategies mitigate the tragedy of the commons? *Journal of Economic Dynamics and Control, 31*(10), 3723–3740.

Chapter 4
Production-Based Pollution vs. Deforestation: Optimal Policy with State-Independent and -Dependent Environmental Absorption Efficiency Restoration Process

Fouad El Ouardighi, Evgeni Khmelnitsky, and Marc Leandri

Abstract An important yet largely unexamined issue is how the interaction between deforestation and pollution affects economic and environmental sustainability. This article seeks to bridge the gap by introducing a dynamic model of pollution accumulation where polluting emissions can be mitigated and the absorption efficiency of pollution sinks can be restored. We assume that emissions are due to a production activity, and we include deforestation both as an additional source of emissions and as a cause of the exhaustion of environmental absorption efficiency. To account for the fact that the switching of natural sinks to a pollution source can be either possible, and in such a case even reversible, or impossible, we consider that restoration efforts can be either independent from or dependent on environmental absorption efficiency, i.e., state-independent versus state-dependent restoration efforts. We determine (1) whether production or deforestation is the most detrimental from environmental and social welfare perspectives, and (2) how state-dependent restoration process affects pollution accumulation and deforestation policies and the related environmental and social welfare consequences.

Keywords Optimal pollution · Deforestation · Environmental absorption efficiency · Restoration process · History dependence

F. El Ouardighi (✉)
ESSEC Business School, Cergy Pontoise Cedex, France
e-mail: elouardighi@essec.fr

E. Khmelnitsky
Tel Aviv University, Ramat Aviv, Israel

M. Leandri
CEMOTEV – UVSQ-Université Paris Saclay, Guyancourt, France

F. El Ouardighi (ed.), *Essays on Pollution Control in Economics and Management Science*, Lecture Notes in Operations Research,
https://doi.org/10.1007/978-3-031-78227-5_4

Introduction

It is well known that deforestation reduces the carbon storage efficiency in the biosphere, thereby causing an increase in the stock of carbon persisting in the atmosphere (e.g., Canadell & Raupach, 2008). However, it is less known that, due to rapidly increased oxidation of soil organic matter created by land use change, deforestation is also a significant carbon source (Baccini et al., 2012; Harris et al., 2012; Houghton et al., 2012). According to Canadell et al. (2007), deforestation was responsible for up to 1.5 billion metric tons of carbon emissions between 2000 and 2007 (Fig. 4.1), which represents 17% of global emissions. Since the carbon accumulation has in turn a destructive impact on the assimilative capacity of oceans, deforestation contributes to switching the biosphere from a carbon sink to a source of carbon in the long run (Cox et al., 2000; Cramer et al., 2001; Joos et al., 2001; Lenton et al., 2006; Piao et al., 2008).

This phenomenon is already patent in the case of tropical forests that have partly switched from sinks to sources of carbon (Baccini et al., 2017). As noted by Gramling (2017), "tropical forests contribute more carbon dioxide to the atmosphere than they remove."

Although the impact of deforestation on climate change is at the heart of international environmental policies, as in the multilateral REDD+ vehicle (Angelsen & Rudel, 2013), the crucial role that deforestation plays in the dynamics of pollution is most often overlooked in the environmental economics literature. This literature has mostly treated pollution control and deforestation as two separate issues so that the interaction between deforestation and pollution and its impact on the economic and environmental sustainability remains yet largely unexamined. One of the main reasons is that possible degradation of environmental absorption efficiency has not been considered in pollution control models for decades. Since the seminal paper by Keeler et al. (1972), pollution control models have aimed to determine an optimal emissions time-path in the presence of negative externalities of the pollution stock, and the related shadow price-based tax scheme necessary to

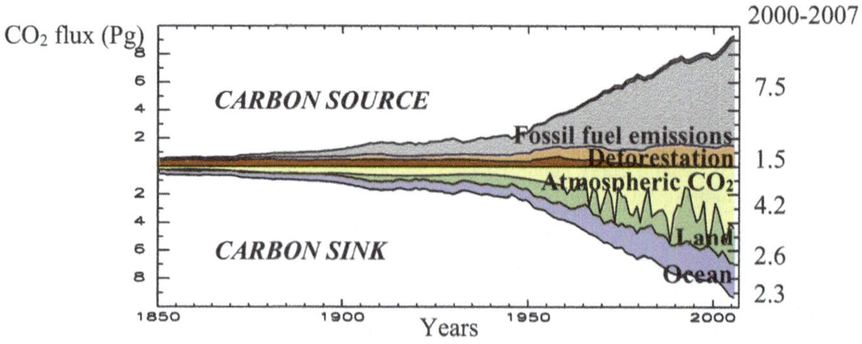

Fig. 4.1 Anthropogenic perturbation of the global carbon budget (Canadell et al., 2007)

implement this time-path. Among the various issues raised by these models, one strand of the literature in the wake of Forster's (1973) intuition has questioned the assumption of constant environmental absorption efficiency. Various authors have considered the possibility for its degradation and possible irreversible exhaustion by formulating it as a function of the pollution stock (Tahvonen & Withagen, 1996; Prieur, 2009; Hediger, 2009; Ayong Le Kama et al., 2014). These improvements did not however capture the effect of inertia on the carbon absorption efficiency degradation process (Leandri & Tidball, 2019), hence the need to model environmental absorption efficiency as a state variable in its own right, either for pollution flows (Leandri, 2009) or the pollution stock (El Ouardighi et al., 2014). As these last models grasp more realistically the dynamics of environmental absorption efficiency, particularly for greenhouse gases accumulation, they can better assess whether the resulting optimal emissions time-path are *environmentally sustainable*, i.e., whether they preserve a positive absorption efficiency in the long run (Leandri & Tidball, 2019). Moreover, they have in turn allowed to considering the restoration of environmental absorption efficiency as another control variable in these models (El Ouardighi et al., 2014, 2016, 2017, 2018).

In contrast, the issue of deforestation has emerged from the need of a trade-off between agriculture and forestry activities in forest-owners' countries (Southgate, 1990; Barbier & Burgess, 1997), which involves a negative externality dimension from a global perspective (Van Soest, 1998; Van Soest & Lensink, 2000) and the need for sustainable international mechanisms for conservation policies (Stähler, 1996; Fredj et al., 2006). A distinction is further suggested between forest owners that draw revenues from timber and agricultural use of deforested land, and a non-forest-owner group that pollutes and suffers the negative externality of carbon accumulation (Andrés-Domenech et al., 2015). Relatedly, conditions for jointly profitable emissions abatement and/or deforestation net reduction are analyzed.

The present article further bridges the gap between these two streams of literature. Our approach differs from the existing studies in that we incorporate a number of essential, empirically established mutual influences between deforestation and pollution accumulation (see Table 4.1 for the essential features of our approach covered by the existing research in pollution accumulation and/or deforestation), that is:

- Deforestation reduces the efficiency of carbon sinks (e.g., Canadell & Raupach, 2008)
- Deforestation generates polluting emissions (e.g., Baccini et al., 2012),
- Pollution accumulation damages the efficiency of carbon sinks (e.g., Raupach et al., 2014),
- The biosphere can switch from carbon sink to a source (e.g., Canadell et al., 2007),
- Restoration of carbon sinks can suffer from inertia (e.g., Liebsch et al., 2008).

Given that pollution accumulation and deforestation both accelerate and intensify global warming and constitute major threats for the earth's future potential of carbon biosequestration, the inclusion of these important mutual influences may enable

Table 4.1 Basic features of research on pollution accumulation and deforestation

Research	Deforestation decreases carbon sinks	Deforestation generates polluting emissions	Destructive impact of pollution on carbon sinks	Possible switch from carbon sink to a source	Restoring carbon sinks can suffer from inertia
Southgate (1990)	–	–	+	–	–
Stähler (1996)	–	–	+	–	–
Tahvonen and Salo (1996)	–	–	+	–	–
Tahvonen and Withagen (1996)	–	–	+	–	–
Barbier and Burgess (1997)	–	–	–	–	–
Van Soest (1998)	+	+	–	–	–
Van Soest and Lensink (2000)	–	–	–	–	–
Sohngen and Mendelsohn (2003)	+	–	–	–	+
Fredj et al. (2006)	–	–	–	–	–
Hediger (2009)	–	–	+	–	–
Leandri (2009)	–	–	+	–	–
Prieur (2009)	–	–	+	–	–
Boucekkine et al. (2013)	–	–	+	–	–
El Ouardighi et al. (2014)	–	–	+	+	–
Moser et al. (2014)	–	–	+	–	–
Andrés-Domenech et al. (2015)	+	–	–	–	–
El Ouardighi et al. (2016)	–	–	+	+	–
El Ouardighi et al. (2017)	–	–	+	+	–
El Ouardighi et al. (2018)	–	–	+	+	–
Leandri and Tidball (2019)	–	–	+	–	–
This paper	+	+	+	+	+

more accurate emissions and deforestation policies than what can be found in the existing literature.

In this regard, this article introduces a dynamic model of pollution accumulation where polluting emissions can be mitigated and the absorption efficiency of pollution sinks can be restored. In the case of greenhouse gases (GHG) accumulation, the absorption efficiency of carbon sinks stems from the aggregated properties of land, notably forests, and oceans to absorb carbon emissions (see Fig. 4.1).[1] The terrestrial absorption efficiency is mechanically reduced by deforestation but also by climate change that can turn tropical forests (Baccini et al., 2017), permafrost and peatlands (Schuur et al., 2015) into carbon sources. The absorption efficiency of oceans is degraded by feedback loops between climate and marine carbon cycle, through either surface water warming, water stratification or thermohaline currents modification (Le Queré et al., 2007; Raupach et al., 2014).

Our second contribution resides in the improved modeling of the environmental absorption efficiency restoration process. In the case of GHG accumulation, efforts to restore this crucial ecosystem service consist mainly in promoting afforestation and reforestation and maintaining other terrestrial carbon sinks such as peatlands (Amesburry et al., 2019). Geo-engineering solutions specifically dedicated to mitigating the negative effects of carbon accumulation on the oceans' carbon absorption, are currently explored (Gattuso et al., 2018; Renforth and Henderson, 2018). Their net impact as well as their possible secondary effects must still be assessed before including them as actual restoration levers. Meanwhile, the potential for additional tree cover on the planet has been estimated to 0.9 billion hectares that would increase by 25% the current atmospheric carbon pool (Bastin et al., 2019). However, biological studies have shown that in the case of reforestation, replanting trees does not allow for a full recuperation of the initial absorption efficiency of forests and that the more severe the depletion of raw forest vegetation the slower the succession time and the more incomplete the recuperation (Liebsch et al., 2008). According to Martin et al. (2013), the carbon pool capacity of a secondary forest "could be expected to be 77–81% of those of undisturbed forests approximately 80 years after disturbance." Therefore, the actual result of restoration spending on the global carbon absorption efficiency can be either independent (the most basic case of isolated afforestation) or dependent on the current state of the forest under restoration. Each assumption has nontrivial consequences on the resulting optimal paths as we will show later. Indeed, if the restoration results are assumed to be state-independent, as they have been until now in the literature (El Ouardighi et al., 2014, 2016, 2017, 2018), it dissociates the actual restoration achieved by a given effort and the current state of the absorption efficiency, thus ignoring the ecological conditions in play in the restoration process. It particularly implies that if the natural carbon sinks have turned to sources, they can be restored back to sinks, which is a rather optimistic take on carbon potential recuperation. Meanwhile, the state-dependent case is better fitted to reflect how the

[1] A similar concept of sink efficiency was introduced by Gloor et al. (2010) and operationalized by Raupach et al. (2014) as the carbon uptake rate by land and ocean sinks.

history of absorption efficiency degradation can influence the effectiveness of restoration spending since the carbon potential added by a given effort will depend on the current conditions of the global absorption efficiency, or at least of its forest component. Moreover, under this assumption it is structurally impossible to turn exhausted carbon sinks to sinks again. It is thus a more conservative assumption that could provide a stronger case for actual environmental policies. Hence our proposition to consider alternatively both hypothesis and compare the subsequent results to capture the effect of restoration *inertia* on optimal pollution/restoration paths.

The dynamic tradeoff between revenues drawn from production of economic goods and deforestation, on the one hand, and costs incurred by pollution externalities and restoration efforts, on the other hand, are analyzed in the setup of a dynamic first-best central planner model. State-independent *versus* state-dependent restoration configurations are considered and their impact on the nature of long-term equilibrium are compared. It is notably shown how key parameters as production-based versus deforestation-based revenues and the discounting rate affect the optimal policy-mix. The environmental sustainability of these optimal policies, i.e. the long-term maintenance of aggregated carbons sinks against a transient shift to carbon sources, are also scrutinized. Furthermore, the behaviors of the optimal paths in the case of state-dependent restoration process highlight the existence of history dependency and the importance of Skiba thresholds (Grass et al., 2008), and thus strengthen our take on the importance of initial conditions in optimal pollution control and deforestation/restoration policies.

The following issues are investigated:

1. Is production-based pollution or deforestation more detrimental from environmental and social welfare perspectives?
2. How does state-dependent restoration process affect pollution accumulation and deforestation policies and the related environmental and social welfare consequences?

To address these issues, we characterize an optimal policy successively for state-independent and state-dependent restoration process and respectively compare in each case its behavior and sustainability with the respective optimal policy mix of mono-activity solutions (production-only and deforestation-only). Depending on the incentive structure in place, this comparison will shed light on the respective impact in the long run of each activity, as well as on the leverage effect that state-dependency induces on the restoration process.

The article is organized as follows. In the next section, we formulate an optimal control model where a social planner seeks to determine the optimal policy in terms of production-based emissions, deforestation rate and restoration efforts. Third section characterizes the general solution of the problem. Fourth section analyzes the model with numerical means, and fifth section concludes the article.

Model Formulation

Our model is an extended version of the model developed in El Ouardighi et al. (2014, 2016). In a dynamic continuous framework, the social planner seeks to maximize intertemporal social welfare by weighing benefits from polluting economic activities, the costs of restoration efforts and environmental damages caused by stock pollution accumulation subject to endogenous variations of the environmental absorption efficiency.

We consider two welfare-relevant activities: forest exploitation and production of economic goods. Both activities, i.e., deforestation and production, generate polluting emissions. The emissions rate resulting from production, that is, production-based emissions, is denoted by $u(t) \geq 0$. On the other hand, the deforestation rate, denoted by $v(t) \geq 0$, instantaneously generates polluting emissions at a proportional rate (e.g., Baccini et al., 2012; Harris et al., 2012; Houghton et al., 2012; Canadell et al., 2007), that is, $\alpha v(t)$, where $\alpha > 0$.

Therefore, the evolution of the pollution stock, denoted by $P(t) \geq 0$, is described as:

$$\dot{P}(t) = u(t) + \alpha v(t) - A(t)P(t) \qquad P(0) = P_0 \geq 0 \tag{4.1}$$

In Eq. 4.1, the pollution stock decreases at an environmental absorption efficiency rate $A(t)$, which obeys the transition equation:

$$\dot{A}(t) = w(t)A(t)^{\beta} - v(t) - \gamma P(t) \qquad A(0) = A_0 > 0 \tag{4.2}$$

where the units of α and $v(t)$ are scaled to fit both Eqs. 4.1 and 4.2. The environmental absorption efficiency, $A(t)$, reflects the aggregated properties of oceans and forests as pollution sinks. Other studies where the environmental absorption efficiency is modeled as a state variable are El Ouardighi et al. (2014, 2016) and El Ouardighi et al. (2017, 2018). According to Eq. 4.2, environmental absorption efficiency decreases with both deforestation and the destructive impact of the pollution stock on oceans. The parameter $\gamma > 0$ inversely reflects the internal capacity of absorption efficiency, that is, oceans, to resist the destructive impact of pollution. It is clear that this simple linear degradation mechanism does not capture all the complex ecological processes at stake in the differentiated reactions of oceans', peatlands', and forests' carbon uptake to accumulated GES. But our main concern is to make sure that the pollution stock feedback loops are accounted for without overwhelming technical complexity.

Further, the environmental absorption efficiency increases with forest restoration efforts, denoted by $w(t) \geq 0$. This assumption is consistent with, e.g., Canadell and Raupach (2008), who assert that reforestation policies can increase the potential for carbon biosequestration. However, to account for the possibility of *inertial* restoration process in Eq. 4.2, restoration efforts can be either independent from or linearly dependent on the magnitude of absorption efficiency. The restoration efforts' state-dependency is such that β takes the value 0 if the restoration process suffers from inertia, or 1 if it is impossible. Thus, the actual impact of restoration efforts $w(t)$ on the environmental absorption efficiency will amount to $w(t)A(t)^{\beta}$ with, as explained above, β capturing whether the restoration process is state-independent ($\beta = 0$) or linearly state-dependent ($\beta = 1$). We aggregate in $A(t)$ the various carbon sinks of the biosphere, assuming that if $\beta = 0$, restoration can increase $A(t)$, through afforestation for example, independently of the current level of $A(t)$. Particularly, even if $A(t)$ is negative, it can be restored back to a positive value. Conversely, if $\beta = 1$, the restoration achieved will depend upon the current state of environmental absorption efficiency and thus on its past evolution. Under this condition, restoration efforts can reflect a densification policy within an existing forest: the carbon sequestration potential is increased by different species mix or lengthening the rotation period (Sohngen & Mendelsohn, 2003) but depends on the actual size and state of the forest already in place. In that case, if the forest has turned to a source, i.e., if $A(t)$ is negative, restoration cannot raise it back to a positive level, i.e., the shift is irreversible. To avoid diverging solution in the case of state-dependent restoration efforts, we assume a strictly positive initial absorption efficiency, that is, $A_0 > 0$. Overall, expressions (Eqs. 4.1 and 4.2) describe the joint dynamics of the pollution stock and environmental absorption efficiency.

We now define an objective criterion. Revenues are supposed to be drawn from production-based emissions and deforestation, whereas costs are incurred from pollution negative externalities and restoration efforts. We assume an increasing concave revenue function for the production-based emissions and use a standard specification (see Wirl, 2007), that is, $u(t)(a - u(t)/2)$, where $a \geq 0$ is the revenue-maximizing production level, if any. The revenues drawn from deforestation are related to the amount of wood retrieved from clear-felling, if any, and to land use change (urbanization, agriculture, ranching, etc.). For simplicity, we omit revenues from selective logging, and consider that the revenues are drawn from deforestation. In all cases, we assume that deforestation generates only short-term rather than long-lasting revenues. This is in contrast with the optimistic assumption of Andrés-Domenech et al. (2015) that revenues can be drawn from deforested land over a significant time interval, i.e., 50 years. In fact, deforestation is a main cause of reduced water cycling (and rainfall) and of local climate change, including major occurrence of droughts (e.g., Sheil & Murdiyarso, 2009; Makarieva et al., 2014), which makes deforested lands more eligible for ephemeral rather than long-lasting beneficial agricultural exploitation and economic development (Rodrigues et al., 2009).

In this regard, as we did for production-based emissions, we choose to assume an increasing concave revenue function for deforestation, that is, $v(t)(b - v(t)/2)$, where $b \geq 0$ is the revenue-maximizing deforestation level, if any. One merit of this prudent assumption is that it lets us avoid treating deforestation as a privileged source of revenues. The tipping point $2b$ interprets as an upper bound for the deforestation rate above which the magnitude of deforestation is so extreme that it engenders costs rather than revenues. Further, the negative externalities of pollution are valued as an increasing convex function of the pollution stock (e.g., Michel & Rotillon, 1995), that is, $cP(t)^2/2$, $c > 0$. Finally, due to decreasing returns, the restoration efforts generate an increasing quadratic cost, $dw(t)^2/2$, $d > 0$, (see El Ouardighi et al., 2016). Without loss of generality, we assume that the cost coefficient of restoration efforts, d, is normalized to 1, $d = 1$, which implies that d can be interpreted relative to c, the cost coefficient of the pollution stock.

Assuming an infinite planning horizon, and denoting the discounting rate by $r > 0$, the optimal control problem writes:

$$\text{Max } W = \int_0^\infty e^{-rt}\left[u(t)(a - u(t)/2) + v(t)(b - v(t)/2) - cP(t)^2/2 - w(t)^2/2\right]\mathrm{d}t \tag{4.3}$$

under the constraints Eqs. 4.1–4.2 and $u(t) \geq 0$, $v(t) \geq 0$, $w(t) \geq 0$.

Overall, the time paths that can be generated by the problem decision defined by Eq. 4.3 under the transition Eqs. 4.1–4.2, are tightly dependent upon the value of β. In the case where restoration efforts are independent from the environmental absorption efficiency, i.e., $\beta = 0$, a negative value of $A(t)$ is possible over a finite time interval. However, in the case where restoration efforts depend upon the environmental absorption efficiency, i.e., $\beta = 1$, it results that $A(t) \geq 0$ for all t.

Optimality Conditions and Stability

Skipping the time index for convenience, the current-value Hamiltonian is:

$$H = u(a - u/2) + v(b - v/2) - cP^2/2 - w^2/2 + \lambda(u + \alpha v - AP) + \varphi\left(wA^\beta - v - \gamma P\right) \tag{4.4}$$

and the extended Hamiltonian is:

$$L(P, A, u, v, w, \lambda, \varphi, \rho_1, \rho_2, \rho_3) := H(P, A, u, v, w, \lambda, \varphi) + \rho_1 u + \rho_2 v + \rho_3 w \tag{4.5}$$

where $\lambda(t)$ and $\varphi(t)$ are costate variables associated with the pollution stock, and the absorption rate, and ρ_1, ρ_2, ρ_3 are Lagrange multipliers corresponding to the nonnegativity constraints imposed on u, v and w, respectively.

Then an optimal solution $(P^*, A^*, u^*, v^*, w^*)$ satisfies the Hamiltonian maximizing condition:

$$(u^*, v^*, w^*) = Argmax_{u \geq 0, v \geq 0, w \geq 0}\, H(P^*, A^*, u\ v, w, \lambda, \varphi) \tag{4.6}$$

and the costates $\lambda(t)$ and $\varphi(t)$ satisfy the adjoint equations:

Table 4.2 Qualification constraint

Arcs	1	2	3	4	5	6	7	8
ρ_1	0	0	–	0	–	–	0	–
ρ_2	0	–	0	0	–	0	–	–
ρ_3	0	0	0	–	0	–	–	–

$$\dot{\lambda} = r\lambda - H_P(P^*, A^*, u^*, v^*, w^*, \lambda, \varphi) \tag{4.7}$$

$$\dot{\varphi} = r\varphi - H_A(P^*, A^*, u^*, v^*, w^*, \lambda, \varphi) \tag{4.8}$$

Lemma 1. The Hamiltonian is concave with respect to the control vector $(u\ v, w)$ and guarantees a (local) maximum of the Hamiltonian.

Proof See Appendix A1.

Therefore, the Hamiltonian maximizing condition (Eq. 4.6) allows us to distinguish eight different arcs, which are presented in the following table (Table 4.2).

The relevant arcs under study for optimal environmental policy will be arcs 1, 2 and 3. Indeed, arcs 4, 6 and 7 display nil restoration efforts and positive emissions, through either or both production and deforestation. According to Eq. 4.2, they will eventually result in diverging paths. They might thus be the initial or an intermediate part of the optimal path but none of them can be followed over an infinite time horizon. Engaging in restoration efforts will in fact always be necessary at some point along our optimal policy mix. As for arcs 5 and 8, they are characterized by no emissions from any source (i.e., uncontrolled model). As such, they are not relevant for actual environmental policies and could only reflect pre-industrial states. Let us then focus on the three arcs with restoration that can appear at any stage of the optimal path, that is, the production-based emissions and deforestation arc, the production-based emissions-only arc, and the deforestation-only arc. These arcs are the only candidates for the final stage when the path converges to equilibrium. In economic terms, we can thus assert that the optimal policy-mix will always include restoration but can rely either on production only (arc 2), deforestation only (arc 3) or production and deforestation combined (arc 1).

For each of arcs 1, 2, and 3, explicit functions $(u^\circ, v^\circ, w^\circ, \rho_1, \rho_2, \rho_3)$ can be determined such that the costate Eqs. 4.7 and 4.8 write:

$$\dot{\lambda} = (r + A)\lambda + \gamma\varphi + cP \tag{4.9}$$

$$\dot{\varphi} = r\varphi + \lambda P - \beta\varphi w^\circ A^{\beta-1} \tag{4.10}$$

The properties of the three arcs are successively analyzed in Appendix A2.

Proposition 1. Along arc 1, the optimal restoration effort increases linearly with the production-based emission rate and decreases linearly with the deforestation rate, for any value of β.

Proof See Appendix A3.

Proposition 2. Given a set of parameter values $(a, b, c, r, \alpha, \beta, \gamma)$, the number of interior steady states, if they exist, is determined by the number of feasible solutions to:

$$\varphi_\infty\left(r - \beta\varphi_\infty A_\infty^{2\beta-1}\right) + \lambda_\infty P_\infty = 0 \tag{4.11}$$

where:

$$\varphi_\infty = \frac{c(b - \alpha a) + [\gamma(a + \alpha b) + bA_\infty](r + A_\infty)}{c + \gamma[\alpha r + \gamma(1 + \alpha^2)] + (r + 2\alpha\gamma + A_\infty)A_\infty + [c(1 + \alpha^2) + (r + A_\infty)A_\infty]A_\infty^{2\beta}}$$

$$\lambda_\infty = -\frac{ac + \gamma[\gamma(a + \alpha b) + bA_\infty] + c(a + \alpha b)A_\infty^{2\beta}}{c + \gamma[\alpha r + \gamma(1 + \alpha^2)] + (r + 2\alpha\gamma + A_\infty)A_\infty + [c(1 + \alpha^2) + (r + A_\infty)A_\infty]A_\infty^{2\beta}}$$

$$P_\infty = \frac{\gamma(\alpha a - b) + \left[a + (a + \alpha b)A_\infty^{2\beta}\right](r + A_\infty)}{c + \gamma[\alpha r + \gamma(1 + \alpha^2)] + (r + 2\alpha\gamma + A_\infty)A_\infty + [c(1 + \alpha^2) + (r + A_\infty)A_\infty]A_\infty^{2\beta}}$$

where the subscript '∞' denotes the steady state.

Proof See Appendix A4.

Clearly, the number of feasible, interior steady states cannot be established analytically. Therefore, we resort to numerical means in the following section.

Numerical Analysis and Economic Implications

To determine which among production and deforestation is the most detrimental in terms of welfare and environmental sustainability depending on whether the exhaustion of environmental absorption efficiency is possible or not, we compare the time paths related to three arcs: production-based emissions-deforestation, production-based emissions-only and deforestation-only.

For each case, the parameter space is divided into regions, and each region is characterized by the number of feasible steady states and by the transient behavior in the neighborhood of each steady state. In this regard, we select the values for the parameters as shown in Table 4.3. By varying parameters a and b, we show that solutions including (locally) stable equilibria with monotonic or oscillating convergence and Skiba behavior (Grass et al., 2008) are possible.

The set of parameter values related to a unique equilibrium with monotonic convergence in Table 4.3 corresponds to the base case. It reflects a situation characterized by relative patience of the regulator with the discounting rate $r = 0.05$, limited destructive impact of pollution on environmental absorption

Table 4.3 Parameter values and related solutions

Description	r	a	b	c	γ	α	β
Unique equilibrium with monotonic convergence (Figs. 4.4 and 4.11)	0.05	0.4	0.1	0.001	0.05	0.15	0, 1
Unique equilibrium with oscillating convergence (Fig. 4.7)	*	0.087	1	*	*	*	0
Skiba with two equilibria and monotonic convergence (Fig. 4.14)	*	0.053	0.1	0.01	*	*	1

efficiency with $\gamma = 0.05$, intermediate marginal incentives for production and deforestation with $a = 0.4 > b = 0.1$, relatively low marginal pollution cost (i.e., lower than marginal restoration cost) with $c = 0.001$, and a typical value for the marginal emissions induced by deforestation with $\dot{P}_v = \alpha = 0.15$ (see Houghton et al. (2012)). Further, the unique equilibrium with oscillating convergence differs from the base case only because of a very low marginal incentive for production and a very high marginal incentive for deforestation with $a = 0.087 < b = 1$, and state-independent restoration efforts ($\beta = 0$). Regarding the case of Skiba behavior with two equilibria, it departs from the base case because of a very low marginal incentive for production ($a = 0.053 < b = 0.1$), a relatively higher marginal pollution cost ($c = 0.01$), and linearly state-dependent restoration efforts ($\beta = 1$).To compare the transient paths associated with the different cases and to assess their sensitivity to the initial conditions, we assume two initial values for each state variable, that is, small and large (i.e., smaller and larger than the base case steady state) values for the initial environmental absorption efficiency with $A_0 = (0.1, 2.5)$, and for the pollution stock with $P_0 = (0.05, 3)$. Although our numerical analysis includes *idyllic initial conditions* with initially small pollution stock and large absorption efficiency as a benchmark, i.e., $(A_0, P_0) = (2.5, 0.05)$, we are mainly interested in *relatively unfavorable* and *favorable* initial conditions, that is, $(A_0, P_0) = (0.1, 3)$ and $(A_0, P_0) = (2.5, 3)$.

For the numerical resolution, a time decomposition method (Maimon et al., 1998) is used to approximate the optimal paths (see Appendix A5). We successively study the cases with state-independent, and linearly state-dependent restoration process.

State-Independent Restoration Process

In the case where restoration efforts are state-independent and can thus reverse the switching of natural sinks into a pollution source, the value of the sum of the principal minors of order 2 of the Jacobian matrix minus the squared discounting rate, $K_{|\beta=0}$, is negative for any feasible steady state, which rules out the occurrence of a limit-cycle (see details in Appendix A2). Therefore, the steady states, if they exist and are feasible, are categorized into saddle node, saddle focus, or unstable equilibrium.

Figure 4.2 shows the bifurcation diagram in the (b, a) parameter space for $\gamma = 0.05$, $\alpha = 0.15$, $r = 0.05$, and three values of c, $c = (0.001, 0.01, 0.1)$.

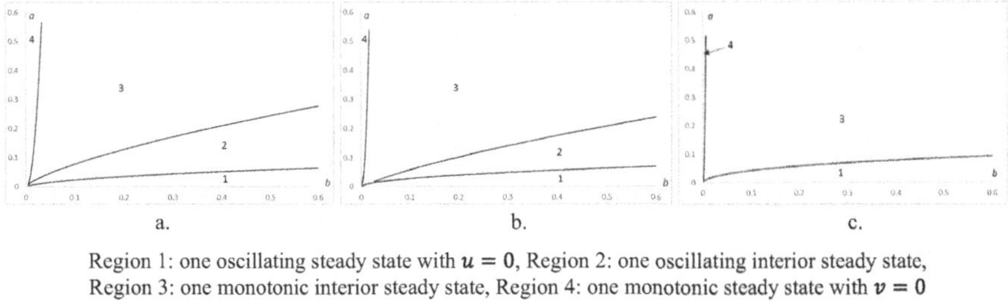

Fig. 4.2 Bifurcation diagram in the (b, a) parameter space with $c = (0.001, 0.01, 0.1)$. (**a**) $c = 0.001$, (**b**) $c = 0.01$, (**c**) $c = 0.1$. Region 1: one oscillating steady state with $u = 0$, Region 2: one oscillating interior steady state, Region 3: one monotonic interior steady state, Region 4: one monotonic steady state with $v = 0$

Figure 4.2a, b predict that a significantly unbalanced structure of incentives for production and deforestation (low *a vs.* high *b* and vice versa), leads in the long run to a unique (locally stable) steady state characterized by specialization of activity with higher revenues and withdrawal from the least remunerating activity. In the case of a prevalent incentive for deforestation (intermediate *a vs.* high *b*), convergence to the unique (locally stable) steady state is oscillatory. Nonetheless, these patterns are affected by a higher pollution cost coefficient. It is interesting to study how these patterns are affected by the magnitude of the environmental damage factor, *c*. First, as the marginal environmental damage increases, by a factor 10 (Fig. 4.2b) then by a factor 100 (Fig. 4.2c), Region 2 respectively decreases and then disappears, in favor of Region 3, and to a lesser extent Region 1. Oscillations are thus eliminated by a large enough environmental damage when it comes to interior solutions. Second, a higher *c* increases slightly Region 1 and reduces Region 4. This means that when a unit of stock pollutant imposes a higher environmental damage, the specialization in deforestation is reinforced for wider range of (b, a) combinations while the specialization in production is less likely. The economic interpretation of this shift relies on the fact that an increase in *c* reduces drastically the restoration cost relatively to the marginal environmental damage, as *d* has been normalized to 1 with respect to *c*. Since we are dealing with state-independent restoration in this subsection, this implies that increasing or maintaining the environmental absorption efficiency through restoration becomes automatically relatively cheaper. Consequently, restoration efforts can be engaged to compensate the detrimental effort of deforestation on the environmental absorption efficiency, and we will show later that this compensation is systematic. Considering that deforestation adds significantly less pollution to the accumulated stock than production given that the deforestation emission factor is $\alpha = 0.15 < 1$, the trade-off between deforestation and production is now in favor of the former.

For the base case values ($a = 0.4$, $b = 0.1$, $c = 0.001$, $\gamma = 0.05$, $\alpha = 0.15$, $r = 0.05$), convergence to the unique locally stable steady-state is monotonic (Fig. 4.4). Table 4.4 reports the values of state and control variables at the steady state with both production and deforestation (interior solution) and with either

Table 4.4 Steady-state values with both production and deforestation and with either production or deforestation

	A_∞	P_∞	u_∞	v_∞	w_∞
Interior solution	0.664	0.602	0.395	0.035	0.065
Corner solution with $v(t) = 0$	0.721	0.551	0.398	–	0.028
Corner solution with $u(t) = 0$	0.063	0.108	–	0.046	0.051

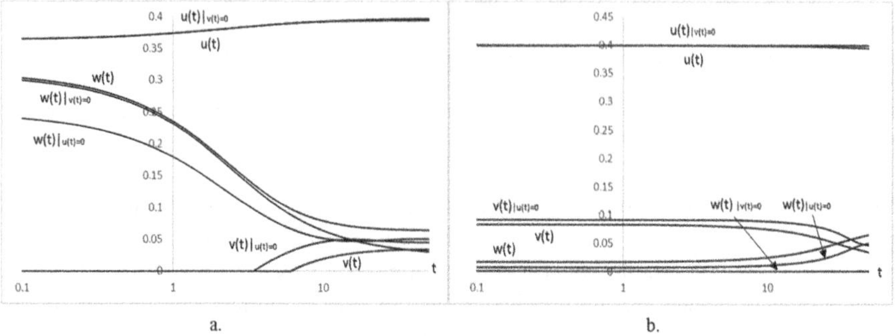

Fig. 4.3 Control policies from (**a**) relatively unfavorable and (**b**) favorable initial conditions (log. time scale)

production or deforestation (corner solutions). We observe that the steady-state production-based emissions are almost unaffected by eradication of deforestation. In contrast, the steady-state restoration efforts are much lower without deforestation than in the converse case. Nevertheless, the steady-state values of absorption efficiency and pollution stock are respectively higher and lower than in the case with deforestation. In other words, environmental sustainability improves with eradication of deforestation. In contrast, eradication of production-based emissions results in both lower environmental absorption efficiency and pollution stock in the long run, with a slightly higher deforestation rate and slightly lower restoration efforts. Overall, deforestation requires more restoration efforts than do production-based emissions in the long run.

Figure 4.4 shows the convergence to the interior steady state (black dot) from various initial conditions. The influence of production and deforestation is illustrated by comparing the interior steady state with the steady state that involves only production-based emissions (unfilled dot) or only deforestation (unfilled star). When starting from unfavorable initial conditions, the optimal path consists in first slightly increasing the absorption efficiency, and then rapidly decreasing the pollution stock while increasing the former until the steady state is reached. This is achieved with initially high and decreasing restoration efforts, and later on with initially intermediate and slightly increasing production-based emissions as shown in Fig. 4.3a. As for deforestation, it is postponed until the absorption efficiency is sufficiently restored to resist its impact. Even the deforestation-only path delays actual deforestation until sufficient restoration has been carried out.

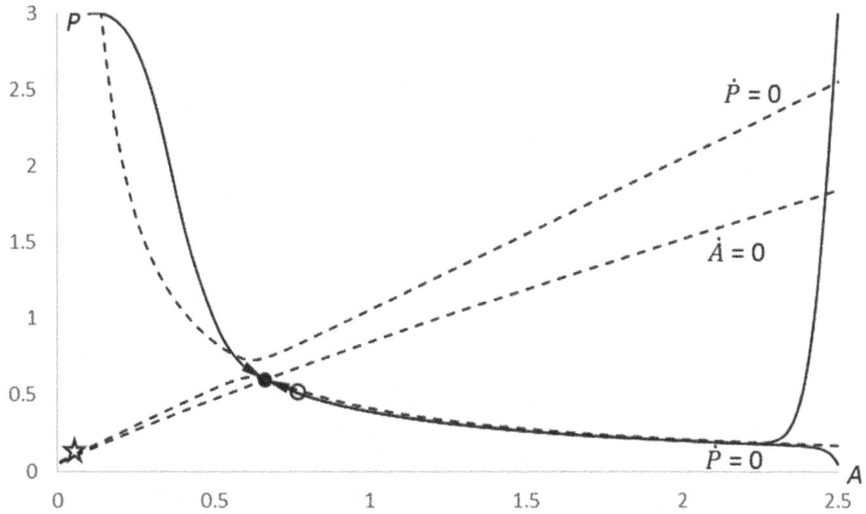

Fig. 4.4 Phase diagram in the state space with monotonic convergence (filled dot: interior solution, unfilled dot: corner solution with $v = 0$, unfilled star: corner solution with $u = 0$)

The policy implications of these results are twofold. If the environmental absorption efficiency is already low, it is optimal to restore it at the earlier stage of the path, even if the restoration costs are high relatively to the environmental damage factor. This is true including for the production-only path, even if it converges towards a lower steady state restoration level than the other two. Moreover, under such initial conditions it is always preferable to delay deforestation until the absorption efficiency has been at least partially restored. This result can be interpreted as a form of consistency between environmental sustainability and economic optimality but it holds under the assumption of this subsection that restoration is state-independent. If it were not the case, restoring the environmental absorption efficiency from a very low initial level could prove extremely costly.

From favorable initial conditions (Fig. 4.3), the optimal path consists first in rapidly decreasing the pollution stock and slowly decreasing the absorption efficiency, and then rapidly decreasing the latter and slowly increasing the former until the steady state is reached. This is implemented with initially low and increasing restoration efforts, and later on with on the one side initially high and slightly decreasing production and on the other side on initially high and decreasing deforestation, as shown in Fig. 4.3b. In that case, the initial absorption efficiency is sufficiently high to endure immediate deforestation, in the policy-mix as well as in the deforestation-only solution. (i.e., greater restoration efforts coincide with lower deforestation, and vice versa).

We note also the level of restoration efforts that is much higher in the interior solution and the deforestation-only case than in the production-only case. This result reflects the fact that in a production-only solution the environmental absorption efficiency will only be harmed by the feedback mechanisms (γ) while a policy involving deforestation will diminish the absorption efficiency much more rapidly. Consequently, heavier long-term restoration efforts are needed in the optimal policy-mix and in the deforestation-only solution. However, if the former succeeds in maintaining a significant level of environmental absorption efficiency, the latter ends up with an almost depleted absorption, while the production-only solution maintains the highest absorption efficiency (see Fig. 4.4). With the set of parameters of this subsection, it appears that deforestation is the activity that threatens most the environmental sustainability along an economic path as it impairs significantly and indefinitely the environmental absorption efficiency. In Fig. 4.4, the convergence towards A_∞ from low initial absorption efficiency levels highlights the crucial role of restoration to avoid the irreversible depletion of this essential ecosystem service (Leandri, 2009).

As shown in Fig. 4.5, both under intermediate (high A_0 and P_0) and unfavorable initial conditions, the cumulative welfare associated to the deforestation-only solution is drastically lower than the welfare of the interior and the production-only paths. This obviously reflects the superior profitability of production relatively to deforestation set by the parameters, but it also translates the need to engage in additional costly restoration efforts if deforestation comes into in play. Although it also includes some deforestation, in the case of the interior solution this additional restoration cost is largely compensated through the exploitation at full rate of the production activity. As expected, the social welfare is lower along any paths when the initial conditions are less favorable, but we can note that the welfare differential between the optimal interior solution and the production-only solution is stronger for intermediate initial conditions. Indeed, in the short-term production only yields more welfare as the restoration efforts are lower (and the emissions are similar). However,

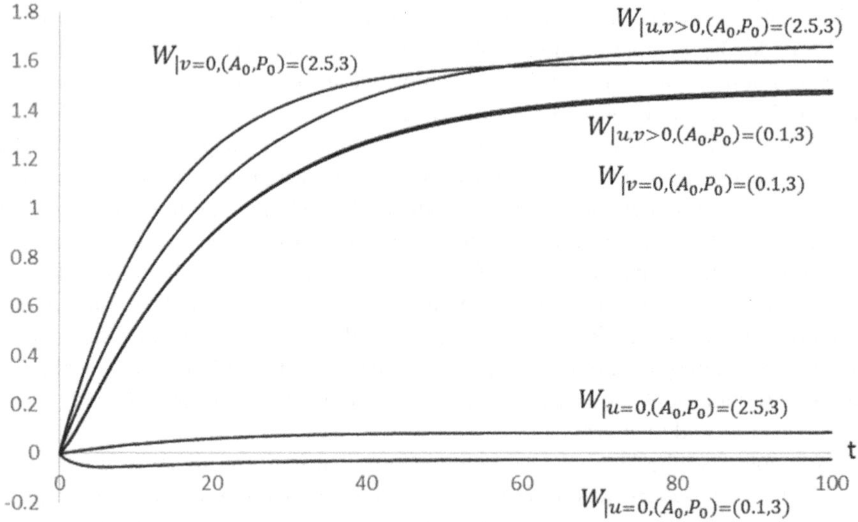

Fig. 4.5 Cumulative utility over time from relatively unfavorable and favorable initial conditions

after some time, the additional profit from deforestation kicks in without threatening environmental sustainability and the interior solution's cumulative welfare exceeds the production-only solution. It is remarkable that under unfavorable initial conditions, both interior and production-only solutions achieve the same welfare, through a similar production/restoration initial mix until the interior solution activates deforestation (see Fig. 4.3a). Considering that the production-only path converges towards a more environmentally sustainable steady state (higher A_∞ in Table 4.4) and imposes less negative externalities (lower P_∞), we can conclude that in this base case scenario (production more profitable and restoration costs relatively high) if the initial ecological conditions are unfavorable, the social planner should exclude deforestation from its policy-mix and resort to a production-only solution. This interesting result illustrates how (unfavorable) initial conditions can shape the content of an optimal policy when feedback effects are taken into account.

Figure 4.6 shows the sensitivity of the steady state with respect to the parameters c, r and γ. For $c = 0.001$ and r going from 0.01 to 0.1, the steady-state values of absorption efficiency and pollution stock are respectively lower and higher, as Fig. 4.6a illustrates. Figure 4.6b shows a similar dependence for $c = 0.1$. The steady-state absorption efficiency and pollution stock respectively decrease and increase monotonically with a higher discounting rate. Since the absorption efficiency is a service that allows to keeping pollution accumulation under control in the long run, its importance diminishes as the discounting rate increases. This sensitivity is less acute when the environmental damage factor is much higher. Indeed, despite a myopic social planner, future damages will nonetheless be taken into consideration when c is very high, thus limiting the increased (decrease) in accumulated pollution (absorption efficiency). The impact of γ on this discounting rate sensitivity is worth noting: when γ is equal to 0.05 as we assumed until now, i.e., when the pollution stock is less destructive to the absorption efficiency, the negative impact of a greater discounting rate is weaker, in the low c case (Fig. 4.6a). Keeping in mind that a low c captures a relatively higher restoration cost, this result can be seen as a reaction to this higher cost. If the pollution stock is more destructive than expected to the absorption efficiency (i.e., $\gamma = 0.2$), then loosing periodically a significant part of the absorption efficiency through these feedbacks will prove much more costly

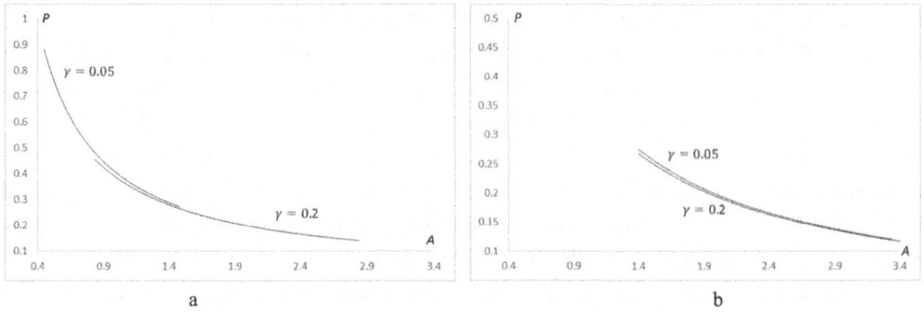

Fig. 4.6 Locus of the steady-state values for $r = (0.01, 0.1)$ and $c = (0.001, 0.1)$. (**a**) $c = 0.001$, (**b**) $c = 0.1$

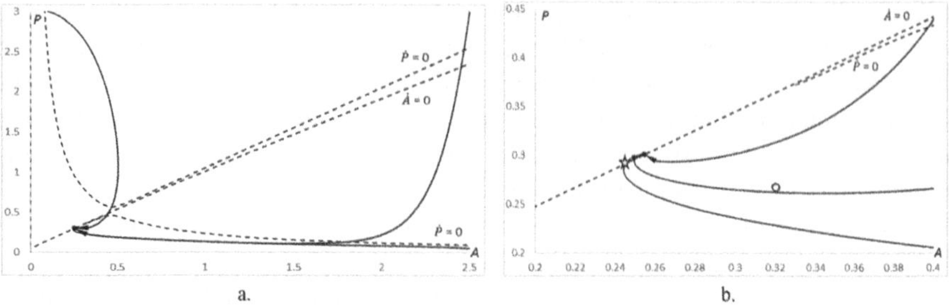

Fig. 4.7 Phase-portrait diagram in the state space with oscillating convergence (filled dot: transiently interior solution, unfilled dot: corner solution with $v = 0$, unfilled star: corner solution with u =0). (**a**) Phase-portrait diagram, (**b**) Zoom in the vicinity of the steady point

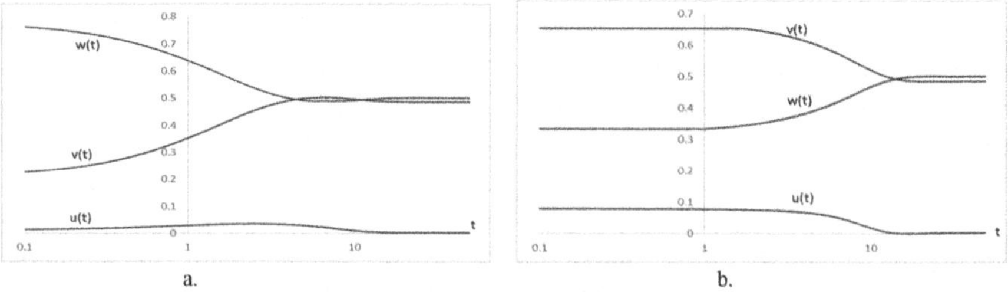

Fig. 4.8 Control policies from (**a**) relatively unfavorable and (**b**) favorable initial conditions

restoration-wise. As a result, the pollution stock will not be so drastically raised under an increased discounting rate. However, when restoration costs are lower (greater c), the value of γ does not affect the discounting rate sensitivity (Fig. 4.6b). This sensitivity with respect to γ is of crucial importance for environmental policies. Indeed, it warns us against an underestimation of γ that would justify an optimal pollution path threatening the environmental sustainability with low A and imposing high damages through a high P. Given the major uncertainties that still characterize the scientific understanding of these feedback loops, especially in the case of oceans (Leandri & Tidball, 2019), it is crucial to dedicate monitoring efforts to better grasp the order of magnitude of γ and avoid over-optimistic parameters.

Figure 4.7 shows that oscillating convergence is obtained for $a = 0.087$ and $b = 1$, all things being equal. Due to the near-zero incentive for production, the steady state without production-based emissions is now very close to the transiently interior steady state (Fig. 4.7b). In contrast, the eradication of deforestation increases absorption efficiency and reduces pollution. This result suggests that environmental sustainability is always worse with deforestation alone than with production-based emissions alone, regardless of the incentive structures for the two activities.

Figure 4.8 confirms that deforestation and restoration efforts are mutual substitutes, and suggests that production-based emissions should be limited both in magnitude and duration for unfavorable initial conditions. In the long run, restoration efforts should always be greater than deforestation.

Linearly State-Dependent Restoration Process

In the case of linearly state-dependent restoration process, $\beta = 1$, and the sum of the principal minors of order 2 of the Jacobian matrix minus the squared discounting rate, $K_{|\beta\,=\,1}$, is negative for all combinations of parameters, which rules out the occurrence of a limit-cycle (see details in Appendix A2). Figure 4.9 shows the dependence of $K_{|\beta\,=\,1}$ on b for $\gamma = 0.05$, $\alpha = 0.15$, $r = 0.05$, $a = (0.01,1)$ and $c = (0.001, 0.01, 0.1)$. Here also, for $a = 0.01$, the three curves corresponding to the three values of c cannot be distinguished.

On the other hand, Fig. 4.10 shows the bifurcation diagram in the (b, a) parameter space for steady states corresponding respectively to interior and corner solutions. The number of (locally) stable steady states is at most 2, with all stable paths being monotonic.

For the base case parameter values ($a = 0.4$, $b = 0.1$, $c = 0.001$, $\gamma = 0.05$, $\alpha = 0.15$, $r = 0.05$), there is a unique steady state with monotonic convergence (Fig. 4.11). Table 4.5 summarizes the value of the steady state and control variables with both production-based emissions and deforestation and with either production-based emissions or deforestation. Under linear state-dependent restoration process, we observe that the absorption efficiency in the interior solution is significantly greater than with the production-only corner solution. Long-run environmental sustainability is improved by the inclusion of deforestation in the optimal policy. This feature is distinct from the state-independent restoration case, where the inclusion of deforestation in the solution actually reduced the absorption efficiency (see Table 4.5). The reason is that the restoration efforts are now greater in the former than in the latter case at the steady state, which allows superior capitalization. These

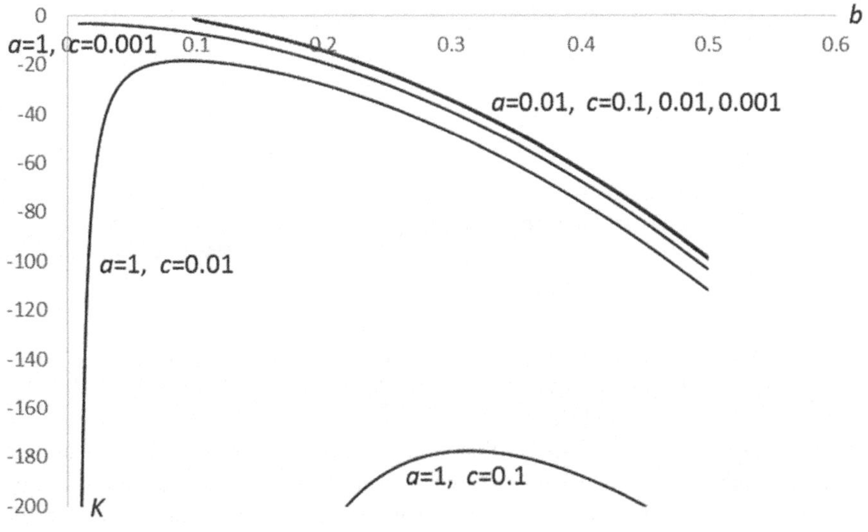

Fig. 4.9 Parameter $K_{|\beta\,=\,1}$

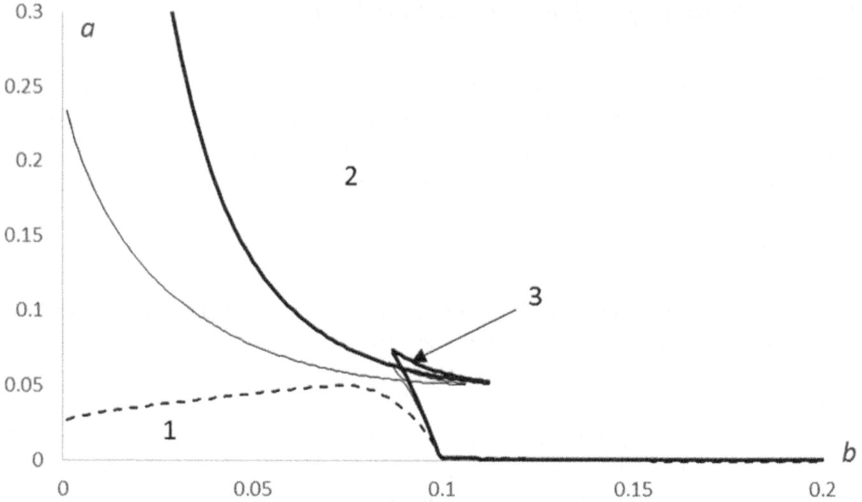

Fig. 4.10 Bifurcation diagram in the (b, a) parameter space. Bold, thin, and dashed lines for $c = (0.001, 0.01, 0.1)$. Region 1: no interior steady state, Region 2: one interior steady state with a monotonic path, Region 3: two interior steady states with monotonic paths

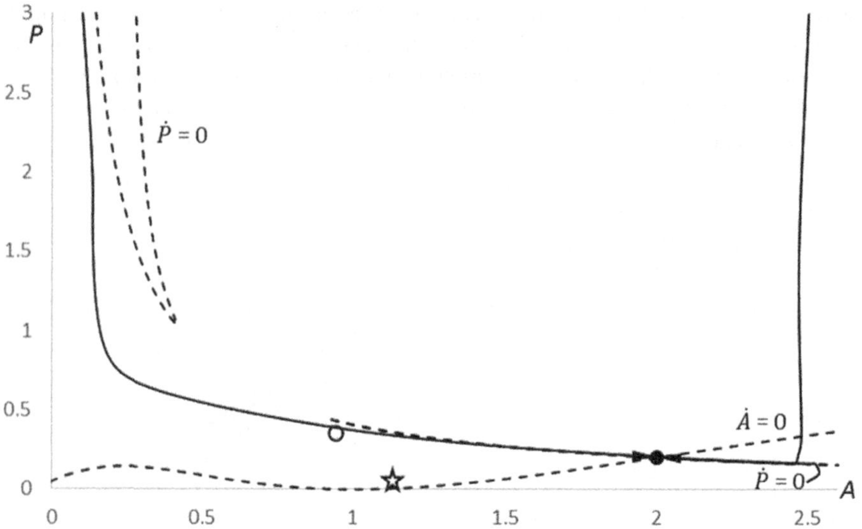

Fig. 4.11 Phase diagram in the state space with monotonic convergence (filled dot: interior solution, unfilled dot: corner solution with $v = 0$, unfilled star: corner solution with $u = 0$)

results can be explained by the fact that in the context of state-dependent restoration efforts, a high absorption efficiency (greater than 1) increases the actual restoration impact of a given restoration effort. The greater the current absorption efficiency is, the greater the effectiveness of the restoration efforts. As a result, the optimal interior

Table 4.5 Steady-state values with both production and deforestation and with either production or deforestation

	A_∞	P_∞	u_∞	v_∞	w_∞
Interior solution	2.001	0.205	0.399	0.078	0.044
Corner solution with $v(t) = 0$	0.957	0.416	0.398	–	0.022
Corner solution with $u(t) = 0$	1.128	0.0074	–	0.056	0.050

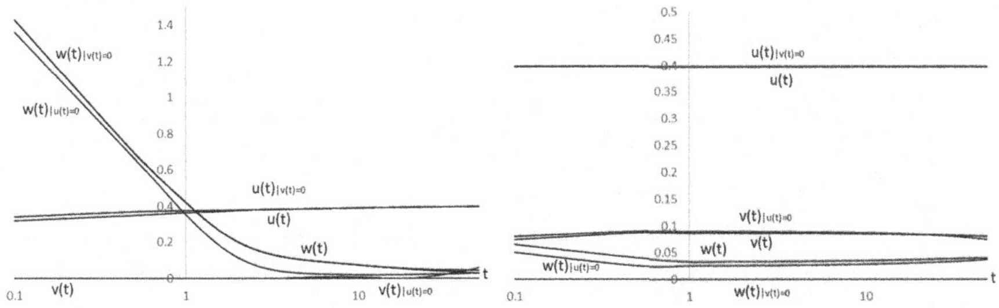

Fig. 4.12 Control policies from relatively unfavorable and favorable initial conditions (log. time scale)

solution can allow significantly more deforestation than in the state-independent restoration process case (roughly twice as much) and compensate for the subsequent loss in absorption efficiency with even less restoration efforts (0.044 *vs.* 0.065), since the latter are now much more effective.

Figure 4.12 confirms that production-based emissions are almost insensitive to deforestation. Restoration efforts are now lesser in the case of deforestation-only in the long run, except in the case of production-based emissions-only. From relatively unfavorable initial conditions, deforestation without production-based emissions is always near zero. Also, it is necessary to apply very high restoration efforts from the beginning that can decrease gradually as the absorption efficiency rise enough to increase the effectiveness of restoration spending. In contrast, from relatively favorable initial conditions, restoration without deforestation is always very low. Also, under the latter conditions, the optimal sequence is less obvious than in the cases of state-independent restoration efforts.

In Fig. 4.13, welfare is now much lower without deforestation from relatively favorable initial conditions, and unchanged from relatively unfavorable initial conditions. Therefore, welfare is more sensitive to deforestation eradication than to initial conditions. Given that environmental sustainability is greater with than without deforestation, the eradication of deforestation is not economically advisable here. The robustness of this result is supported by our prudent assumption of short-term revenues drawn from deforestation. Finally, from relatively favorable (unfavorable) initial conditions, welfare is slightly greater (dramatically lower) without production-based emissions than without deforestation. This is due to the fact that from favorable conditions with state-dependent restoration a deforestation only path is made more profitable since restoration efforts are more effective in the long run,

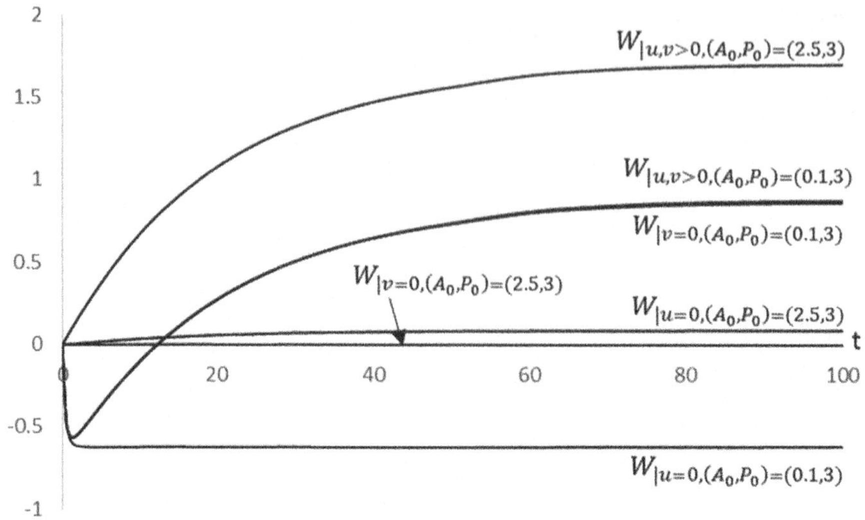

Fig. 4.13 Cumulative utility over time from relatively unfavorable and favorable initial conditions

Table 4.6 Steady-state values with both production and deforestation and with either production or deforestation

		A_∞	P_∞	u_∞	v_∞	w_∞
Interior solution	First steady state	0.315	0.119	0.037	0.003	0.030
	Second steady state	1.409	0.044	0.052	0.066	0.048
Corner solution with $v(t) = 0$		0.272	0.130	0.035	–	0.024
Corner solution with $u(t) = 0$		1.129	0.007	–	0.056	0.050

once we have settled at a higher than one absorption efficiency at the steady state. Conversely, from unfavorable conditions, the state-dependent property makes restoration much less effective, and the adverse effects of deforestation on absorption efficiency affects thus much more the social welfare, in part because they have to be compensated with heavy restoration spending at the beginning.

To conclude this analysis, we focus on a particularly interesting set of parameter values which reflects a scenario where production is significantly less profitable than deforestation, and that give rise to peculiar dynamics. For the set of parameter values such that $a = 0.053$, $b = 0.1$, $c = 0.01, \gamma = 0.05$, $\alpha = 0.15$, $r = 0.05$, there are two steady states for the interior solution (Table 4.6).

Figure 4.14 shows the steady states and the Skiba threshold (dotted line) that divides the state plane into two regions, each corresponding to the set of initial states that lead the system to either the left- or the right-hand steady-state value. The paths that start on the left-hand side of the Skiba threshold converge monotonically to $\left(A_\infty^1, P_\infty^1\right) = (0.315, 0.119)$, and those that start on the right-hand side converge monotonically to the more environmentally sustainable steady state $\left(A_\infty^2, P_\infty^2\right) = (1.409, 0.044)$.

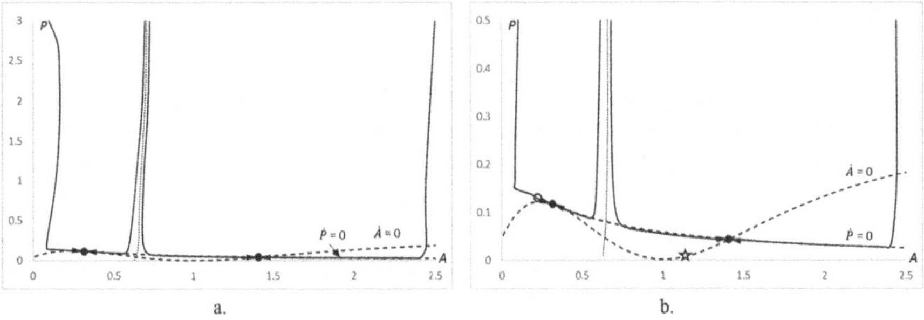

Fig. 4.14 Phase-portrait diagram in the state space with Skiba threshold (filled dot: interior solution, unfilled dot: corner solution with $v = 0$, unfilled star: corner solution with $u = 0$). (**a**) Phase-portrait diagram, (**b**) Zoom in the vicinity of the steady states

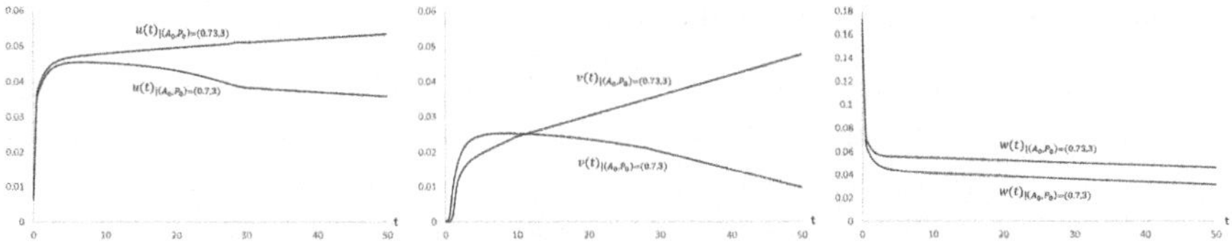

Fig. 4.15 Control policies from the left- and right-hand sides of the Skiba threshold

In a setting where deforestation revenues are significantly higher than production's, the optimal policy will depend on the initial conditions, thus reflecting history dependency. An economy starting from already deteriorated initial conditions, as on the left of the Skiba threshold in Fig. 4.15, will settle at the less desirable steady state $\left(A_{\infty}^{1}, P_{\infty}^{1}\right)$ whereas in presence of more favorable initial conditions a more sustainable steady state $\left(A_{\infty}^{2}, P_{\infty}^{2}\right)$ will be attained. This particular property does not arise under a state-independent restoration process, it is specific to the linear state-dependent restoration mechanism when deforestation is more profitable than production. Indeed, in that case, the economic tradeoff will be in favor of deforestation despite its more detrimental effect on absorption efficiency. But depending on the initial level of absorption efficiency, the leverage effect highlighted previously will go one way or the other. If A_0 is already low, so will be the effectiveness of restoration efforts, and it will thus be optimal to implement deforestation with little restoration to compensate (Fig. 4.15). In this case, the only way to neutralize history dependency is to impose eradication of production-based emissions rather than deforestation, because this results in a steady state on the right-hand side of the Skiba threshold. Conversely, if A_0 is already large enough, highly effective restoration spending will allow to draw benefits from a higher level of deforestation (and production), while maintaining a sustainable level of absorption efficiency through higher and more effective restoration (Fig. 4.15).

Conclusions

This article extends the strand of literature on pollution accumulation by introducing deforestation and its impact on environmental absorption efficiency, and by allowing different restoration mechanisms. The objective is to determine an optimal policy mix between production, deforestation, and restoration of environmental absorption efficiency. In this respect, we suggest a first-best model where production and deforestation are the main pollution sources. Although the environmental absorption efficiency is negatively affected by both deforestation and pollution accumulation, it can be improved by restoration efforts, which can be either independent from or linearly dependent on the magnitude of pollution sinks. The economic tradeoff involves instantaneous revenues from production-based emissions and deforestation, and costs of pollution externalities and restoration efforts over an infinite time horizon.

Our conclusions are summarized below.

- Significantly unbalanced revenues from production and deforestation tend to lead to specialization in an activity with higher revenues and withdrawal from an activity with lower revenues.
- Restoration is used along the optimal path to compensate for the degradation of environmental absorption efficiency entailed directly by deforestation and indirectly by pollution accumulation in general.
- In general, switching from a policy mix of production and deforestation to deforestation only leads to a higher level of deforestation. However, giving up deforestation does not lead to a significantly higher level of production-based emissions.
- In all cases, the optimal policy consists of *first restoring, then deforesting* from initially low environmental absorption efficiency, and *first deforesting, then restoring* from initially high environmental absorption efficiency.
- If the restoration mechanism is state-independent, the eradication of deforestation improves long-run environmental sustainability regardless of the structure of revenues from the two activities. In this case, however, stopping deforestation is economically justified only by initially low environmental absorption efficiency.
- If the restoration mechanism is linearly state-dependent, the effectiveness of restoration efforts is enhanced, which results in a positive influence on long-run environmental sustainability.
- If the restoration mechanism is linearly state-independent, stopping deforestation is neither environmentally nor economically desirable, as greater effectiveness of restoration efforts can be achieved through the leverage effect to compensate for the detrimental impact of deforestation on environmental absorption efficiency.
- If the restoration mechanism is linearly state-dependent, long-run environmental sustainability is greater along a deforestation-only path than along a production-only path, due to the leverage effect mentioned beforehand, regardless of the structure of revenues from the two activities.

– In the case of a low revenue-maximizing production level and an intermediate revenue-maximizing deforestation level, introducing linear state-dependency of restoration efforts leads to a Skiba point situation. As a result, unfavorable initial conditions will lead to a long-run solution with low environmental absorption efficiency and high pollution stock. In this scenario, the optimal policy displays a strong history-dependence that can be overcome by a deforestation-only path ensuring environmental sustainability with a more desirable steady state. This feature is quite noteworthy as it shows that under this specific incentive structure, there are initial situations from which it is not possible to revert to a sustainable long-run solution. In doing so, our model echoes the concern for the *actual* state of "initial" conditions when it comes to dwelling on theoretical models to enlighten actual environmental policies.

Most economic studies consider stopping deforestation the most effective way to improve both environmental sustainability and social welfare (e.g., Van Soest & Lensink, 2000; Fredj et al. 2006; Stern, 2006, 2015), yet our results, which are based on the prudent assumption that deforestation is a source of ephemeral revenues only, suggest that dropping deforestation from the policy mix is neither necessary nor sufficient. Instead, we conclude that a policy mix combining sustained levels of deforestation and significant restoration efforts can prove more efficient, both economically and environmentally, than a solution exclusively based on production. In addition, by permitting us to consider the two kinds (state-independent and dependent) of restoration mechanism, our model sheds light on their importance on the optimal environmental policy at stake and thus calls for a more precise approach to environmental restoration, beyond the topical "*let's plant more trees*" perspective. Finally, the state-independency property generates over-confident results not only because, having disregarded that restoration of pollution sinks can suffer from inertia, it suggests that pollution sources be restored to sinks, but also because it systematically leads to overestimating the social welfare, especially under unfavorable initial conditions. This property is not compatible with the precaution principle and should be dropped in future research.

In this paper, the possibility of nonlinear degradation of the absorption efficiency has not been explored. An accurate representation of a nonlinear degradation mechanism in the absorption efficiency dynamics could be a concave-convex specification, as the one used for the damage cost function in Moser et al. (2014). This specification, which would involve an inertial effect on the degradation mechanism side, might result in history dependency and a multiple optimal long-run solution. We leave it as an extension for future research. Another interesting extension could consider the impact on the optimal policies of alternative revenue specification for production and deforestation, for example, convex-concave functions, that would account for a wider range of revenue profiles for these activities. Finally, our results could be further generalized by accounting for the relative impact of forest owners' and non-owners' non-cooperative strategies on welfare and environmental sustainability. An important issue related to this context is whether the onus for the restoration of pollution sinks should fall on forest owners or non-owners.

Acknowledgements The authors acknowledge helpful comments from one anonymous referee. This research was supported by ESSEC Business School (France) and Tel Aviv University (Israel).

Appendix

A1 The Legendre-Clebsch condition holds for Eq. 4.4 since the Hessian is negative definite, that is:

$$\begin{bmatrix} H_{uu} & H_{uv} & H_{uw} \\ H_{vu} & H_{vv} & H_{vw} \\ H_{wu} & H_{wv} & H_{ww} \end{bmatrix} = \begin{bmatrix} -1 & 0 & 0 \\ 0 & -1 & 0 \\ 0 & 0 & -1 \end{bmatrix}$$

The Hamiltonian is therefore concave with respect to (u, v, w), which guarantees a (local) maximum.

A2 The candidate arc segments of the optimal path are described in the following way.

Production-Based Emissions and Deforestation Arc

In this case, Eq. 4.7 yields $(u^*, v^*, w^*) = (u\,^\circ\,(\lambda), v\,^\circ\,(\lambda, \varphi), w\,^\circ\,(A^*, \varphi))$ and $\rho_1 = 0$, $\rho_2 = 0$, $\rho_3 = 0$, with:

$$\left(u^\circ \quad v^\circ \quad w^\circ\right)' := \left(a+\lambda \quad b+\lambda\alpha-\varphi \quad \varphi A^\beta\right)' \tag{4.12}$$

being the solution of $H_u(P, A, u^\circ, v^\circ, w^\circ, \lambda, \varphi) = 0$, $H_v(P, A, u^\circ, v^\circ, w^\circ, \lambda, \varphi) = 0$and $H_w(P, A, u^\circ, v^\circ, w^\circ, \lambda, \varphi) = 0$, respectively.

Plugging the expressions of u^*, $v^*$and w^* in Eqs. 4.1–4.2 and 4.9–4.10, respectively, gives:

$$\dot{P} = a + \left(1+\alpha^2\right)\lambda + \alpha(b-\varphi) - AP \tag{4.13}$$

$$\dot{A} = \left(1 + A^{2\beta}\right)\varphi - b - \lambda\alpha - \gamma P \tag{4.14}$$

$$\dot{\lambda} = (r+A)\lambda + \gamma\varphi + cP \tag{4.15}$$

$$\dot{\varphi} = \left(r - \beta\varphi A^{2\beta-1}\right)\varphi + \lambda P \tag{4.16}$$

To analyze the behavior of the canonical system (Eqs. 4.13–4.16) in the neighborhood of the steady state, if it exists, we compute the Jacobian matrix:

$$J = \begin{bmatrix} -A & -P & 1+\alpha^2 & -\alpha \\ -\gamma & 2\beta\varphi A^{2\beta-1} & -\alpha & 1+A^{2\beta} \\ c & \lambda & r+A & \gamma \\ \lambda & -\beta(2\beta-1)\varphi^2 A^{2\beta-2} & P & r-\beta\varphi A^{2\beta-1} \end{bmatrix} \quad (4.17)$$

where (P, A, λ, φ) are evaluated at their steady-state values, and whose determinant is given by:

$$\begin{aligned} |J| = {} & \beta\varphi^2 A^{2\beta-2}\Big\{(1-2\beta)\left[(\alpha\gamma+A)(r+\alpha\gamma+A)+\gamma^2+c\right] \\ & +(1+2\beta)A^{2\beta}\left[A(r+A)+c\left(1+\alpha^2\right)\right]\Big\} \\ & -2\beta\varphi A^{2\beta-1}\Big\{r\left[A(r+A)+c\left(1+\alpha^2\right)\right]-(r+2A)(\alpha\lambda+\gamma P) \\ & -2\left[\gamma\lambda\left(1+\alpha^2\right)-\alpha cP\right]\Big\}-\alpha^2\lambda^2A^{2\beta}+\left(1+A^{2\beta}\right)\left[cP^2-\lambda(\lambda+rP+2AP)\right] \\ & -r\lambda\left[\gamma\left(1+\alpha^2\right)+\alpha A\right]+P\{\alpha(rc-2\gamma\lambda)+\gamma[\gamma P-r(2r+A)]\} \end{aligned}$$

and the sum of the principal minors of J of order 2 minus the squared discounting rate is:

$$\begin{aligned} K = {} & -A(r+A)-c\left(1+\alpha^2\right)+2(\alpha\lambda-\gamma P) \\ & +\beta\varphi A^{2\beta-1}\left\{2r+\varphi A^{-1}\left[2\beta-1-(2\beta+1)A^{2\beta}\right]\right\} \end{aligned}$$

In the case of state-independent restoration efforts, $K_{|\beta=0} < 0$, which rules out the possibility of limit cycles (Dockner & Feichtinger, 1991). For linearly state-dependent restoration efforts, the sign of $K_{|\beta=1} = -A(r+A) - c(1+\alpha^2) + 2(\alpha\lambda - \gamma P) + \varphi[2rA + \varphi(1 - A^2)]$ is not clear.

Production-Based Emissions-Only Arc

In this case, the control constraint $v(t) \geq 0$ is active. Using Eq. 4.5, the maximizing condition (Eq. 4.6) yields $(u^*, v^*, w^*) = (u^\circ(\lambda), 0, w^\circ(A^*, \varphi))$ and $\rho_1 = 0$, $\rho_2 \leq 0$, $\rho_3 = 0$, with:

$$(u^\circ \quad w^\circ)' := \left(a+\lambda \quad \varphi A^\beta\right)' \text{ and } \rho_2 = \varphi - b - \lambda\alpha \quad (4.18)$$

being the solution of $L_u(P, A, u^\circ, 0, w^\circ, \lambda, \varphi, 0, \rho_2, 0) = 0$, $L_v(P, A, u^\circ, 0, w^\circ, \lambda, \varphi, 0, \rho_2, 0) = 0$and $L_w(P, A, u^\circ, 0, w^\circ, \lambda, \varphi, 0, \rho_2, 0) = 0$, respectively. Plugging the expressions of u^* and w^* in Eqs. 4.1–4.2 and 4.9–4.10, respectively, gives the canonical system:

$$\dot{P} = a + \lambda - AP \tag{4.19}$$

$$\dot{A} = \varphi A^{2\beta} - \gamma P \tag{4.20}$$

$$\dot{\lambda} = (r + A)\lambda + \gamma\varphi + cP \tag{4.21}$$

$$\dot{\varphi} = \left(r - \beta\varphi A^{2\beta-1}\right)\varphi + \lambda P \tag{4.22}$$

The steady state, if it exists, is obtained by solving Eq. 4.11 where:

$$(\varphi_\infty \quad \lambda_\infty \quad A_\infty)^t = \left(\frac{\gamma a(r+A_\infty)}{\Phi} \quad -\frac{a\left(\gamma^2 + cA_\infty^{2\beta}\right)}{\Phi} \quad \frac{a(r+A_\infty)A_\infty^{2\beta}}{\Phi}\right)^t$$

with $\Phi = A_\infty^{2\beta}[c + A_\infty(r + A_\infty)] + \gamma^2$. It can be shown that the resolution of the system (Eqs. 4.19–4.22) for the case of state-independent restoration efforts ($\beta = 0$), leads to a steady state that is a saddle point with either monotonic or spiraling convergence. On the other hand, if $\beta = 1$, we obtain $K_{|\beta=1} = -A[r + A(1 + 3\varphi^2)] - 2\gamma P - c + 2r\varphi A$. Therefore, the possibility of limit cycles cannot be ruled out for linearly state-dependent restoration efforts.

Deforestation-Only Arc

In this case, the control constraint $u(t) \geq 0$ is active. Using Eq. 4.5, the maximizing condition (Eq. 4.6) yields $(u^*, v^*, w^*) = (0, v^\circ(\lambda, \varphi), w^\circ(A^*, \varphi))$ and $\rho_1 \leq 0, \rho_2 = 0, \rho_3 = 0$, with:

$$(v^\circ \quad w^\circ)^t := \left(b + \lambda\alpha - \varphi \quad \varphi A^\beta\right)^t \text{ and } \rho_1 = -\lambda - a \tag{4.23}$$

being the solution of $L_u(P, A, 0, v^\circ, w^\circ, \lambda, \varphi, \rho_1, 0, 0) = 0$, $L_v(P, A, 0, v^\circ, w^\circ, \lambda, \varphi, \rho_1, 0, 0) = 0$and $L_w(P, A, 0, v^\circ, w^\circ, \lambda, \varphi, \rho_1, 0, 0) = 0$, respectively. Plugging the expressions of v^* and w^* in Eqs. 4.1–4.2 and 4.9–4.10, respectively, gives the canonical system:

$$\dot{P} = \alpha(b + \lambda\alpha - \varphi) - AP \tag{4.24}$$

$$\dot{A} = \varphi\left(1 + A^{2\beta}\right) - b - \lambda\alpha - \gamma P \tag{4.25}$$

$$\dot{\lambda} = (r + A)\lambda + \gamma\varphi + cP \tag{4.26}$$

$$\dot{\varphi} = \left(r - \beta\varphi A^{2\beta-1}\right)\varphi + \lambda P \tag{4.27}$$

The steady state, if it exists, is obtained by solving Eq. 4.11 where:

$$(\varphi_\infty \quad \lambda_\infty \quad A_\infty)^t = \left(\frac{b(r+A_\infty)(\alpha\gamma+A_\infty)}{\Psi} \quad -\frac{b\left[\gamma(\alpha\gamma+A_\infty)+\alpha c A_\infty^{2\beta}\right]}{\Psi} \quad \frac{\alpha b(r+A_\infty)A_\infty^{2\beta}}{\Psi} \right)^t$$

with $\Psi = (A_\infty + \alpha\gamma)(r + A_\infty + \alpha\gamma) + A_\infty^{2\beta}[c\alpha^2 + A_\infty(r + A_\infty)]$. If $\beta = 0$, there can be no limit cycle. If $\beta = 1$, $K_{|\beta = 1} = -A[r + A(1 + 3\varphi^2)] - \alpha^2 c + 2(\alpha\lambda - \gamma P) + \varphi(\varphi + 2rA)$, whichl implies that limit cycles are possible for linearly state-dependent restoration efforts.

A3 Using Eq. 4.12 gives:

$$w^\circ = A^\beta(\alpha u^\circ - v^\circ + b - \alpha a)$$

A4 The steady state, if it exists, is obtained by solving Eqs. 4.19–4.22 to zero. Equations 4.19–4.21 are linear in φ, λ and P, and after substituting their solutions into Eq. 4.22, we get Eq. 4.11.

A5 The time decomposition method used for the numerical resolution follows the main steps below:

Step 1. Set a large enough time horizon $t \epsilon [0, T]$. Set the initial states $P(0)$ and $A(0)$ at the given values. Set the terminal values of the costate variables at the steady-state values, $\lambda(T) = \lambda_\infty$ and $\varphi(T) = \varphi_\infty$.
Step 2. Guess feasible control functions, $u(t)$, $v(t)$ and $w(t)$.
Step 3. Integrate the state system from left to right.
Step 4. Integrate the costate system from right to left.
Step 5. If the optimality conditions are satisfied (the Hamiltonian is maximized at each t) with a required tolerance, stop. Otherwise, go to step 6.
Step 6. At each t, where the Hamiltonian is not maximized, change $u(t)$, $v(t)$ and $w(t)$ to $u(t) + \delta u(t)$, $v(t) + \delta v(t)$ and $w(t) + \delta w(t)$ where $\delta u(t)$, $\delta v(t)$ and $\delta w(t)$ are small enough positive/negative increments to make the Hamiltonian rise.
Step 7. Go to Step 3.

References

Amesbury, M. J., Gallego-Sala, A., & Loisel, J. (2019). Peatlands as prolific carbon sinks. *Nature Geoscience, 12*, 880–881.

Andrés-Domenech, P., Martín-Herrán, G., & Zaccour, G. (2015). Cooperation for sustainable forest management: An empirical differential game approach. *Ecological Economics, 117*, 118–128.

Angelsen, A., & Rudel, T. K. (2013). Designing and implementing effective REDD+ policies: A forest transition approach. *Review of Environmental Economics and Policy, 7*(1), 91–113.

Ayong Le Kama, A., Pommeret, A., & Prieur, F. (2014). Optimal emission policy under the risk of irreversible pollution. *Journal of Public Economic Theory, 16*(6), 959–980.

Baccini, A., Goetz, S. J., Walker, W. S., Laporte, N. T., Sun, M., Sulla-Menashe, D., Hackler, J., Beck, P. S. A., Dubayah, R., Friedl, M. A., Samanta, S., & Houghton, R. A. (2012). Estimated carbon dioxide emissions from tropical deforestation improved by carbon-density maps. *Nature Climate Change, 2*(3), 182–185.

Baccini, A., Walker, W., Carvalho, L., Farina, M., Sulla-Menashe, D., & Houghton, R. A. (2017). Tropical forests are a net carbon source based on aboveground measurements of gain and loss. *Science, 358*(6360), 230–234.

Barbier, E. B., & Burgess, J. C. (1997). The economics of forest land use options. *Land Economics, 73*(2), 174–195.

Bastin, J. F., Finegold, Y., Garcia, C., Mollicone, D., Rezende, M., Routh, D., & Crowther, T. W. (2019). The global tree restoration potential. *Science, 365*(6448), 76–79.

Boucekkine, R., Pommeret, A., & Prieur, F. (2013). Optimal regime switching and threshold effects. *Journal of Economic Dynamics and Control, 37*(2), 2979–2997.

Canadell, J. G., Le Quéré, C., Raupach, M. R., Field, C. B., Buitenhuis, E. T., Ciais, P., Conway, T. J., Gillett, N. P., Houghton, R. A., & Marland, G. (2007). Contributions to accelerating atmospheric CO_2 growth from economic activity, carbon intensity, and efficiency of natural sinks. *Proceedings of the National Academy of Sciences, 104*(47), 18866–18870.

Canadell, J. G., & Raupach, M. R. (2008). Managing forests for climate change mitigation. *Science, 320*(5882), 1456–1457.

Cox, P. M., Betts, R. A., Jones, C., Spall, S. A., & Totterdell, I. (2000). Acceleration of global warming due to carbon-cycle feedbacks in a coupled climate model. *Nature, 408*, 184–187.

Cramer, W., Bondeau, A., Woodward, F. I., Prentice, I. C., Betts, R. A., Brovkin, V., Cox, P. M., Fisher, V., Foley, J. A., Friend, A. D., Kucharik, C., Lomas, M. R., Ramankutty, N., Sitch, S., Smith, B., White, A., & Young-Molling, C. (2001). Global response of terrestrial ecosystem structure and function to CO2 and climate change: results from six dynamic global vegetation models. *Global Change Biology, 7*(4), 357–373.

Dockner, E., & Feichtinger, G. (1991). On the optimality of limit cycles in dynamic economic systems. *Journal of Economics, 53*(1), 31–50.

El Ouardighi, F., Benchekroun, H., & Grass, D. (2014). Controlling pollution and environmental absorption capacity. *Annals of Operations Research, 220*(1), 1–23.

El Ouardighi, F., Benchekroun, H., & Grass, D. (2016). Self-regenerating environmental absorption efficiency and the *Soylent Green* scenario. *Annals of Operations Research, 238*(1), 179–198.

El Ouardighi, F., Kogan, K., Gnecco, G., & Sanguineti, M. (2017). Commitment-based equilibrium environmental strategies under time-dependent absorption efficiency. *Group Decision and Negotiation, 27*(2), 235–249.

El Ouardighi, F., Kogan, K., Gnecco, G., & Sanguineti, M. (2018). Transboundary pollution control and environmental absorption efficiency management. *Annals of Operations Research*, 1–29. https://doi.org/10.1007/s10479-018-2927-7

Forster, B. (1973). Optimal consumption planning in a polluted environment. *Economic Record, 49*(4), 534–545.

Fredj, K., Martín-Herrán, G., & Zaccour, G. (2006). Incentive mechanisms to enforce sustainable forest exploitation. *Environmental Modeling Assessment, 11*(2), 145–156.

Gattuso, J.-P., Magnan, A. K., Bopp, L., Cheung, W. W. L., Duarte, C. M., Hinke, J., Mcleod, E., Micheli, F., Oschlies, A., Williamson, P., Billé, R., Chalastani, V. I., Gates, R. D., Irisson, J.-O., Middelburg, J. J., Pörtner, H. O., & Rau, G. H. (2018). Ocean solutions to address climate change and its effects on marine ecosystems. *Frontiers in Marine Science, 5*(337), 1–18.

Gloor, M., Sarmiento, J. L., & Gruber, N. (2010). What can be learned about carbon cycle climate feedbacks from the CO2 airborne fraction? *Atmospheric Chemistry and Physics, 10*(16), 7739–7751.

Gramling, C. (2017, September 28). Tropical forests have flipped from sponges to sources of carbon dioxide. *Science News*.

Grass, D., Caulkins, J. P., Feichtinger, G., Tragler, G., & Behrens, D. A. (2008). *Optimal control of nonlinear processes with applications in drugs, corruption, and terror*. Springer.
Harris, N. L., Brown, S., Hagen, S. C., Saatchi, S. S., Petrova, S., Salas, W., Hansen, M. C., Potapov, P. V., & Lotsch, A. (2012). Baseline map of carbon emissions from deforestation in tropical regions. *Science, 336*(6088), 1573–1576.
Hediger, W. (2009). Sustainable development with stock pollution. *Environment and Development Economics, 14*(6), 759–780.
Houghton, R. A., House, J. I., Pongratz, J., van der Werf, G. R., DeFries, R. S., Hansen, M. C., Le Quéré, C., & Ramankutty, N. (2012). Carbon emissions from land use and land-cover change. *Biogeosciences, 9*, 5125–5142.
Joos, F., Prentice, I. C., Sitch, S., Meyer, R., Hooss, G., Plattner, G.-K., Gerber, S., & Hasselmann, K. (2001). Global warming feedbacks on terrestrial carbon uptake under the Intergovernmental Panel on Climate Change (IPCC) emission scenarios. *Global Biogeochemical Cycles, 15*(4), 891–907.
Keeler, E., Spence, M., & Zeckhauser, R. (1972). The optimal control of pollution. *Journal of Economic Theory, 4*, 19–34.
Le Quéré, C., Rödenbeck, C., Buitenhuis, E. T., Conway, T. J., Langenfelds, R., Gomez, A., Labuschagne, C., Ramonet, M., Nakazawa, T., Metzl, N., Gillett, N., Heimann, M. (2007). Saturation of the southern ocean CO2 sink due to recent climate change. *Science, 316*(5832), 1735–1738.
Leandri, M. (2009). The shadow price of assimilative capacity in optimal flow pollution control. *Ecological Economics, 68*(4), 1020–1031.
Leandri, M., & Tidball, M. (2019). Assessing the sustainability of optimal pollution paths in a world with inertia. *Environmental Modeling & Assessment, 24*(2), 249–263.
Lenton, T. M., Williamson, M. S., Edwards, N. R., Marsh, R., Price, A. R., Ridgwell, A. J., Shepherd, J. G., Cox, S. J., & The GENIET. (2006). Millennial timescale carbon cycle and climate change in an efficient Earth system model. *Climate Dynamics, 26*(7/8), 687–711.
Liebsch, D., Marques, M. C., & Goldenberg, R. (2008). How long does the Atlantic Rain Forest take to recover after a disturbance? Changes in species composition and ecological features during secondary succession. *Biological Conservation, 141*(6), 1717–1725.
Makarieva, A. M., Gorshkov, V. G., Sheil, D., Nobre, A. D., Bunyard, P., & Li, B.-L. (2014). Why does air passage over forest yield more rain? Examining the coupling between rainfall, pressure, and atmospheric moisture content. *Journal of Hydrometeorology, 15*(1), 411–426.
Maimon, O., Khmelnitsky, E., & Kogan, K. (1998). *Optimal flow control in manufacturing systems: Production planning and scheduling*. Kluwer Academic.
Martin, P. A., Newton, A. C., & Bullock, J. M. (2013). Carbon pools recover more quickly than plant biodiversity in tropical secondary forests. *Proceedings of the Royal Society B: Biological Sciences, 280*(1773), 22–36.
Michel, P., & Rotillon, G. (1995). Disutility of pollution and endogenous growth. *Environmental and Resource Economics, 6*(3), 279–300.
Moser, E., Seidl, A., & Feichtinger, G. (2014). History-dependence in production-pollution-trade-off models: A multi-stage approach. *Annals of Operation Research, 222*, 457–481.
Piao, S. L., Ciais, P., Friedlingstein, P., Peylin, P., Reichstein, M., Luyssaert, S., Margolis, H., Fang, J. Y., Barr, A., Chen, A. P., Grelle, A., Hollinger, D. Y., Laurila, T., Lindroth, A., Richardson, A. D., & Vesala, T. (2008). Net carbon dioxide losses of northern ecosystems in response to autumn warming. *Nature, 451*, 49–52.
Prieur, F. (2009). The environmental Kuznets curve in a world of irreversibility. *Economic Theory, 40*(1), 57–90.
Raupach, M. R., Gloor, M., Sarmiento, J. L., Canadell, J. G., Frölicher, T. L., Gasser, T., Houghton, R. A., Le Quéré, C., & Trudinger, C. M. (2014). The declining uptake rate of atmospheric CO_2 by land and ocean sinks. *Biogeosciences, 11*(13), 3453–3475.
Renforth, P., & Henderson, G. (2018). Assessing ocean alkalinity for carbon sequestration. *Reviews of Geophysics, 55*(3), 636–674.

Rodrigues, A. S. L., Ewers, R. M., Parry, L., Souza, C., Jr., Veríssimo, A., & Balmford, A. (2009). Boom-and-bust development patterns across the Amazon deforestation frontier. *Science, 324*(5933), 1435–1437.

Schuur, E. A., McGuire, A. D., Schädel, C., Grosse, G., Harden, J. W., Hayes, D. J., & Natali, S. M. (2015). Climate change and the permafrost carbon feedback. *Nature, 520*(7546), 171–179.

Sheil, D., & Murdiyarso, D. (2009). How forests attract rain: an examination of a new hypothesis. *BioScience, 59*(4), 341–347.

Sohngen, B., & Mendelsohn, R. (2003). An optimal control model of forest carbon sequestration. *American Journal of Agricultural Economics, 85*(2), 448–457.

Southgate, D. (1990). The causes of land degradation along 'spontaneous' expanding agricultural frontiers. *Land Economics, 66*(1), 93–101.

Stähler, F. (1996). On international compensations for environmental stocks. *Environmental Resource Economics, 8*(1), 1–13.

Stern, N. (2006). *Stern review report on the economics of climate change*. HM Treasury.

Stern, N. (2015). *Why are we waiting? The logic, urgency, and promise of tackling climate change*. MIT Press.

Tahvonen, O., & Withagen, C. (1996). Optimality of irreversible pollution accumulation. *Journal of Economic Dynamics and Control, 20*(9), 1775–1795.

Tahvonen, O., & Salo, S. (1996). Nonconvexities in optimal pollution accumulation. *Journal of Environmental Economics and Management, 31*(2), 160–177.

Van Soest, D. (1998). *Tropical deforestation: An economic perspective*. Labyrinth Publications.

Van Soest, D., & Lensink, R. (2000). Foreign transfers and tropical deforestation: What terms of conditionality? *American Journal of Agricultural Economics, 82*(2), 389–399.

Wirl, F. (2007). Do multiple Nash equilibria in Markov strategies mitigate the tragedy of the commons? *Journal of Economic Dynamics and Control, 31*(10), 3723–3740.

Part II
Pollution Control in Management Science

Chapter 5
Pollution Accumulation and Abatement Policy in a Supply Chain

Fouad El Ouardighi, Jeong Eun Sim, and Bowon Kim

Abstract This paper seeks to determine how the double marginalization phenomenon affects the tradeoff between polluting emissions and abatement activities related to pollution accumulation in a supply chain composed of one manufacturer and one retailer. The environmental consequence of this inefficiency, which emerges in a noncooperative vertical setting governed by a single-parameter contract, is overlooked in the literature on pollution control. In the setup of a two-stage game, we investigate the impact of double marginalization for noncooperative equilibrium. To check whether there are differences between the dynamic and strategic effects of double marginalization on pollution accumulation, both the open-loop and feedback Nash equilibria are derived over a finite time horizon, with the cooperative solution as a benchmark.

Keywords Double marginalization · Pollution accumulation · Abatement efforts · Supply chain

Introduction

The environmental economics literature has dealt extensively with the horizontal impact of decentralized decision-making on pollution accumulation, which consists of improper internalization of the social cost of pollution by firms or nations [for recent surveys, see Jørgensen et al. (2010), Long (2010), Benchekroun and Long (2011)]. Accordingly, strategy types (Dockner et al., 2000) in pollution dynamic

F. El Ouardighi (✉)
ESSEC Business School, Cergy Pontoise Cedex, France
e-mail: elouardighi@essec.fr

J. E. Sim
Sauder School of Business, University of British Columbia, Vancouver, BC, Canada

B. Kim
KAIST Business School, Seoul, Korea

F. El Ouardighi (ed.), *Essays on Pollution Control in Economics and Management Science*, Lecture Notes in Operations Research,
https://doi.org/10.1007/978-3-031-78227-5_5

game models are strongly influential in that commitment (that is, open-loop Nash equilibrium) strategies result in a lower steady-state pollution stock than contingent (or feedback Nash equilibrium) strategies do.

In contrast, the vertical impact of decentralized decision-making in a bilateral monopoly (supply chain) on pollution accumulation has been under-investigated in both the economics and management literature. It is well-known that poor supply chain coordination creates a common inefficiency: the double marginalization effect is maximized under conventional wholesale price contracts (Spengler, 1950). Accordingly, retailers pay suppliers a fixed transfer price per unit ordered, and mark-ups applied by both firms increase consumer prices, lower consumer demand, and reduce overall profits compared with centralized decision-making (Bresnahan & Reiss, 1985).

Because double marginalization decreases both channel profit and consumer surplus, it should be eliminated from the supply chain. Supply chain coordination has been advocated and can be induced via mechanisms like contracts (e.g., Hu et al., 2013; El Ouardighi, 2014).

How does double marginalization affect pollution accumulation in a supply chain? The general perception is that greater collaboration and integration within a supply chain would also enhance firms' environmental practices (Handfield et al., 1997; Klassen & Vachon, 2003; Lee, 2010; Simpson, 2010; Hollos et al., 2012). Thus, compared with a noncooperative supply chain, the mitigation of double marginalization in a cooperative supply chain intensifies greater abatement efforts by eliminating the underinvestment problem in a decentralized context. However, considering that mitigating double marginalization ultimately increases consumer demand and thus the absolute volume of pollution emissions as a by-product of production activities, it is unclear whether the increase in abatement efforts in a cooperative supply chain driven by the increase in revenues would far outweigh the increase in emission volumes. Put differently, double marginalization purportedly lowers the pollution stock due to lower consumer demand, yet because the under-investment problem typically occurs in a decentralized context, the decentralized supply chain with double marginalization putatively spends less effort on pollution abatement. Therefore, the tradeoff between emissions and abatement is not straightforward.

A primary goal of this paper is to determine how double marginalization affects the tradeoff between polluting emissions and abatement efforts related to pollution accumulation in a supply chain. Particularly, we investigate whether the virtue of supply chain coordination would be still valid from an environmental standpoint. Given the increasing importance of environmental sustainability and the traditional emphasis on supply chain coordination, this linkage deserves a thorough examination. Should regulators encourage supply chain coordination to enhance environmental performance and social welfare? What would an optimal abatement strategy be for profit-maximizing firms, along with optimal emissions in a cooperative or noncooperative supply chain? Although the impact of double marginalization has been widely discussed in the literature, little research has addressed these issues simultaneously. To fill this gap and to respond to Tang and Zhou's (2012) call for future research on vertical interactions between supply chain members in the context of environmentally/socially sustainable operations, we address the following research questions:

- What is the impact of double marginalization in a supply chain from the environmental perspective, specifically on the tradeoff between emissions and abatement decisions?
- From a broader perspective that considers social welfare and cumulative pollution, is the centralized supply chain more beneficial than the decentralized chain?

To investigate this issue, we use a noncooperative two-stage game model of a simple supply chain with one manufacturer that sells a finished good to one retailer, which resells it to a final consumer. As a key component of the problem, pollution is interpreted as a stock variable in which pollutants would be generated per unit output as a by-product of the production process. However, it is reduced by the firm's abatement efforts throughout the planning horizon. Production is determined by price-dependent consumer demand: a make-to-order system is assumed in this paper. We analyze the supply chain's pricing and abatement effort decisions under a wholesale price contract. In this setup, the optimal transfer price is determined at the first stage of the game by the manufacturer, which presumably maximizes its best interest. The optimal contract can be reached by both firms if their resulting payoffs are positive. Due to the cumulative nature of pollution, a differential game of finite duration is played during the second stage of the game. To investigate optimal strategies and their dynamic effects on the environment, a differential game is often adopted in environment-related studies [e.g., Chen and Sheu (2009), Boucekkine et al. (2011), Bertinelli et al. (2014), for a review, see Jørgensen et al. (2010)]. Jørgensen et al. (2010) also explained that a differential game approach is particularly useful to model environmental issues that are intrinsically dynamic and influenced by interactions between firms over time, not just one-shot phenomena. To check whether there are differences between the dynamic and strategic effects of double marginalization on pollution accumulation, both the open-loop and feedback Nash equilibria are derived over finite time horizons, with the cooperative solution as a benchmark.

The paper proceeds as follows. Second section reviews the literature. Third section develops the game model. Fourth, fifth, and sixth sections present the cooperative, open-loop, and feedback Nash equilibrium strategies, respectively. In seventh section, we compare the results. Eighth section provides numerical illustrations. Ninth section concludes the paper.

Literature Review

Most of the research on the environment investigates how a single firm or regulator manages pollution or environment-related issues. It includes studies on the problem of emission technology selection in response to environmental policies (Krass et al., 2013), the optimal pollution abatement effort or emission banking strategy with production-inventory dynamics (Li, 2013), and optimal environmental policies (Milliman & Prince, 1989; Jung et al., 1996). Extending these studies, the impact of horizontal interactions on the environment has been widely explored. In a horizontal context, interactions between firms' or countries' polluting emissions are often investigated (Boucekkine et al., 2011; Bertinelli et al., 2014). Subramanian et al. (2007) examined firms' production, bidding, and emissions abatement strategies when firms compete for scarce emission permits, or even for customer demand. Atasu et al. (2009) studied the economic and environmental impacts of take-back legislation when manufacturers engage in Cournot competition. Using a differential game, Chen and Sheu (2009) found that regulation standards should be gradually raised to enhance product recyclability and promote Extended Product Responsibility of firms competing for sales. Bagnoli and Watts (2003) examined firms' competition for socially responsible consumers and found that a less competitive market promotes more eco-friendly activities by firms.

In a vertical context, however, the impact of double marginalization on environmental performance remains largely underexplored. Recently, a few studies investigated the impact of diverse contract schemes for environment-related issues in a supply chain or examined firms' environmental decision-making in a supply chain network. For instance, Giovanni (2014) investigated environmental collaboration between a manufacturer and a retailer in a closed-loop remanufacturing supply chain by implementing a reverse revenue-sharing contract. Zhang et al. (2015) examined how consumer environmental awareness influences order quantity decisions for environmental and traditional products and the economic performance of supply chain members. They also investigated how a return contract coordinates a supply chain, assuming that the level of environmental quality is exogenously given. Cruz (2008) explored product flows, prices, and social responsibility activities (e.g., activities for environmental preservation or quality assurance) in a supply chain network consisting of multiple manufacturers, retailers, and consumers when each entity engages in multi-criteria decision-making behaviors such as profit maximization and minimization of emission and risk. Chung et al. (2013) studied how supply chain network members decide on outsourcing, shipment, distribution, emissions, or pollution mitigation in response to environmental taxes. They found that taxes on the pollution stock should be carefully designed given the game structure.

Although these studies shed light on environmental issues in a supply chain, in-depth research into how double marginalization in a supply chain influences pollution dynamics is lacking. When supply chain centralization is considered in the environmental context, the commonly accepted view is that greater collaboration and integration within a supply chain enhances environmental performance (Handfield et al., 1997; Klassen & Vachon, 2003; Lee, 2010; Simpson, 2010; Hollos et al., 2012). However, given the motivation of profit maximization, little is understood about how firms in a centralized or decentralized supply chain actually behave regarding the trade-off between pollution abatement decisions and pollution emission, and how firms' behavior influences environmental performance and firms' profits and consumer surplus throughout the time horizon.

Model

We consider a supply chain with one manufacturer (upward firm) and one retailer (downward firm) involved in a make-to-order system where no firm can carry an inventory. We assume that the manufacturer and retailer first agree on a wholesale price contract, and supply consumer demand over a finite time horizon. Consumer demand is assumed to depend negatively on price. At the first stage of the game, the manufacturer sets the optimal transfer price that prevails throughout the game. In the second stage, consumer demand generates polluting emissions that accumulate over time. Specifically, pollutants are generated per unit of output as a byproduct of the manufacturer's production. We assume that each unit of goods produced and sold results in one unit of emitted pollutant.

Production activities of manufacturing firms are major sources of pollution, and pollution emissions are a by-product of manufacturing processes (Klassen & Whybark, 1999; Krass et al., 2013; Piot-Lepetit, 2014). The United States Department of State (2010) reports that manufacturing accounts for approximately four-fifths of total energy use in the industrial sector resulting in CO_2 emissions. Emissions from industrial processes such as production activities and transformation of raw materials ranked third highest, followed by emissions from the energy and agriculture sectors. Chung et al. (2013), examining a supply chain network with multiple manufacturers and retailers, also assumed that the costs associated with pollution stock and pollution abatement efforts are incurred only by manufacturers, not retailers. They identified two sources of pollutants at the manufacturer stage; one related to the use of specific production inputs and the other to manufacturing operations and production of goods. We therefore assume that the pollution stock entails a cost incurred by the manufacturer only.[1]

To reduce the pollution stock and related costs, the manufacturer can invest in abatement efforts (e.g., Fowlie, 2010). Given thermodynamic restrictions on the extent to which emissions can be reduced, abatement efforts should be bounded to account for the partial reduction of emissions. We assume that the manufacturer's production process and abatement operations are mutually independent so that the abatement effort is nonproportional to the output.

The environmental absorption capacity of pollution is approximated by zero. This approximation is consistent with recent empirical observations that terrestrial and oceanic natural pollution sinks are becoming much less efficient (Canadell et al., 2007; Le Quéré et al., 2007) and will either become saturated or turn into a pollution source after 2050 (e.g., Zhao & Running, 2010; Cox et al., 2000; Piao et al., 2008). These trends suggest that our approximation of a zero environmental absorption capacity of pollution describes the current situation more accurately than standard assumptions based on linear or nonlinear environmental absorption capacity of pollution [for a review, see El Ouardighi et al. (2014)]. In practical terms, the supply

[1] We also examined the assumption that both the manufacturer and the retailer emit pollutants and found that our major findings still hold qualitatively under this assumption.

chain cannot benefit from a natural abatement of the pollution stock. As an economic counterpart of such absence of environmental subsidy, the initial pollution stock is normalized to zero to account for the specific part of the supply chain in the pollution accumulation process and thus determine a proper economic imputation of the manufacturer's corresponding pollution cost.

Based on the previous assumptions, the dynamics of the pollution stock, $y(t) \geq 0$, are given by:

$$\dot{y}(t) = D(t) - \varphi v(t), \qquad y(0) = y_0 = 0$$

where $D(t) \geq 0$ is the number of units produced and sold (i.e., consumer demand), $v(t) \geq 0$ is the level of the manufacturer's abatement effort, $\varphi > 0$ is the effectiveness of the manufacturer's pollution abatement effort (i.e., given the manufacturer's investment in the pollution abatement, how much pollutants would be actually reduced), and $t \in [0, T)$ is for time. The value of φ depends on the firm's capability or the maturity of available environmental technologies/facilities in the industry. The constraint $v(t) \leq D(t)/\varphi$ should be satisfied so that current emissions remain nonnegative, which implies that $\dot{y}(t) \geq 0$, $\forall t \geq 0$.

Consumer demand is defined as:

$$D(t) \equiv \alpha - \beta p_2(t) \geq 0$$

where $0 \leq p_2(t) \leq \alpha/\beta$ is the retailer's consumer price, and α and β are positive constants.[2] Price-dependent linear demand is one of the most commonly used demand specifications (Krishnamoorthy et al., 2010), long used in the economic and management literature as a basic demand function to investigate a problem of double marginalization and vertical coordination (e.g., Bresnahan & Reiss, 1985; Carr & Karmarkar, 2005; Cui et al., 2007; Mishra et al., 2009; Dellarocas, 2012). It is also commonly applied in environment-related studies (e.g., Majumder & Groenevelt, 2001; Krass et al., 2013).

Each firm's objective is to maximize its cumulative profits over the planning period. We assume a fixed and finite planning horizon, $T < \infty$, and omit the discounting of future profits.

The manufacturer's instantaneous gross profit is given by$(p_1 - c)(\alpha - \beta p_2(t))$, where $p_1 > 0$ is the (constant) transfer price to the retailer, and $c \geq 0$ is the manufacturer's unit production cost. For simplicity, c is normalized to zero, without loss of generality.

The negative externalities of the pollution stock are valued as an increasing convex cost function of pollution, that is, $fy(t)^2/2$, where $f > 0$ is the cost coefficient

[2]Isoelastic demand functions are usual alternatives to linear functions (e.g., Xu & Hopp, 2006; Krishnamoorthy et al., 2010; Krass et al., 2013). For the robustness of the results, we also examine isoelastic demand $D(t) = ap_2(t)^{-b}$, where $a > 0$ is the market size and $b > 1$ is the price elasticity of demand. The results show that our major findings and implications hold qualitatively under the isoelastic demand function as well.

associated with the manufacturer's pollution. This quadratic form of the cost associated with the pollution stock is commonly accepted in the literature on pollution control (e.g., Michel & Rotillon, 1995; Chung et al., 2013; Bertinelli et al., 2014; Li, 2014; El Ouardighi et al., 2014). It is also of practical relevance because manufacturers producing high pollution face pressures not only from the institutional environment (e.g., taxes or fines imposed by regulators, protests from environmental groups) but also from shareholders. A quadratic pollution cost function means that the damages due to the pollution stock entail a social cost that increases more than proportionally with the pollution stock. In practice, it is often difficult to assess exhaustively the social cost of pollution damages that should be imposed on polluters. Nevertheless, the current trend for regulators is to include in the social cost of pollution—though approximately—damages like rising sea levels, the increased number of Katrina-level hurricanes, or sustained droughts (Wald, 2013). Recently, six companies in China were imposed fines which amount to 26 million dollars in total for their accumulated pollution (Wong, 2014). The fine was the biggest of any Chinese environmental pollution case, signaling that the agencies in China regard pollution as a top concern. On the other hand, the Carbon Disclosure Project (CDP) works with hundreds of institutional investors along with governments and international institutions (e.g., UN and EU) to disclose the greenhouse gas emissions of major companies. CDP disclosed the emission performance of over 4500 companies in 2013 and specifically identified "high emitting" companies in each industry sector. Because investors are often averse to long-term investment risks, they might avoid companies with low environmental performance. Klassen and McLaughlin (1996) also found a link between firms' environmental crises and major negative stock returns. They explained that firms with weak environmental performance might incur environmental spills, crises, and liabilities, triggering negative market reactions. Therefore, firms with huge emissions often impose a tremendous cost from the damage to reputation and from the loss in stock value that far exceeds the environmental taxes (Hamilton, 1995; Pierce, 1998).

Finally, we suppose that abatement efforts are subject to diminishing returns and that their cost is an increasing quadratic function for the manufacturer, that is, $ev(t)^2/2$, where $e > 0$ is the cost coefficient associated with the manufacturer's investment in pollution abatement. It encompasses resources (e.g., capital, human resources, time) a manufacturer needs to introduce and operate environment-friendly technology/facility/equipment, innovate in operating processes, train employees, and set up superior management systems/infrastructure for emission reduction. The literature often assumes that investment in pollution abatement has a rising marginal cost, that is, reducing additional pollution becomes increasingly difficult due to the need for more advanced technology or more costly investment (Ni et al., 2010; Liu et al., 2012; Chung et al., 2013; Bertinelli et al., 2014). For simplicity, we assume that $e = 1$, without loss of generality; that is, the cost coefficient of pollution abatement (e) is normalized to one. A similar assumption is made by Bertinelli et al. (2014). It implies that f, the cost coefficient associated with the pollution stock, could be interpreted relative to e, the cost coefficient of the pollution abatement effort.

Table 5.1 Model parameters and variables

$y(t)$	Cumulative pollution at t
$v(t)$	Firm's pollution abatement effort at t
$p_2(t)$	Retailer's sales price (consumer price) at t
$D(t)$	Consumer demand at t
e	Cost coefficient associated with firm's pollution abatement effort
f	Cost coefficient associated with firm's pollution stock
p_1	Transfer price
α	Potential market size
β	Coefficient in the demand function associated with the sales price
φ	Effectiveness of the manufacturer's pollution abatement effort
v_{cum}	Firm's cumulative pollution abatement effort over the entire period
D_{cum}	Cumulative consumer demand over the entire period
J, J_m, J_r	Cumulative profit of the supply chain, the manufacturer, and the retailer, respectively over the entire period

The retailer's instantaneous net profit is given by the difference between its gross sales revenue and its purchasing cost from the manufacturer that is, $(p_2(t) - p_1)(\alpha - \beta p_2(t))$.

The differential game model is then given as:

$$\underset{p_1, v(t)}{\text{Max}} J_m = \int_0^T \left[p_1(\alpha - \beta p_2(t)) - ev(t)^2/2 - fy(t)^2/2\right] dt \tag{5.1}$$

$$\underset{p_2(t)}{\text{Max}} J_r = \int_0^T [(p_2(t) - p_1)(\alpha - \beta p_2(t))] dt \tag{5.2}$$

subject to:

$$\dot{y}(t) = \alpha - \beta p_2(t) - \varphi v(t), \qquad y(0) = y_0 = 0 \tag{5.3}$$

where $0 \leq v(t) \leq (\alpha - \beta p_2(t))/\varphi$ and $0 \leq p_2(t) \leq \alpha/\beta$, $\forall t$. As will be shown below, the control constraints are never binding. We therefore do not explicitly derive the related optimality conditions. Table 5.1 summarizes the parameters and variables of the model.

A particular feature of the game (Eqs. 5.1–5.2–5.3) is that the manufacturer's problem is linear-quadratic and the retailer's problem has the linear-state property. This feature makes it difficult to predict whether the strategy types (open-loop and feedback Nash equilibrium strategies) will have different outcomes. Therefore, both strategy types are analyzed.

Cooperative Equilibrium

This section analyzes the cooperative setting where the manufacturer and the retailer centralize their decisions to maximize their joint profits, as in the case of vertical integration.

Skipping the time index for convenience, the cooperative objective criterion is:

$$J = J_m + J_r = \int_0^T \left[p_2(\alpha - \beta p_2) - v^2/2 - fy^2/2\right] dt \tag{5.4}$$

The Hamiltonian of the cooperative equilibrium is:

$$H = p_2(\alpha - \beta p_2) - \frac{v^2}{2} - \frac{fy^2}{2} + \eta(\alpha - \beta p_2 - \varphi v) \tag{5.5}$$

where $\eta \equiv \eta(t)$ is the costate variable associated with the pollution stock, that evolves as:

$$\dot{\eta} = fy \qquad \eta(T) = 0 \tag{5.6}$$

Assuming interior solutions, the necessary conditions for optimality are:

$$H_v = -v - \varphi\eta = 0 \Rightarrow v = -\varphi\eta \tag{5.7}$$

$$H_{p_2} = \alpha - 2\beta p_2 - \beta\eta = 0 \Rightarrow p_2 = \frac{\alpha - \beta\eta}{2\beta} \tag{5.8}$$

The Hamiltonian is concave in (v, p_2) because the Hessian matrix $\begin{bmatrix} -2\beta & 0 \\ 0 & -1 \end{bmatrix}$ is definite negative and therefore the Legendre-Clebsch is satisfied (Grass et al., 2008). It can be easily shown that the maximized Hamiltonian, that is:

$$H^0 = \frac{1}{2}\left[(\alpha + \beta\eta)^2/2\beta + \varphi^2\eta^2 - fy^2\right] \tag{5.9}$$

is concave in y, which implies that the necessary conditions are also sufficient for optimality.

From Eqs. 5.7 and 5.8, we get:

$$p_2(t) = \frac{1}{2}\left(\frac{\alpha}{\beta} + \frac{v(t)}{\varphi}\right) \tag{5.10}$$

which implies that greater abatement efforts should result in fewer emissions *via* lower demand.

We use the superscript of "*c*" to denote the equilibrium strategies in the cooperative setting.

Theorem 1. The cooperative consumer price and abatement effort are given by:

$$p_2{}^c(t) = \frac{\alpha}{(\beta + 2\varphi^2)}\left[\frac{\beta + \varphi^2}{\beta} - \frac{\text{ch}(\theta t)}{2\text{ch}(\theta T)}\right] \tag{5.11}$$

$$v^c(t) = \frac{\varphi\alpha}{(\beta + 2\varphi^2)}\left[1 - \frac{\text{ch}(\theta t)}{\text{ch}(\theta T)}\right] \tag{5.12}$$

where $\theta \equiv \sqrt{f(\beta + 2\varphi^2)/2}$.

Accordingly, the cooperative pollution stock is:

$$y^c(t) = \frac{\text{sh}(\theta t)\alpha}{2\theta\text{ch}(\theta T)} \tag{5.13}$$

Proof See A1.

Because sufficiency is granted, the cooperative equilibrium is unique. It can be easily shown that both the control and state variables in Eqs. 5.11–5.13 have feasible values for any $T > 0$.

From Eq. 5.11 we get $\dot{p}_2{}^c(t) < 0$ and $\ddot{p}_2{}^c(t) < 0$. That is, the initial consumer price is high:

$$p_2{}^c(0) = \frac{\alpha}{(\beta + 2\varphi^2)}\left[\frac{\beta + \varphi^2}{\beta} - \frac{1}{2ch(\theta T)}\right] < \frac{\alpha}{\beta}$$

and diminishes at a decreasing rate over time until $p_2{}^c(T) = \alpha/2\beta$. Accordingly, sales grow at an increasing rate over time.

Similarly, from Eq. 5.12, we get $\dot{v}^c(t) < 0$ and $\ddot{v}^c(t) < 0$, which imply that the abatement effort is initially high, that is:

$$v^c(0) = \frac{\varphi\alpha}{(\beta + 2\varphi^2)}\left[1 - \frac{1}{\text{ch}(\theta T)}\right]$$

and concavely decreases over time until it vanishes at the final period, that is, $v^c(T) = 0$. Note that the consumer price pattern is consistent with that of the abatement effort.

Finally, from Eq. 5.13, we get:

$$\dot{y}^c(t) = \text{ch}(\theta t)\alpha/2\text{ch}(\theta T) \tag{5.14}$$

which is always strictly positive. This ensures that $0 \leq v^c(t) < D^c(t)/\varphi$, $\forall t \geq 0$, that is, at any time $t \geq 0$, the polluting emissions are nonnegative. At the final period, we get:

$$y^c(T) = \mathrm{th}(\theta T)\alpha/2\theta$$

Corollary 1. The cooperative payoff is given by:

$$J^c = \frac{1}{2(\beta + 2\varphi^2)}\left[\frac{\varphi^2\alpha^2 T}{\beta} + \frac{(e^{2\theta T} - 1)\alpha^2}{2\theta(e^{2\theta T} + 1)}\right] \tag{5.15}$$

Proof See A2.

It can be easily shown that the cooperative payoff in Eq. 5.15 is strictly positive.

Open-Loop Nash Equilibrium Strategies

This section identifies an open-loop Nash equilibrium (OLNE). We assume that because information on pollution is disregarded or unavailable, firms' strategies depend on time only.

The manufacturer's Hamiltonian is given as:

$$H_m = p_1(\alpha - \beta p_2) - \frac{v^2}{2} - \frac{fy^2}{2} + \lambda(\alpha - \beta p_2 - \varphi v) \tag{5.16}$$

where $\lambda \equiv \lambda(t)$ is the manufacturer's costate variable associated with the pollution stock, which evolves according to:

$$\dot{\lambda} = fy \qquad \lambda(T) = 0 \tag{5.17}$$

Assuming an interior solution, the manufacturer's necessary condition for optimality is:

$$\frac{\partial H_m}{\partial v} = -v - \lambda\varphi = 0 \Rightarrow v = -\lambda\varphi \tag{5.18}$$

The manufacturer's Hamiltonian is concave in v. Further, the maximized Hamiltonian:

$$H_m{}^0 = (p_1 + \lambda)(\alpha - \beta p_2) + \frac{1}{2}\left(\varphi^2\lambda^2 - fy^2\right) \tag{5.19}$$

is concave in y, and the necessary conditions for optimality are therefore also sufficient.

The retailer's Hamiltonian is:

$$H_r = (p_2 - p_1)(\alpha - \beta p_2) + \mu(\alpha - \beta p_2 - \varphi v) \tag{5.20}$$

where $\mu \equiv \mu(t)$ is the retailer's costate variable related to the pollution stock, given by:

$$\dot{\mu} = 0 \qquad \mu(T) = 0 \tag{5.21}$$

Given Eq. 5.21, the retailer's costate variable associated with the pollution stock is zero throughout the planning horizon.

Assuming an interior solution, the retailer's necessary condition for optimality is:

$$\frac{\partial H_r}{\partial p_2} = \alpha - 2\beta p_2 + \beta(p_1 - \mu) = 0 \;\Rightarrow p_2 = \frac{\alpha + \beta p_1}{2\beta} \tag{5.22}$$

The retailer's Hamiltonian is concave in p_2. Further, sufficiency is granted. We use the superscript of "*ol*" to denote the equilibrium strategies in the open-loop Nash setting.

Theorem 2. The manufacturer's OLNE transfer price and abatement effort and the retailer's OLNE consumer price are given by:

$$p_1^{ol} = \frac{\left[\varphi T\sqrt{f}(\beta + 2\varphi^2)\left(e^{2\varphi T\sqrt{f}} + 1\right) - \beta\left(e^{2\varphi T\sqrt{f}} - 1\right)\right]\alpha}{\beta\left[\varphi T\sqrt{f}(\beta + 4\varphi^2)\left(e^{2\varphi T\sqrt{f}} + 1\right) - \beta\left(e^{2\varphi T\sqrt{f}} - 1\right)\right]} \tag{5.23}$$

$$v^{ol}(t) = \frac{\varphi^2 T\sqrt{f}\left(e^{2\varphi T\sqrt{f}} + 1\right)\alpha}{\varphi T\sqrt{f}(\beta + 4\varphi^2)\left(e^{2\varphi T\sqrt{f}} + 1\right) - \beta\left(e^{2\varphi T\sqrt{f}} - 1\right)}\left[1 - \frac{\mathrm{ch}(\varphi t\sqrt{f})}{\mathrm{ch}(\varphi T\sqrt{f})}\right] \tag{5.24}$$

$$p_2^{ol}(t) = \frac{\left[\varphi T\sqrt{f}(\beta + 3\varphi^2)\left(e^{2\varphi T\sqrt{f}} + 1\right) - \beta\left(e^{2\varphi T\sqrt{f}} - 1\right)\right]\alpha}{\beta\left[\varphi T\sqrt{f}(\beta + 4\varphi^2)\left(e^{2\varphi T\sqrt{f}} + 1\right) - \beta\left(e^{2\varphi T\sqrt{f}} - 1\right)\right]} \tag{5.25}$$

Accordingly, the OLNE pollution stock is:

$$y^{ol}(t)=\frac{\varphi^2 T\mathrm{sh}(\varphi t\sqrt{f})\left(e^{2\varphi T\sqrt{f}}+1\right)\alpha}{\mathrm{ch}(\varphi T\sqrt{f})\left[\varphi T\sqrt{f}(\beta+4\varphi^2)\left(e^{2\varphi T\sqrt{f}}+1\right)-\beta\left(e^{2\varphi T\sqrt{f}}-1\right)\right]} \tag{5.26}$$

Proof See A3.

Due to the sufficiency condition, the OLNE is unique. Because $1+\varphi T\sqrt{f}+e^{2\varphi T\sqrt{f}}(\varphi T\sqrt{f}-1)>0$, the control and state variables in Eqs. 5.30–5.33 all have feasible values.

From Eq. 5.25, the retailer's consumer price is constant over time, strictly lower than α/β, and positively affected by both the manufacturer's transfer price and abatement effort. Consumer demand is also constant, given by:

$$D^{ol}(t)=\frac{\varphi^3 T\sqrt{f}\left(e^{2\varphi T\sqrt{f}}+1\right)\alpha}{\varphi T\sqrt{f}(\beta+4\varphi^2)\left(e^{2\varphi T\sqrt{f}}+1\right)-\beta\left(e^{2\varphi T\sqrt{f}}-1\right)}$$

Hence, the instantaneous manufacturer's gross profit and retailer's net profit are constant.

From Eq. 5.24, we get $\dot{v}^{ol}(t)<0$ and $\ddot{v}^{ol}(t)<0$, which imply a high initial abatement effort, that is:

$$v^{ol}(0)=\frac{\varphi^2 T\sqrt{f}\left(e^{2\varphi T\sqrt{f}}+1\right)\alpha}{\varphi T\sqrt{f}(\beta+4\varphi^2)\left(e^{2\varphi T\sqrt{f}}+1\right)-\beta\left(e^{2\varphi T\sqrt{f}}-1\right)}\left[1-\frac{1}{\mathrm{ch}(\varphi T\sqrt{f})}\right]$$

and decreasing at an increasing rate over time until it is zero at the end of the game.

Finally, from Eq. 5.26, we get:

$$\dot{y}^{ol}(t)=\frac{\varphi^3 T\sqrt{f}\mathrm{ch}(\varphi t\sqrt{f})\left(e^{2\varphi T\sqrt{f}}+1\right)\alpha}{\mathrm{ch}(\varphi T\sqrt{f})\left[\varphi T\sqrt{f}(\beta+4\varphi^2)\left(e^{2\varphi T\sqrt{f}}+1\right)-\beta\left(e^{2\varphi T\sqrt{f}}-1\right)\right]}>0$$

which ensures that $0\leq v^{ol}<(\alpha-\beta p_2^{ol})/\varphi$. At the final period, we get:

$$y^{ol}(T)=\frac{\varphi^2 T\mathrm{sh}(\varphi T\sqrt{f})\left(e^{2\varphi T\sqrt{f}}+1\right)\alpha}{\mathrm{ch}(\varphi T\sqrt{f})\left[\varphi T\sqrt{f}(\beta+4\varphi^2)\left(e^{2\varphi T\sqrt{f}}+1\right)-\beta\left(e^{2\varphi T\sqrt{f}}-1\right)\right]}$$

Lemma 1. The game admits a unique open-loop Stackelberg equilibrium, that is a Nash equilibrium.

Proof See A4.

In other words, the game admits a unique open-loop equilibrium irrespective of whether moves are simultaneous or sequential.

Corollary 2. The manufacturer's OLNE payoff is:

$$J_m^{ol} = \frac{\varphi^3\alpha^2T^2\sqrt{f}e^{\varphi T\sqrt{f}}\mathrm{ch}(\varphi T\sqrt{f})}{\beta\left[\varphi T\sqrt{f}(4\varphi^2+\beta)\left(e^{2\varphi T\sqrt{f}}+1\right)-\beta\left(e^{2\varphi T\sqrt{f}}-1\right)\right]} \tag{5.27}$$

and the retailer's OLNE payoff is:

$$J_r^{ol} = \frac{\varphi^6 fT^3\left(e^{2\varphi T\sqrt{f}}+1\right)^2\alpha^2}{\beta\left[\varphi T\sqrt{f}(\beta+4\varphi^2)\left(e^{2\varphi T\sqrt{f}}+1\right)-\beta\left(e^{2\varphi T\sqrt{f}}-1\right)\right]^2} \tag{5.28}$$

Proof See A5.

Because $1+\varphi T\sqrt{f}+e^{2\varphi T\sqrt{f}}(\varphi T\sqrt{f}-1)>0$, the manufacturer's and retailer's OLNE payoff are both positive. Therefore, the wholesale price contract is agreeable for both firms.

Feedback Nash Equilibrium Strategies

If information on pollution stock is available to both firms throughout the game, they can condition their actions on the current value of this stock, and on time. In this case, a standard approach is to look for a feedback Nash equilibrium (FBNE). We use the superscript of "*fb*" to denote the equilibrium strategies in the open-loop Nash setting.

Theorem 3. The manufacturer's FBNE transfer price and abatement effort and the retailer's FBNE consumer price are given by:

$$p_1^{fb}(t) = \frac{[\varphi T\sqrt{f}(\beta+2\varphi^2)-\beta\mathrm{th}(\varphi T\sqrt{f})]\alpha}{\beta[\varphi T\sqrt{f}(\beta+4\varphi^2)-\beta\mathrm{th}(\varphi T\sqrt{f})]} \tag{5.29}$$

$$v^{fb}(t) = \frac{2\varphi^2T\sqrt{f}\mathrm{sech}(\varphi T\sqrt{f})\mathrm{sh}[\varphi\sqrt{f}(T-t)/2]\mathrm{sh}[\varphi\sqrt{f}(T+t)/2]\alpha}{\varphi T\sqrt{f}(\beta+4\varphi^2)-\beta\mathrm{th}(\varphi T\sqrt{f})} \tag{5.30}$$

$$p_2^{fb}(t) = \frac{[\varphi T\sqrt{f}(\beta + 3\varphi^2) - \beta \mathrm{th}(\varphi T\sqrt{f})]\alpha}{\beta[\varphi T\sqrt{f}(\beta + 4\varphi^2) - \beta \mathrm{th}(\varphi T\sqrt{f})]} \tag{5.31}$$

Accordingly, the FBNE pollution stock is:

$$y^{fb}(t) = \frac{\varphi^2 T \mathrm{sech}(\varphi T\sqrt{f})\mathrm{sh}(\varphi t\sqrt{f})\alpha}{\varphi T\sqrt{f}(\beta + 4\varphi^2) - \beta \mathrm{th}(\varphi T\sqrt{f})} \tag{5.32}$$

Proof See A6.

Note that the FBNE is unique. Because $\varphi T\sqrt{f}(\beta + 4\varphi^2) - \beta \mathrm{th}(\varphi T\sqrt{f}) > 0$, the control and state variables in Eqs. 5.29–5.32 have feasible values. From Eq. 5.31, the retailer's consumer price is constant over time, strictly lower than α/β, which implies constant demand, equal to:

$$D^{fb}(t) = \frac{\varphi^3 T\sqrt{f}\alpha}{\varphi T\sqrt{f}(\beta + 4\varphi^2) - \beta \mathrm{th}(\varphi T\sqrt{f})}$$

and constant instantaneous manufacturer's gross profit and retailer's net profit over time. It can be easily shown from Eq. 5.32 that $\dot{y}^{fb}(t) > 0$, which ensures that $0 \leq v^{fb} < \left(\alpha - \beta p_2^{fb}\right)/\varphi$.

Lemma 2. The game admits a unique subgame perfect Stackelberg equilibrium, that is a Nash equilibrium.

Proof See A7.

Thus, the game admits a unique and subgame perfect equilibrium irrespective of whether moves are simultaneous or sequential.

Corollary 3. Assuming that the game starts at $t = 0$, the manufacturer's FBNE payoff is:

$$J_m^{fb} = \frac{\varphi^3 T^2\sqrt{f}\,\mathrm{ch}(\varphi T\sqrt{f})\mathrm{sech}(\varphi T\sqrt{f})^2[\varphi T\sqrt{f}(\beta + 4\varphi^2)\mathrm{ch}(\varphi T\sqrt{f}) - \beta \mathrm{sh}(\varphi T\sqrt{f})]\alpha^2}{2\beta[\varphi T\sqrt{f}(\beta + 4\varphi^2) - \beta \mathrm{th}(\varphi T\sqrt{f})]^2} \tag{5.33}$$

and the retailer's FBNE payoff is:

$$J_r^{fb} = \frac{\varphi^6 f T^3 \alpha^2}{\beta[\varphi T\sqrt{f}(\beta + 4\varphi^2) - \beta \mathrm{th}(\varphi T\sqrt{f})]^2} \tag{5.34}$$

Proof See A8.

Because $\varphi T\sqrt{f} > \text{th}(\varphi T\sqrt{f})$ in Eqs. 5.33–5.34, the manufacturer's and retailer's payoffs are both positive. Hence, the wholesale price contract can be agreed on by both firms also in FBNE.

Comparative Analysis

In this section, we use the previous analytical results to determine the main differences between cooperative and noncooperative equilibria in terms of control and state variables.

Theorem 4. The OLNE strategy is subgame perfect.

Proof See A9.

Because p_1, p_2, v and y are equivalent in the OLNE and FBNE models, it holds that $D^{ol}(t) = D^{fb}(t)$, $J_m{}^{ol} = J_m{}^{fb}$ and $J_r{}^{ol} = J_r{}^{fb}$. The equivalence between OLNE and FBNE is due to the linear state property of the retailer's problem. Lemma 2 and Theorem 4 show that the unique open-loop Stackelberg equilibrium of the game is also subgame perfect.

Below, we use either open-loop or feedback solutions to compare cooperative and noncooperative equilibria, respectively denoted by the respective superscripts c and nc.

Proposition 1. For any given values of α and f, there exists a threshold $\widetilde{\varphi} > 0$ ($\widetilde{\beta} > 0$) such that for any $\varphi > \widetilde{\varphi}$ ($\beta < \widetilde{\beta}$), it holds that:

$$v^c(t) \geq v^{nc}(t) \tag{5.35}$$

Proof See A10.

Proposition 1 implies that manufacturer's abatement effort is greater in a cooperative case than in a noncooperative case if the effectiveness of the abatement effort is sufficiently large or demand sensitivity to price is weak. In a more profitable context, firms can more easily invest in abatement efforts. This is why the cooperative supply chain participates in the pollution abatement more actively than in the noncooperative case. In the opposite context, the noncooperative supply chain might heighten the manufacturer's abatement effort, which increases environmental virtue.

Proposition 2. For sufficient marginal effectiveness of the manufacturer's abatement effort, it holds that:

$$p_2{}^c(t) < p_2{}^{nc}(t), \forall\ t \in [0, T] \tag{5.36}$$

Proof See A11.

From Eqs. 5.11 and 5.25, it is obvious that:

$$p_2{}^c(T) = \frac{\alpha}{2\beta} < \frac{[\varphi T\sqrt{f}(\beta + 3\varphi^2) - \beta\,\mathrm{th}(\varphi T\sqrt{f})]\alpha}{\beta[\varphi T\sqrt{f}(\beta + 4\varphi^2) - \beta\,\mathrm{th}(\varphi T\sqrt{f})]} = p_2{}^{nc}(T)$$

However, the following inequality:

$$p_2{}^c(0) = \frac{\alpha}{(\beta + 2\varphi^2)}\left[\frac{\beta + \varphi^2}{\beta} - \frac{1}{2\,\mathrm{ch}(\theta T)}\right] < \frac{[\varphi T\sqrt{f}(\beta + 3\varphi^2) - \beta\,\mathrm{th}(\varphi T\sqrt{f})]\alpha}{\beta[\varphi T\sqrt{f}(\beta + 4\varphi^2) - \beta\,\mathrm{th}(\varphi T\sqrt{f})]} = p_2{}^{nc}(0)$$

holds whenever $\varphi \geq \sqrt{\beta}/2$. This means that consumer demand in a cooperative case is greater than in a noncooperative case throughout the planning horizon if the manufacturer's effectiveness of abatement effort is great enough to avoid an excessively high pollution cost. It follows that polluting emissions may be lower during an initial time interval and higher over the subsequent time interval in the cooperative case than in the noncooperative case.

Proposition 3. It holds that:

$$y^c(T) > y^{nc}(T) \tag{5.37}$$

Proof See A12.

Hence, the cooperative case generates a greater pollution stock than the noncooperative case. Although an increase in revenues driven by mitigation of double marginalization in a cooperative supply chain may result in greater abatement efforts, the efforts do not outweigh the increase in emissions and thus lower the pollution stock. In other words, a cooperative supply chain might choose to sacrifice environmental performance rather than increase abatement effort. In a noncooperative supply chain, there is a lesser burden of abatement effort due to relatively small polluting emissions.

Numerical Illustrations

In this section, we illustrate the optimal paths of the cooperative and noncooperative equilibria numerically. The parameter values used are in Table 5.2.

The planning horizon is supposed to be short enough with $T = 12$ months. The potential market size is $\alpha = 100$. We use two values for demand sensitivity to price

Table 5.2 Parameter values

T	α	β	φ	f
12	100	[1,10]	[0.1,10]	[0.01, 0.1]

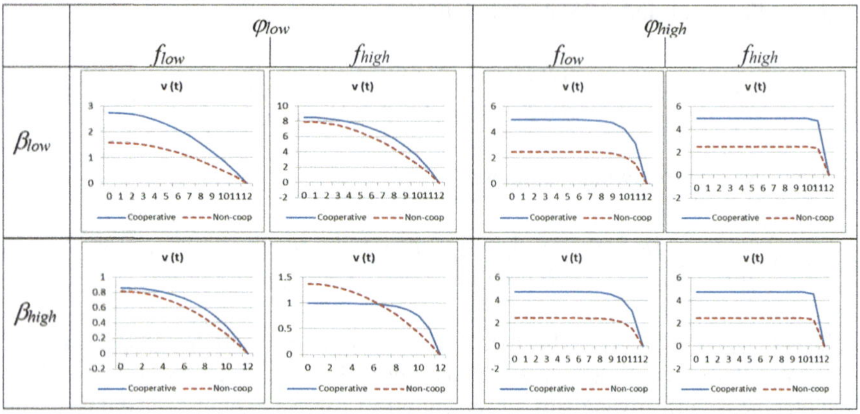

Fig. 5.1 Cooperative and noncooperative manufacturer's abatement effort over time

($\beta_{low} = 1$,$\beta_{high} = 10$), marginal effectiveness of manufacturer's abatement effort ($\varphi_{low} = 0.1$,$\varphi_{high} = 10$) and pollution cost coefficient ($f_{low} = 0.01, f_{high} = 0.1$).The pollution cost coefficient values are lower than the abatement effort cost coefficient ($e = 1$) because in the converse case, the manufacturer would maximally invest in the abatement effort.

Impact of Double Marginalization on Pollution Abatement Effort

Figure 5.1 confirms that $v^c(t) > v^{nc}(t)$ holds over the whole game except under high β and low φ for a high pollution cost coefficient.

In a cooperative game, more efforts are generally allocated to pollution abatement because the supply chain's costs and revenues are both affected, while only the manufacturer's costs and revenues are affected in a noncooperative game. However, an increase in abatement efforts is less beneficial to a cooperative supply chain if its marginal ability to reduce pollution is low and there is little room to raise the consumer price due to high price sensitivity of demand. Therefore, greater initial abatement efforts (while the pollution stock is still relatively low) would raise the consumer price, which would reduce demand.

From Eqs. 5.12 and 5.24, the cumulative abatement efforts for the entire planning horizon in the cooperative and noncooperative cases are:

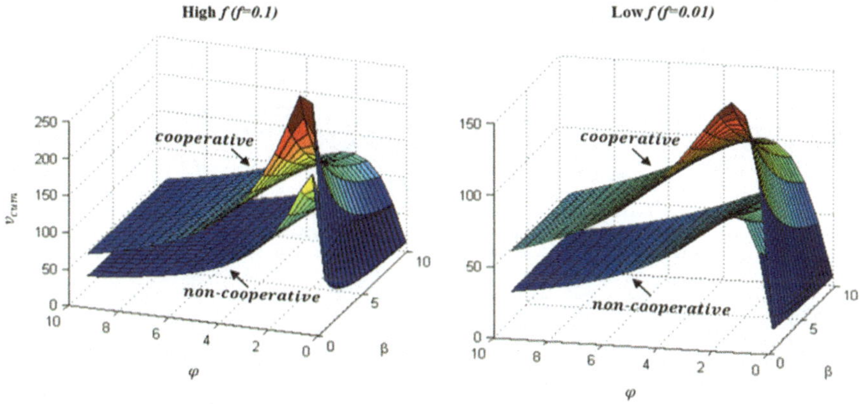

Fig. 5.2 Cumulative abatement effort of cooperative and noncooperative manufacturer

$$v_{cum}^{c}=\int_0^T v^c(t)\mathrm{d}t=\frac{\varphi[\theta T-\mathrm{th}(\theta T)]\alpha}{\theta(\beta+2\varphi^2)},\quad v_{cum}^{nc}=\int_0^T v^{nc}(t)\mathrm{d}t=\frac{\varphi T[\varphi T\sqrt{f}-\mathrm{th}(\varphi T\sqrt{f})]\alpha}{(\beta+4\varphi^2)\varphi T\sqrt{f}-\beta\,\mathrm{th}(\varphi T\sqrt{f})}$$

Whether firms' pollution abatement effort at each period is greater in the cooperative case than in the noncooperative case depends on parameter values such as demand sensitivity to price and marginal effectiveness of abatement efforts, yet numerical results show that the cumulative abatement efforts over the whole period are greater in the cooperative than in the noncooperative case; up to twice that of the noncooperative manufacturer (Fig. 5.2).

This result corroborates the findings that cooperation within a supply chain is related to extensive environmental efforts (Klassen & Vachon, 2003; Vachon & Klassen, 2008; Simpson, 2010; Hollos et al., 2012).

Impact of Double Marginalization on Price and Demand

Figure 5.3 shows that if the marginal effectiveness of abatement effort (φ) is sufficiently high, $p_2{}^c(t) < p_2{}^{nc}(t)$ holds $\forall\ t$, which implies that $D^c(t) > D^{nc}(t)$, resulting in more emissions in the cooperative than in the noncooperative case. Conversely, these inequalities do not always hold, notably in the case of a large pollution cost coefficient. As suggested in Eq. 5.10, the consumer price patterns are consistent with those of abatement efforts in the cooperative case.

While double marginalization usually leads to higher consumer prices, we show that consumer prices in the cooperative case might not be strictly larger than in the noncooperative case for a given time interval. Specifically, when the abatement effort is ineffective, a cooperative supply chain often imposes similar or even higher consumer prices than a noncooperative supply chain in the initial periods. The

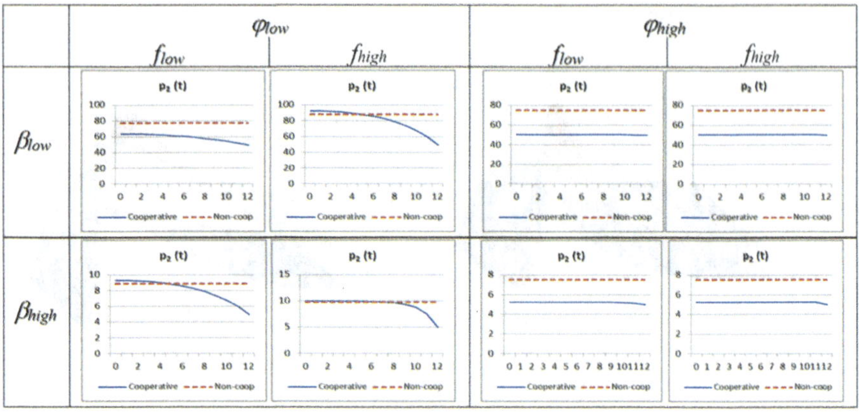

Fig. 5.3 Cooperative and noncooperative retailer's consumer price over time

intuition behind this result is that low abatement effectiveness increases total abatement cost to maintain low pollution stock. This increases the consumer price in both cooperative and noncooperative supply chains. However, in the cooperative case, the increase in consumer price is often concentrated in the initial periods, while consumer price decreases sharply in the later periods.

Numerical results show that cumulative consumer demand, given by:

$$D^c_{cum} = \int_0^T D^c(t)\mathrm{d}t = \left[\frac{\varphi^2 T}{\beta + 2\varphi^2} + \frac{\beta f \,\mathrm{th}(\theta T)}{4\theta^3}\right] \quad D^{nc}_{cum} = \int_0^T D^{nc}(t)\mathrm{d}t = \frac{\varphi^3 T\sqrt{f}\alpha T}{(\beta + 4\varphi^2)\varphi T\sqrt{f} - \beta\,\mathrm{th}(\varphi T\sqrt{f})}$$

is always greater in the cooperative than in the noncooperative case. This is because the cooperative price is significantly lower than the noncooperative one during the later periods.

Figure 5.4 shows that regardless of the pollution cost coefficient, the retailer's profit margin, $p_2^{nc}(t) - p_1^{nc}(t)$, is maximized (minimized) when both demand sensitivity to price is low (high) and abatement effort effectiveness is high (low). For high demand sensitivity to price and low marginal effectiveness of abatement effort, a large pollution cost coefficient involves more abatement effort in the noncooperative case than in the cooperative one. Hence, this configuration mitigates the double marginalization effect.

Impact of Double Marginalization on Pollution Emissions

At the final period, $y^c(T) > y^{nc}(T)$ holds although $v^c(t) > v^{nc}(t)$ tends to hold especially for low β and high φ, and $v^c_{cum} > v^{nc}_{cum}$ holds (Fig. 5.5a).

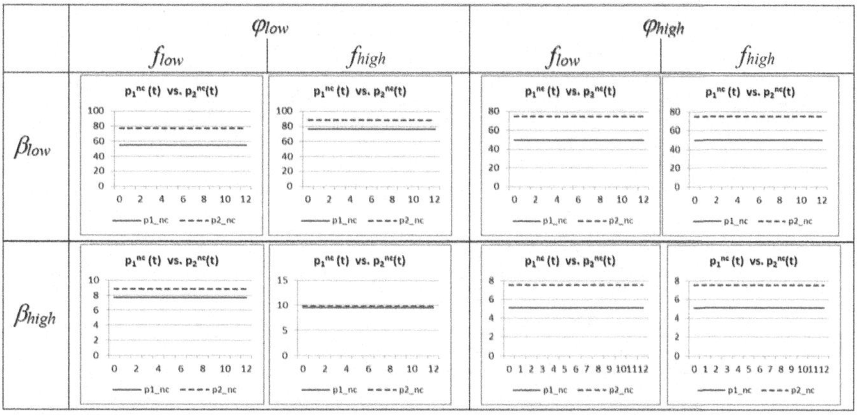

Fig. 5.4 Noncooperative manufacturer's transfer price and retailer's profit margins over time

Stated differently, the cooperative supply chain results in lower environmental performance than the noncooperative supply chain, although the pollution abatement effort is generally greater due to an increase in revenue. This result implies that the abatement effort in the cooperative supply chain would not suffice, despite its high volume, to outweigh high emissions.

This compromise in environmental performance is less apparent in the noncooperative supply chain because the required level of abatement effort is lesser due to low emissions. In addition, in the cooperative case, consumer demand is higher during the later periods. Because the abatement effort decreases and vanishes by the final period, the pollution stock in the later periods is more significantly influenced by consumer demand than by the abatement effort. Conversely, the pollution stock is influenced more by the abatement effort in the initial periods, and thus the cooperative supply chain might be a more eco-friendly configuration during those periods.

Figure 5.5b shows that, in terms of cumulative output, the cooperative supply chain still results in a higher pollution stock per unit of cumulative consumer demand than the noncooperative supply chain, confirming the lower environmental performance of the cooperative supply chain. This result continues to hold, regardless of pollution cost coefficient, marginal effectiveness of abatement effort and demand sensitivity to price, and suggests that the cooperative supply chain makes in average less per unit abatement efforts than the noncooperative supply chain.

Impact of Double Marginalization on Firms' Payoffs

Finally, Fig. 5.6 compares firms' payoffs in the cooperative and noncooperative settings. The Nash bargaining scheme (e.g., Dockner et al., 2000) is used to compute the manufacturer's and retailer's cooperative payoffs as follows:

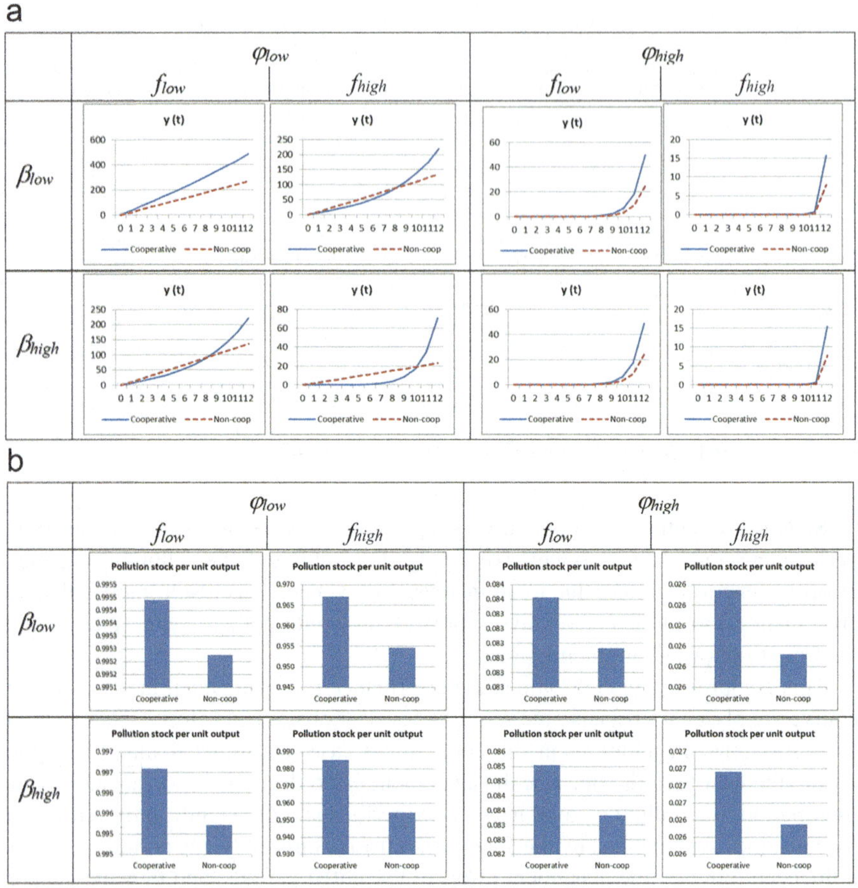

Fig. 5.5 (**a**) Cooperative and noncooperative pollution stock over time. (**b**) Cooperative and noncooperative pollution stock per unit cumulative consumer demand $\left(\text{i.e., } \frac{y(T)}{D_{cum}}\right)$

$$J_m{}^c = J_m{}^{nc} + (J^c - J_m{}^{nc} - J_r{}^{nc})/2, \qquad J_r{}^c = J_r{}^{nc} + (J^c - J_m{}^{nc} - J_r{}^{nc})/2$$

As expected, the manufacturer and the retailer earn higher payoffs in the cooperative than in the noncooperative case. The relative difference between the cooperative and the noncooperative payoffs is greatest in the case of high price sensitivity of demand and low effectiveness of manufacturer's abatement effort under a high pollution cost coefficient.

Overall, we get the following rankings from Fig. 5.6:

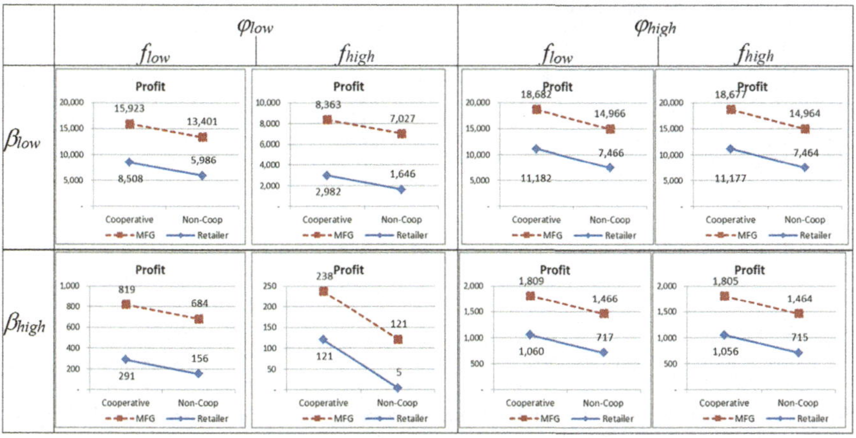

Fig. 5.6 Cooperative and noncooperative manufacturer's and retailer's payoffs

$$J_m^c|\beta_{low},\ \varphi_{high} > J_m^c|\beta_{low},\ \varphi_{low} > J_m^c|\beta_{high},\ \varphi_{high} > J_m^c|\beta_{high},\ \varphi_{low},$$

$$J_r^c|\beta_{low},\ \varphi_{high} > J_r^c|\beta_{low},\ \varphi_{low} > J_r^c|\beta_{high},\ \varphi_{high} > J_r^c|\beta_{high},\ \varphi_{low}$$

$$J_m^{nc}|\beta_{low},\ \varphi_{high} > J_m^{nc}|\beta_{low},\ \varphi_{low} > J_m^{nc}|\beta_{high},\ \varphi_{high} > J_m^{nc}|\beta_{high},\ \varphi_{low},$$

$$J_r^{nc}|\beta_{low},\ \varphi_{high} > J_r^{nc}|\beta_{low},\ \varphi_{low} > J_r^{nc}|\beta_{high},\ \varphi_{high} > J_r^{nc}|\beta_{high},\ \varphi_{low}$$

For a given pollution cost coefficient, the best configuration of firms' payoffs both for cooperative and noncooperative equilibria is that where the sensitivity of demand to price is low and the manufacturer's effectiveness of abatement effort is high. Conversely, the worst configuration both for cooperative and noncooperative equilibria is where the sensitivity of demand to price is high and the manufacturer's effectiveness of abatement effort is low. The predominant influence of demand sensitivity to price over the marginal effectiveness of abatement effort is obvious in the ranking of the payoffs of the two remaining intermediate configurations.

Impact of Double Marginalization on Eco-friendly Social Welfare

We investigate the impact of double marginalization in a supply chain from a global perspective including both social welfare and the environmental impact. To do so, we introduce a measure of eco-friendly social welfare, defined as the amount of social welfare generated at the expense of the pollution stock. Social welfare for the entire period $t \in [0, T]$ is defined as the sum of firms' total profits and total consumer surplus as follows (Long, 2010), that is:

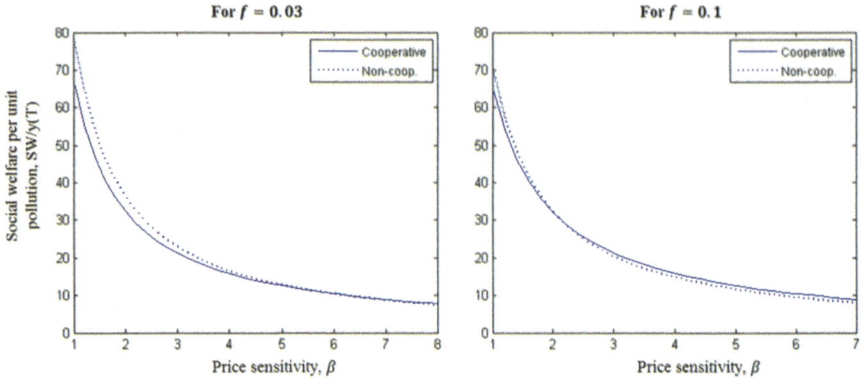

Fig. 5.7 Eco-friendly social welfare depending on price sensitivity ($\alpha = 100,\ \varphi = 0.1,\ T = 12$)

$$\text{Social welfare } (SW) = \text{Total supply chain profit } (J) + \text{Consumer surplus } (CS)$$
$$= \text{Total supply chain profit } (J) + \int_0^T \frac{1}{2\beta}[\alpha - \beta p_2(t)]^2\, dt$$

Figure 5.7 shows the eco-friendly social welfare, $SW/y(T)$, as a function of demand sensitivity β in cooperative and noncooperative settings. The most interesting implication in Fig. 5.7 is that a cooperative supply chain does not always yield higher eco-friendly social welfare than a noncooperative supply chain. Whether a cooperative supply chain generates higher eco-friendly social welfare is influenced by market characteristics and the institutional environment—i.e., demand sensitivity to price (β) and the pollution cost coefficient (f).

Figure 5.7 shows that a cooperative supply chain tends to yield greater eco-friendly social welfare than a noncooperative supply chain as the market becomes more price-sensitive. In contrast, when consumers are not very responsive to price changes, a noncooperative supply chain tends to generate more eco-friendly social welfare than a cooperative supply chain. The intuition behind these results is that the advantage of a large consumer surplus resulting from low prices in the cooperative case tends to be more significant as demand price sensitivity rises. Thus, despite a large pollution stock, the advantage of high social welfare in the cooperative supply chain dominates when price sensitivity is high, leading to higher eco-friendly social welfare than in the noncooperative supply chain.

Conclusions

This paper assessed the impact of decentralized decision-making on pollution accumulation in a supply chain composed of one manufacturer and one retailer under a conventional wholesale price contract. To do so, we suggested a differential game model of vertical pollution control with nonexistent natural abatement of the pollution stock.

We showed that the game allows a unique and subgame perfect equilibrium irrespective of whether moves are simultaneous or sequential, and that the influence of the double marginalization effect on pollution accumulationis unaffected by strategy types, that is, open-loop *vs.* feedback Nash equilibrium strategies. While the pricing strategy is farsighted in the cooperative case, it is myopic in the noncooperative case because the retailer does not internalize the cost of pollution. In the cooperative case, the abatement effort path generally shapes the consumer price path. Further, greater abatement efforts reduce consumer demand (and emissions). In the noncooperative case, abatement efforts are negatively influenced by the transfer price, which in turn positively affects the consumer price. As a result, double marginalization is amplified because lower abatements are concomitant with lower emissions.

This paper contributes to the literature first by bringing the double marginalization effect in a vertical supply chain explicitly to environmental economics. Although the virtue of supply chain coordination has been widely accepted, we show that this statement is only partially correct when an environmental aspect is considered. The cooperative supply chain results in lower sales prices and consequently higher consumer demand in general. Also, it indeed leads to higher cumulative abatement efforts as is commonly believed, driven by an increase in revenue. However, these efforts are not sufficient to offset the greater emissions as by-products from higher production activities in the centralized supply chain because abatement efforts often require substantial investments in environment-friendly technologies, facilities and processes (Krass et al., 2013). As a result, cooperative supply chain emits more cumulative pollution and pollution per unit output than a decentralized chain and compromises on environmental performance until it increases the supply chain profit. In the noncooperative supply chain, however, this problem is less apparent because the level of required abatement effort is lower due to low polluting emissions. In addition, abatement efforts might be higher in the noncooperative supply chain during the initial periods under particular configurations involving high price-sensitive demand or low effectiveness of abatement efforts. Stated differently, although the noncooperative supply chain ruled by a wholesale price contract and its corollary, the double marginalization effect, have been commonly recognized as inefficient, they have environmental virtue because they not only reduce polluting emissions *via* lower production, but also heighten manufacturer's abatement efforts.

Next, we introduced a notion of eco-friendly social welfare, measured as social welfare per unit of pollution stock, to explore the impact of double marginalization from a broader perspective including supply chain profit, consumer surplus and environmental performance. We find that the elimination of double marginalization in the cooperative supply chain generates higher eco-friendly social welfare only in some cases. When demand is highly sensitive to the sales price, a cooperative supply chain often generates greater eco-friendly social welfare than a noncooperative chain. However, a decentralized supply chain and the presence of double marginalization tend to be more socially desirable in the converse case.

We also provide several insights for regulators. First, the general perception that a centralized supply chain is more socially desirable should be reexamined if an environmental aspect is considered. Which supply chain configuration (i.e., centralized *vs.* decentralized) should be promoted varies depending on the priorities. If regulators intend to maximize social welfare only (i.e., firms' profits and consumer surplus), they should facilitate a centralized supply chain. Yet if their main goal is to reduce emissions, then regulators should encourage a decentralized supply chain rather than supply chain coordination, because the elimination of double marginalization in a centralized supply chain results in greater cumulative emissions and higher emissions per unit output despite higher abatement efforts. This could be particularly relevant to industries where pollution is the most urgent issue and significant emission reduction is crucial. Moreover, if the regulators consider broader aspects of the entire economic system, including not only the environment but also both firms' profits and consumer surplus, the optimal supply chain configuration would vary depending on the market characteristics. Specifically, regulators should encourage supply chain coordination to increase eco-friendly social welfare only when the market is price-sensitive enough. Otherwise, it might be more effective to focus on maintaining a double marginalization effect in a supply chain.

Our findings also imply that the benefits of supply chain coordination might vary depending on the firm's objective. If the goal of the firm is purely profit-seeking, managers should try to move toward a centralized supply chain. Yet if the profit-maximizing firm also cares about long-term environmental sustainability, then managers should be aware of a downside of supply chain coordination: lower environmental performance. Managers should learn how to overcome this environmental issue in a coordinated supply chain and find ways to balance short-term profit maximization and long-term environmental sustainability.

This paper has demonstrated the importance of double marginalization on pollution accumulation. For future research, our vertical pollution game can be extended to account for price competition in a bilateral duopoly. This important topic is high on our research agenda.

Acknowledgments The authors are grateful to Steffen Jørgensen and Georges Zaccour for constructive suggestions on an early draft.

Appendix

A1 Plugging the expressions from Eqs. 5.7–5.8 in Eq. 5.3 and using Eq. 5.6, the two-point boundary value problem (TPBVP) is:

$$\dot{y} = \left[\alpha + \left(\beta + 2\varphi^2\right)\eta\right]/2 \tag{5.38}$$

$$\dot{\eta} = fy \tag{5.39}$$

Deriving Eq. 5.39 with respect to time and using Eq. 5.38 gives:

$$\ddot{\eta} - f(\beta + 2\varphi^2)\eta/2 - f\alpha/2 = 0 \tag{5.40}$$

We derive the complementary solutions of Eq. 5.40, and solve the auxiliary equation:

$$m_1^2 - f(\beta + 2\varphi^2)/2 = 0$$

which gives: $m_{11} = \sqrt{f(\beta + 2\varphi^2)/2}$ and $m_{12} = -\sqrt{f(\beta + 2\varphi^2)/2}$. We derive a particular solution of the differential equation, that is, $Q_1 = -\alpha/(2\varphi^2 + \beta)$. Therefore, a general solution of Eq. 5.40 is:

$$\eta(t) = K_{11}e^{m_{11}t} + K_{12}e^{m_{12}t} - \alpha/\left(2\varphi^2 + \beta\right) \tag{5.41}$$

which, using Eqs. 5.7–5.8, respectively gives:

$$v(t) = \varphi\left[\alpha/\left(2\varphi^2 + \beta\right) - K_{11}e^{m_{11}t} - K_{12}e^{m_{12}t}\right] \tag{5.42}$$

$$p_2(t) = \frac{1}{2}\left[\alpha/\beta - K_{11}e^{m_{11}t} - K_{12}e^{m_{12}t} + \alpha/\left(2\varphi^2 + \beta\right)\right] \tag{5.43}$$

which, using Eq. 5.39, finally yields:

$$y(t) = \left(m_{11}K_{11}e^{m_{11}t} + m_{12}K_{12}e^{m_{12}t}\right)/f \tag{5.44}$$

To determine K_{11} and K_{12}, we use $y(0) = 0$ and $\eta(T) = 0$ to solve the system:

$$y(0) = (m_{11}K_{11} + m_{12}K_{12})/f = 0$$

$$\eta(T) = K_{11}e^{m_{11}T} + K_{12}e^{m_{12}T} - \frac{\alpha}{2\varphi^2 + \beta} = 0$$

which gives $K_{11} = -\frac{m_{12}\alpha}{(\beta + 2\varphi^2)(m_{11}e^{m_{12}T} - m_{12}e^{m_{11}T})}$ and $K_{12} = \frac{m_{11}\alpha}{(\beta + 2\varphi^2)(m_{11}e^{m_{12}T} - m_{12}e^{m_{11}T})}$. Plugging the expressions of K_{11}, K_{12}, m_{11} and m_{12}, in Eqs. 5.42, 5.43, and 5.44, gives Eqs. 5.11, 5.12, and 5.13. □

A2 Plugging the expressions from Eqs. 5.11, 5.12, and 5.13 in Eq. 5.4 and solving gives Eq. 5.15. □

A3 As usual in a two-stage game, one solves the game backward. We first analyze the second stage to derive the manufacturer's abatement effort and the pollution stock, and then turn back to the first stage to determine the manufacturer's transfer price. Plugging the expressions from Eqs. 5.18 and 5.22 in Eq. 5.3 and using Eq. 5.17, the TPBVP is:

$$\dot{\lambda} = fy \tag{5.45}$$

$$\dot{y} = (\alpha - \beta p_1)/2 + \varphi^2 \lambda \tag{5.46}$$

Using Eqs. 5.45–5.46, we get:

$$\ddot{\lambda} - f\varphi^2 \lambda - f(\alpha - \beta p_1)/2 = 0 \tag{5.47}$$

Using similar methods to the cooperative case, the general solution of Eq. 5.47 is:

$$\lambda(t) = K_{21} e^{\varphi t \sqrt{f}} + K_{22} e^{-\varphi t \sqrt{f}} - (\alpha - \beta p_1)/2\varphi^2 \tag{5.48}$$

which, using Eqs. 5.45 and 5.18, gives:

$$v^{ol}(t) = (\alpha - \beta p_1)/2\varphi - \varphi\left(K_{21} e^{\varphi t \sqrt{f}} + K_{22} e^{-\varphi t \sqrt{f}}\right) \tag{5.49}$$

and:

$$y(t) = \varphi\left(K_{21} e^{\varphi t \sqrt{f}} - K_{22} e^{-\varphi t \sqrt{f}}\right) / \sqrt{f} \tag{5.50}$$

Using $y(0) = 0$ and $\lambda(T) = 0$, we get: $K_{21} = K_{22} = \frac{\alpha - \beta p_1}{4\varphi^2\left(e^{\varphi T\sqrt{f}} + e^{-\varphi T\sqrt{f}}\right)}$. Plugging the expressions of K_{21} and K_{22} respectively in Eqs. 5.49 and 5.50 gives:

$$v^{ol}(t) = \frac{(\alpha - \beta p_1)}{2\varphi}\left[1 - \frac{\text{ch}(\varphi t\sqrt{f})}{\text{ch}(\varphi T\sqrt{f})}\right] \tag{5.51}$$

and:

$$y^{ol}(t) = \frac{(\alpha - \beta p_1)\text{sh}(\varphi t\sqrt{f})}{2\varphi\sqrt{f}\text{ch}(\varphi T\sqrt{f})} \tag{5.52}$$

Plugging Eqs. 5.22, 5.51, and 5.52 in Eq. 5.1 and solving gives the manufacturer's payoff:

$$J_m = \frac{1}{8\varphi^2}\left\{\frac{\left(e^{2\varphi T\sqrt{f}} - 1\right)(\alpha - \beta p_1)}{\varphi\sqrt{f}\left(e^{2\varphi T\sqrt{f}} + 1\right)} - \left[\alpha - (\beta + 4\varphi^2)p_1\right]T\right\}(\alpha - \beta p_1) \tag{5.53}$$

Deriving the right-hand side of Eq. 5.53 with respect to p_1 and solving for p_1 gives Eq. 5.23. To confirm that p_1^{ol} is indeed optimal, we need to prove that $\partial^2 J_m / \partial p_1^{ol2} < 0$:

$$\frac{\partial^2 J_m}{\partial p_1^{ol2}} = -\frac{\beta\left[\varphi\sqrt{f}(\beta+4\varphi^2)\left(e^{2\varphi T\sqrt{f}}+1\right)T-\beta\left(e^{2\varphi T\sqrt{f}}-1\right)\right]}{4\varphi^3\sqrt{f}\left(e^{2\varphi T\sqrt{f}}+1\right)}$$

As $\varphi T\sqrt{f}>0$, we only need to show that $1+\varphi T\sqrt{f}+e^{2\varphi T\sqrt{f}}(\varphi T\sqrt{f}-1)>0$ for $\partial^2 J_m/\partial p_1^{ol2}<0$ to hold. Let $x=\varphi T\sqrt{f}$, $f(x)=1+\varphi T\sqrt{f}+e^{2\varphi T\sqrt{f}}(\varphi T\sqrt{f}-1)=1+x+e^{2x}(x-1)$. Then, $f_x = 1 - 2e^{2x} + e^{2x} + 2xe^{2x} = 1 - e^{2x} + 2xe^{2x}$ and $f_{xx} = - 2e^{2x} + 2e^{2x} + 4xe^{2x} = 4xe^{2x}$. At $x=0$, we have $f(x)=0$ and $f_x=0$. As $f_{xx}>0$ for $x>0, f_x>0$ holds for $x>0$. Thus, we can conclude that $f(x) > 0$ for all $x > 0$. Then, $1+\varphi T\sqrt{f}+e^{2\varphi T\sqrt{f}}(\varphi T\sqrt{f}-1)>0$ and $\partial^2 J_m/\partial p_1^{ol2}<0$ for all $\varphi T\sqrt{f}>0$. Finally, plugging the right-hand side of Eq. 5.23 in Eqs. 5.53, 5.22, and 5.52, respectively, gives Eqs. 5.24, 5.25, and 5.26. □

A4 Assuming that the manufacturer is the leader and the retailer the follower, the retailer's necessary condition in Eq. 5.22 does not contain the retailer's costate variable, which therefore does not affect the manufacturer's Hamiltonian in Eq. 5.16, and confirms that the unique open-loop Stackelberg equilibrium is a Nash equilibrium. Conversely, assuming that the retailer is the leader and the manufacturer the follower, the manufacturer's necessary condition in Eq. 5.18 and the costate equation in Eq. 5.17 lead us to rewrite the retailer's Hamiltonian as:

$$H_r=(p_2-p_1)(\alpha-\beta p_2)+\mu\left(\alpha-\beta p_2+\varphi^2\lambda\right)+\psi f y$$

where $\mu\equiv\mu(t)$ and $\psi\equiv\psi(t)$ are the retailer's costate variable associated with the pollution stock and the manufacturer's costate variable, respectively, which evolve according to:

$$\dot{\mu}=-\psi f \qquad \mu(T)=0$$
$$\dot{\psi}=-\varphi^2\mu \qquad \psi(T)=0$$

Solving this differential system with the related transversality conditions, the retailer's costate variables associated with the pollution stock and the manufacturer's costate variable are both zero over the whole game, which confirms that the unique open-loop Stackelberg equilibrium is a Nash equilibrium also in this case. □

A5 Plugging the expression from Eq. 5.23 in Eqs. 5.53 and 5.22, and solving gives Eqs. 5.27 and 5.28. □

A6 Here also, the game is solved backward. To identify an FBNE, we first construct the Hamilton-Jacobi-Bellman (HJB) equations of the two players, that is:

$$-\dot{V}_m = \underset{v(.)}{\text{Max}}\left[p_1(\alpha-\beta p_2) - v^2/2 - fy^2/2 + V_y^m(\alpha-\beta p_2-\varphi v)\right] \tag{5.54}$$

$$-\dot{V}_r = \underset{p_2(.)}{\text{Max}}\left[(p_2-p_1)(\alpha-\beta p_2) + V_y^r(\alpha-\beta p_2-\varphi v)\right] \tag{5.55}$$

where V_m and V_r are the manufacturer's and retailer's value functions. We need to establish the existence of bounded and continuously differentiable value functions V_m and V_r, such that there is a unique solution $y \geq 0$ to the HJB equations. Necessary conditions for optimality are:

$$v = -\varphi\frac{\partial V_m}{\partial y} \tag{5.56}$$

$$p_2 = \frac{\alpha+\beta(p_1-\partial V_r/\partial y)}{2\beta} \tag{5.57}$$

Substituting from the necessary conditions in the HJB equations, and assuming that positive controls, yields:

$$\dot{V}_m + \left(p_1+\frac{\partial V_m}{\partial y}\right)\left[\alpha-\beta\left(p_1-\frac{\partial V_r}{\partial y}\right)\right]/2 + \varphi^2\left(\frac{\partial V_m}{\partial y}\right)^2/2 - fy^2/2 = 0 \tag{5.58}$$

$$\dot{V}_r + \left[\alpha-\beta\left(p_1-\frac{\partial V_r}{\partial y}\right)\right]^2/4\beta + \varphi^2\frac{\partial V_r}{\partial y}\frac{\partial V_m}{\partial y} = 0 \tag{5.59}$$

We make the following conjectured functional forms for the value functions:

$$V_m(y,t) = \frac{A(t)y^2}{2} + B(t)y + C(t) \tag{5.60}$$

$$V_r(y,t) = D(t)y + E(t) \tag{5.61}$$

Note that the conjecture for the value function is quadratic for the manufacturer and linear for the retailer. Whenever controls are positive, FBNE efforts for both players are given by:

$$v^{fb}(t) = -\varphi(A(t)y(t) + B(t)) \tag{5.62}$$

$$p_2^{fb}(t) = \frac{\alpha+\beta(p_1-D(t))}{2\beta} \tag{5.63}$$

Deriving the conjectures in Eqs. 5.60–5.61 respectively with respect to time and y, and plugging the resulting expressions in Eqs. 5.62–5.63 gives:

$$\frac{(\dot{A}+\varphi^2A^2-f)}{2}y^2+\left\{\dot{B}+\frac{A[\alpha-\beta(p_1-D)+2\varphi^2B]}{2}\right\}y+\dot{C}$$
$$+\frac{(p_1+B)[\alpha-\beta(p_1-D)]+B^2\varphi^2}{2}=0 \tag{5.64}$$

$$(\dot{D}+\varphi^2AD)y+\dot{E}+\frac{[\alpha-\beta(p_1-D)]^2}{4\beta}+\varphi^2BD=0 \tag{5.65}$$

To satisfy the equations in Eqs. 5.64–5.65, the following set of equations should be met:

$$\dot{A}=f-\varphi^2A^2, \qquad A(T)=0$$
$$\dot{B}=-\frac{A}{2}\left[\alpha-\beta(p_1-D)+2\varphi^2B\right], \qquad B(T)=0$$
$$\dot{C}=-\frac{(p_1+B)[\alpha-\beta(p_1-D)]+B^2\varphi^2}{2}, \qquad C(T)=0$$
$$\dot{D}=-\varphi^2AD, \qquad D(T)=0$$
$$\dot{E}=-\frac{[\alpha-\beta(p_1-D)]^2}{4\beta}-\varphi^2BD, \qquad E(T)=0$$

This system has the following solutions:

$$A(t)=-\frac{\sqrt{f}\,\mathrm{th}[\varphi\sqrt{f}(T-t)]}{\varphi} \tag{5.66}$$

$$B(t)=-(\alpha-\beta p_1)\left\{1-\mathrm{sech}\left[\varphi\sqrt{f}(T-t)\right]\right\}/2\varphi^2 \tag{5.67}$$

$$C(t)=-\frac{(\alpha-\beta p_1)}{8\varphi^2}\left\{\left[\alpha-(\beta+4\varphi^2)p_1\right](T-t)-\frac{\tanh[\varphi\sqrt{f}(T-t)](\alpha-\beta p_1)}{\varphi\sqrt{f}}\right\} \tag{5.68}$$

$$D(t)=0 \tag{5.69}$$

$$E(t)=-(\alpha-\beta p_1)^2(T-t)/4\beta \tag{5.70}$$

Plugging the right-hand side of Eqs. 5.66 and 5.67 in Eq. 5.62, and Eq. 5.69 in Eq. 5.63, respectively, gives:

$$v^{fb}(t)=\sqrt{f}\,\mathrm{th}\left[\varphi\sqrt{f}(T-t)\right]y(t)+\frac{(\alpha-\beta p_1)\{1-\mathrm{sech}[\varphi\sqrt{f}(T-t)]\}}{2\varphi} \tag{5.71}$$

and:

$$p_2^{fb}(t) = \frac{\alpha + \beta p_1}{2\beta} \tag{5.72}$$

It can be shown that $v^{fb}(t) \geq 0, \forall t \in [0, T]$, because $p_1 < \alpha/\beta$. However, plugging Eqs. 5.71 and 5.72 in Eq. 5.3 and solving the resulting expression gives:

$$y^{fb}(t) = \frac{\mathrm{sech}(\varphi T\sqrt{f})(\alpha - \beta p_1)\mathrm{sh}(\varphi t\sqrt{f})}{2\varphi\sqrt{f}} \tag{5.73}$$

which is nonnegative for all $t \in [0, T]$ because $p_1 < \alpha/\beta$. Plugging Eqs. 5.71, 5.72, and 5.73 in Eq. 5.1 and solving the resulting expression gives the manufacturer's payoff:

$$J_m^{fb} = \frac{(\alpha - \beta p_1)}{2}\left\{p_1 T - \frac{[\varphi T\sqrt{f} - \mathrm{th}(\varphi T\sqrt{f})](\alpha - \beta p_1)}{4\varphi^3\sqrt{f}}\right\} \tag{5.74}$$

Deriving the right-hand side of Eq. 5.74 with respect to p_1 and solving for p_1 gives Eq. 5.29. To confirm that p_1^{fb} is indeed optimal, we compute:

$$\frac{\partial^2 J_m}{\partial p_1^{fb2}} = -\frac{\beta[\varphi T\sqrt{f}(\beta + 4\varphi^2) - \beta\,\mathrm{th}(\varphi T\sqrt{f})]}{4\varphi^3\sqrt{f}}$$

Because $\varphi T\sqrt{f} > \mathrm{th}(\varphi T\sqrt{f})$ for all $T > 0$, $\partial^2 J_m/\partial p_1^{fb^2} < 0$ holds. Finally, plugging the right-hand side of Eq. 5.29 in Eqs. 5.72, 5.73, and 5.74 respectively gives Eqs. 5.30, 5.31, and 5.32. □

A7 Assuming that the manufacturer is the leader and the retailer the follower, plugging the retailer's necessary condition in Eq. 5.57 in the manufacturer's HJB equation in Eq. 5.54 does not affect the manufacturer's necessary condition in Eq. 5.56, which confirms that the unique feedback Stackelberg equilibrium is a Nash equilibrium. Conversely, assuming that the retailer is the leader and the manufacturer the follower, we plug the manufacturer's necessary condition in Eq. 5.56 in the retailer's HJB equation in Eq. 5.55 and see that the retailer's necessary condition in Eq. 5.57 is unaffected, which implies that the unique feedback Stackelberg equilibrium is also a Nash equilibrium.□

A8 Plugging the expression from Eq. 5.29 in Eq. 5.74 and the expressions from Eqs. 5.29 and 5.31 in Eq. 5.2, respectively, and solving the resulting expression gives Eqs. 5.33 and 5.34. □

A9 The OLNE strategy is subgame perfect if the following equalities hold, that is:

$$p_1^{ol}(t) = p_1^{fb}(t) \tag{5.75}$$

$$v^{ol}(t) = v^{fb}(t) \tag{5.76}$$

$$p_2^{ol}(t) = p_2^{fb}(t) \tag{5.77}$$

First, we show that $v^{ol}(t) = v^{fb}(t)$ holds for a given p_1. From Eq. 5.51, we have:

$$v^{ol}(t) = \left[1 - \frac{\text{ch}(\varphi t\sqrt{f})}{\text{ch}(\varphi T\sqrt{f})}\right]\frac{(\alpha - \beta p_1)}{2\varphi} \tag{5.78}$$

However, plugging the right-hand side of Eqs. 5.73 in Eq. 5.71 and rearranging gives:

$$v^{fb}(t) = \left\{1 - \text{sech}\left[\varphi\sqrt{f}(T-t)\right] + \text{th}\left[\varphi\sqrt{f}(T-t)\right]\text{sech}\left(\varphi T\sqrt{f}\right)\text{sh}\left(\varphi t\sqrt{f}\right)\right\}\frac{(\alpha - \beta p_1)}{2\varphi} \tag{5.79}$$

Using the first part of the righthand side of Eq. 5.79, we compute:

$$1 - \text{sech}\left[\varphi\sqrt{f}(T-t)\right] + \text{th}\left[\varphi\sqrt{f}(T-t)\right]\text{sech}\left(\varphi T\sqrt{f}\right)\text{sh}\left(\varphi t\sqrt{f}\right) = 1$$
$$- \frac{e^{\varphi t\sqrt{f}} + e^{-\varphi t\sqrt{f}}}{e^{\varphi T\sqrt{f}} + e^{-\varphi T\sqrt{f}}} = 1 - \frac{\text{ch}(\varphi t\sqrt{f})}{\text{ch}(\varphi T\sqrt{f})} \tag{5.80}$$

By Eq. 5.80, $v^{ol}(t) = v^{fb}(t)$ holds for any given p_1. We show that $p_1^{ol}(t) = p_1^{fb}(t)$ holds because:

$$p_1^{ol} = \frac{\left[\varphi T\sqrt{f}(\beta + 2\varphi^2)\left(e^{2\varphi T\sqrt{f}} + 1\right) - \beta\left(e^{2\varphi T\sqrt{f}} - 1\right)\right]\alpha}{\beta\left[\varphi T\sqrt{f}(\beta + 4\varphi^2)\left(e^{2\varphi T\sqrt{f}} + 1\right) - \beta\left(e^{2\varphi T\sqrt{f}} - 1\right)\right]}$$
$$= \frac{[\varphi T\sqrt{f}(\beta + 2\varphi^2) - \beta\text{th}(\varphi T\sqrt{f})]\alpha}{\beta[\varphi T\sqrt{f}(\beta + 4\varphi^2) - \beta\text{th}(\varphi T\sqrt{f})]} = p_1^{fb}(t) \tag{5.81}$$

where $p_1^{fb}(t)$ is given by Eq. 5.29. Therefore, $v^{ol}(t) = v^{fb}(t)$ holds. From Eqs. 5.22 and 5.72, Eq. 5.81 implies that $p_2^{ol}(t) = p_2^{fb}(t)$. Finally, $y^{ol}(t) = y^{fb}(t)$ because the state equations for OLNE and FBNE are equivalent as well as p_1, p_2, v. Hence, the OLNE strategy is subgame perfect. □

A10 Because $\dot{v}^c(t) < 0$ from Eq. 5.12 and $\dot{v}^{ol}(t) < 0$ from Eq. 5.24), and $v^c(T) = v^{ol}(T) = 0$, we compute $v^c(0) - v^{ol}(0)$ to determine the paths of the two control variables, which gives:

$$v(0)-v^{ol}(0)=\varphi\left\{\frac{(e^{\theta T}-1)^2}{(\beta+2\varphi^2)(e^{2\theta T}+1)}-\frac{\varphi T\sqrt{f}\left(e^{\varphi T\sqrt{f}}-1\right)^2}{\left[\varphi T\sqrt{f}(\beta+4\varphi^2)\left(e^{2\varphi T\sqrt{f}}+1\right)-\beta\left(e^{2\varphi T\sqrt{f}}-1\right)\right]}\right\}\alpha$$

Because $\varphi T\sqrt{f}(\beta+4\varphi^2)\left(e^{2\varphi T\sqrt{f}}+1\right)-\beta\left(e^{2\varphi T\sqrt{f}}-1\right)>0$, $v^c(0)>v^{ol}(0)$ holds whenever:

$$\frac{(\beta+4\varphi^2)}{(\beta+2\varphi^2)}\left(\frac{e^{2\varphi T\sqrt{f}}+1}{e^{2\theta T}+1}\right)-\frac{\beta\sqrt{f}}{2\varphi\theta^2 T}\left(\frac{e^{2\varphi T\sqrt{f}}-1}{e^{2\theta T}+1}\right)-\left(\frac{e^{\varphi T\sqrt{f}}-1}{e^{\theta T}-1}\right)^2>0$$

See Table 5.3.

Let $G_v=\left(\frac{\beta+4\varphi^2}{\beta+2\varphi^2}\right)\left(\frac{e^{2\varphi T\sqrt{f}}+1}{e^{2\theta T}+1}\right)-\frac{\beta\sqrt{f}}{2\varphi\theta^2 T}\left(\frac{e^{2\varphi T\sqrt{f}}-1}{e^{2\theta T}+1}\right)-\left(\frac{e^{\varphi T\sqrt{f}}-1}{e^{\theta T}-1}\right)^2$. Then, whether $G_v>0$ holds depends on the parameter values β and φ. We highlight below the role played by the parameters β and φ in the value of G_v. Although the range of β and φ such that $G_v>0$ cannot be obtained explicitly, we can show that $G_v>0$ holds for smaller values of β and/or larger values of φ, other things being equal, as shown in Table 5.3. From the limit value analysis, we conclude that there exists a threshold $\tilde{\beta}>0$ ($\tilde{\varphi}>0$) such that for any $\beta<\tilde{\beta}$ ($\varphi>\tilde{\varphi}$), Eq. 5.35 holds. □

A11 Comparing Eqs. 5.11 and 5.31 shows that Eq. 5.36 holds if:

$$-\frac{\beta\mathrm{ch}(\theta t)}{2\mathrm{ch}(\theta T)}\left[\varphi T\sqrt{f}(\beta+4\varphi^2)-\beta\mathrm{th}\left(\varphi T\sqrt{f}\right)\right]<\varphi^2\left[2\varphi^3 T\sqrt{f}-\beta\mathrm{th}\left(\varphi T\sqrt{f}\right)\right]$$

A sufficient but not necessary condition that satisfies the above inequality is $\varphi\geq\sqrt{\beta}/2$, that is, if the marginal effectiveness of the manufacturer's abatement effort is sufficiently large. □

A12 Let $\beta=2k\varphi^2$, where k is a constant. Then, $\theta=\sqrt{f(\beta+2\varphi^2)/2}=\varphi\sqrt{f(k+1)}$. From Eqs. 5.13 and 5.26, we get:

Table 5.3 Limit value analysis

	$\lim\left(\frac{\beta+4\varphi^2}{\beta+2\varphi^2}\cdot\frac{e^{2\varphi T\sqrt{f}}+1}{e^{2\theta T}+1}\right)$	$\lim\left(\frac{\beta\sqrt{f}}{2\varphi\theta^2 T}\cdot\frac{e^{2\varphi T\sqrt{f}}-1}{e^{2\theta T}+1}\right)$	$\lim\left(\frac{e^{\varphi T\sqrt{f}}-1}{e^{\theta T}-1}\right)^2$	$\lim G_v$
$\varphi\mapsto+\infty$	2	1	1	1
$\varphi\mapsto 0$	$\frac{2}{e^{T\sqrt{2\beta f}}+1}$	$\frac{2}{e^{T\sqrt{2\beta f}}+1}$	0	0
$\beta\mapsto+\infty$	0	0	0	0
$\beta\mapsto 0$	2	0	1	1

$$\frac{y(T)}{y^{ol}(T)}=\frac{\text{th}(\varphi T\sqrt{f(k+1)})}{\text{th}(\varphi T\sqrt{f})}\left\{\frac{k[\varphi T\sqrt{f}-\text{th}(\varphi T\sqrt{f})]}{\varphi T\sqrt{f(k+1)}}+\frac{2}{\sqrt{k+1}}\right\} \tag{5.82}$$

Two cases can be distinguished, that is, either $k<3$ or $k\geq 3$. For $k<3$, it holds that $2/\sqrt{k+1}>1$. Also, $\frac{\text{th}\left(\varphi T\sqrt{f(k+1)}\right)}{\text{th}\left(\varphi T\sqrt{f}\right)}\geq 1$ and $\frac{k\left[\varphi T\sqrt{f}-\text{th}\left(\varphi T\sqrt{f}\right)\right]}{\varphi T\sqrt{f(k+1)}}>0$ hold for any $0<k<3$. Therefore, from Eq. 5.82, $y(T)/y^{ol}(T)>1$ holds for $k<3$. Conversely, if $k\geq 3$, because $\frac{\text{th}\left(\varphi T\sqrt{f(k+1)}\right)}{\text{th}\left(\varphi T\sqrt{f}\right)}>1$, a sufficient (though not necessary) condition for $y(T)/y^{ol}(T)>1$ to hold is that $\frac{\sqrt{k+1}-2}{k}<\frac{\varphi T\sqrt{f}-\text{th}\left(\varphi T\sqrt{f}\right)}{\varphi T\sqrt{f}}$. Let $k^*=\arg\max_{k\in[3,\infty]}\left(\frac{\sqrt{k+1}-2}{k}\right)$, then $\frac{\sqrt{k+1}-2}{k}\Big|_{k^*}=0.134$ and $\lim\limits_{k\mapsto\infty}\left(\frac{\sqrt{k+1}-2}{k}\right)=0$. Define $\gamma=\varphi T\sqrt{f}>1$ and let$\gamma^*=\arg\min_{\gamma\in[1,\infty]}\left(\frac{\varphi T\sqrt{f}-\text{th}\left(\varphi T\sqrt{f}\right)}{\varphi T\sqrt{f}}\right)$. Then $\frac{\varphi T\sqrt{f}-\text{th}\left(\varphi T\sqrt{f}\right)}{\varphi T\sqrt{f}}\Big|_{\gamma^*}=0.238$ and $\lim\limits_{\gamma\mapsto\infty}\left(\frac{\varphi T\sqrt{f}-\text{th}\left(\varphi T\sqrt{f}\right)}{\varphi T\sqrt{f}}\right)=1$. Therefore, the sufficient condition $\frac{\sqrt{k+1}-2}{k}<\frac{\varphi T\sqrt{f}-\text{th}\left(\varphi T\sqrt{f}\right)}{\varphi T\sqrt{f}}$, and thereby $\frac{y(T)}{y^{ol}(T)}>1$, holds for $k\geq 3$ and $\gamma>1$. When $k\geq 3$ and $\gamma\leq 1$, Eq. 5.82 is larger than 1 if $\frac{k}{\sqrt{k+1}}\cdot\frac{[\gamma-\text{th}(\gamma)]}{\gamma}+\frac{2}{\sqrt{k+1}}>\frac{\text{th}(\gamma)}{\text{th}\left(\gamma\sqrt{k+1}\right)}$. If $\gamma=1$, the latter inequality holds for any $k\geq 3$. When $\gamma\mapsto 0$, $\lim\limits_{\gamma\mapsto 0}\left\{\frac{k}{\sqrt{k+1}}\cdot\frac{[\gamma-\text{th}(\gamma)]}{\gamma}+\frac{2}{\sqrt{k+1}}\right\}=\frac{2}{\sqrt{k+1}}$, and $\lim\limits_{\gamma\mapsto 0}\left[\frac{\text{th}(\gamma)}{\text{th}\left(\gamma\sqrt{k+1}\right)}\right]=\frac{1}{\sqrt{k+1}}$ imply that $\frac{k}{\sqrt{k+1}}\cdot\frac{[\gamma-\text{th}(\gamma)]}{\gamma}+\frac{2}{\sqrt{k+1}}>\frac{\text{th}(\gamma)}{\text{th}\left(\gamma\sqrt{k+1}\right)}$ holds. Therefore, $y(T)/y^{ol}(T)>1$ holds when $k\geq 3$ and $\gamma(<1)$ approaches zero. Whether Eq. 5.37 holds analytically is less clear when $k\geq 3$ and $\gamma<1$, than for the case where $k<3$ and for the case where $k\geq 3$ and $\gamma>1$ due to the complexity of Eq. 5.82. However, it can be shown that Eq. 5.37 holds numerically for all (k,γ) because the minimum value of $\frac{\text{th}\left(\gamma\sqrt{k+1}\right)}{\text{th}(\gamma)}\times\left\{\frac{k}{\sqrt{k+1}}\cdot\frac{[\gamma-\text{th}(\gamma)]}{\gamma}+\frac{2}{\sqrt{k+1}}\right\}$ is strictly larger than 1 when numerically computed. Finally, when $k\mapsto\infty$, $\frac{k}{\sqrt{k+1}}\cdot\frac{[\gamma-\text{th}(\gamma)]}{\gamma}+\frac{2}{\sqrt{k+1}}>\frac{\text{th}(\gamma)}{\text{th}\left(\gamma\sqrt{k+1}\right)}$ is satisfied, which implies that $y(T)/y^{ol}(T)>1$ holds. □

References

Atasu, A., Van Wassenhove, L., & Sarvary, M. (2009). Efficient take-back legislation. *Production and Operations Management, 18*(3), 243–258.

Bagnoli, M., & Watts, S. G. (2003). Selling to socially responsible consumers: Competition and the private provision of public goods. *Journal of Economics & Management Strategy, 12*(3), 419–445.

Benchekroun, H., & Long, N. V. (2011). Game theory: Static and dynamic games. In A. A. Batabyal & P. Nijkamp (Eds.), *Research tools in natural resource and environmental economics*. World Scientific Publishing.

Bertinelli, L., Camacho, C., & Zou, B. (2014). Carbon capture and storage and transboundary pollution: A differential game approach. *European Journal of Operational Research, 237*(2), 721–728.

Boucekkine, R., Krawczyk, J. B., & Vallée, T. (2011). Environmental quality versus economic performance: A dynamic game approach. *Optimal Control Applications and Methods, 32*(1), 29–46.

Bresnahan, T. F., & Reiss, P. C. (1985). Dealer and manufacturer margins. *Rand Journal of Economics, 16*(2), 253–268.

Canadell, J. G., Le Quéré, C., Raupach, M. R., Field, C. B., Buitenhuis, E. T., Ciais, P., Conway, T. J., Gillett, N. P., Houghton, R. A., & Marland, G. (2007). Contributions to accelerating atmospheric CO_2 growth from economic activity, carbon intensity, and efficiency of natural sinks. *Proceedings of the National Academy of Sciences, 104*(47), 18866–18870.

Carr, S. M., & Karmarkar, U. S. (2005). Competition in multiechelon assembly supply chains. *Management Science, 51*(1), 45–59.

Chen, Y. J., & Sheu, J.-B. (2009). Environmental-regulation pricing strategies for green supply chain management. *Transportation Research Part E, 45*(5), 667–677.

Chung, S. H., Weaver, R. D., & Friesz, T. L. (2013). Strategic response to pollution taxes in supply chain networks: Dynamic, spatial, and organizational dimensions. *European Journal of Operational Research, 231*(2), 314–327.

Cox, P. M., Betts, R. A., Jones, C., Spall, S. A., & Totterdell, I. (2000). Acceleration of global warming due to carbon-cycle feedbacks in a coupled climate model. *Nature, 408*, 184–187.

Cruz, J. M. (2008). Dynamics of supply chain networks with corporate social responsibility through integrated environmental decision-making. *European Journal of Operational Research, 184*(3), 1005–1031.

Cui, T. H., Raju, J. S., & Zhang, Z. J. (2007). Fairness and channel coordination. *Management Science, 53*(8), 1303–1314.

Dellarocas, C. (2012). Double marginalization in performance-based advertising: Implications and solutions. *Management Science, 58*(6), 1178–1195.

Dockner, E., Jørgensen, S., Van Long, N., & Sorger, G. (2000). *Differential games in economics and management science*. Cambridge University Press.

El Ouardighi, F. (2014). Supply quality management with optimal wholesale price and revenue sharing contracts: A two-stage game approach. *International Journal of Production Economics, 156*(1), 260–268.

El Ouardighi, F., Benchekroun, H., & Grass, D. (2014). Controlling pollution and environmental absorption capacity. *Annals of Operations Research, 220*(1), 111–133.

Fowlie, M. (2010). Emissions trading, electricity restructuring, and investment in pollution abatement. *American Economic Review, 100*(3), 837–869.

Giovanni, P. D. (2014). Environmental collaboration in a closed-loop supply chain with a reverse revenue sharing contract. *Annals of Operations Research, 220*(1), 135–157.

Grass, D., Caulkins, J. P., Feichtinger, G., Tragler, G., & Behrens, D. A. (2008). *Optimal control of nonlinear processes with applications in drugs, corruption, and terror*. Springer.

Hamilton, J. T. (1995). Pollution as news: Media and stock market reactions to the toxics release inventory data. *Journal of Environmental Economics and Management, 28*(1), 98–113.

Handfield, R. B., Walton, S. V., Seegers, L. K., & Melnyk, S. A. (1997). 'Green' value chain practices in the furniture industry. *Journal of Operations Management, 15*(4), 293–315.

Hollos, D., Blome, C., & Foerstl, K. (2012). Does sustainable supplier co-operation affect performance? Examining implications for the triple bottom line. *International Journal of Production Research, 50*(11), 2968–2986.

Hu, X., Caldentey, R., & Vulcano, G. (2013). Revenue sharing in airline alliances. *Management Science, 59*(5), 1177–1195.

Jørgensen, S., Martín-Herrán, G., & Zaccour, G. (2010). Dynamic games in the economics and management of pollution. *Environmental Modelling and Assessment, 15*(6), 433–467.

Jung, C., Krutilla, K., & Boyd, R. (1996). Incentives for advanced pollution abatement technology at the industry level: An evaluation of policy alternatives. *Journal of Environmental Economics and Management, 30*(1), 95–111.

Klassen, R. D., & McLaughlin, C. P. (1996). The impact of environmental management on firm performance. *Management Science, 42*(8), 1199–1214.

Klassen, R. D., & Vachon, S. (2003). Collaboration and evaluation in the supply chain: The impact on plant-level environmental investment. *Production and Operations Management, 12*(3), 336–352.

Klassen, R. D., & Whybark, D. C. (1999). The impact of environmental technologies on manufacturing performance. *Academy of Management Journal, 42*(6), 599–615.

Krass, D., Nedorezov, T., & Ovchinnikov, A. (2013). Environmental taxes and the choice of green technology. *Production and Operations Management, 22*(5), 1035–1055.

Krishnamoorthy, A., Prasad, A., & Sethi, S. P. (2010). Optimal pricing and advertising in a durable-good duopoly. *European Journal of Operational Research, 200*(2), 486–497.

Le Quéré, C., Rödenbeck, C., Buitenhuis, E. T., Conway, T. J., Langenfelds, R., Gomez, A., Labuschagne, C., Ramonet, M., Nakazawa, T., Metzl, N., Gillett, N., & Heimann, M. (2007). Saturation of the Southern ocean CO_2 sink due to recent climate change. *Science, 316*, 1735–1738.

Lee, H. L. (2010). Don't tweak your supply chain – rethink it end to end. *Harvard Business Review, 88*(10), 62–69.

Li, S. (2013). Emission permit banking, pollution abatement and production-inventory control of the firm. *International Journal of Production Economics, 146*(2), 679–685.

Li, S. (2014). A differential game of transboundary industrial pollution with emission permits trading. *Journal of Optimization Theory and Applications, 163*(2), 642–659.

Liu, Z., Anderson, T. D., & Cruz, J. M. (2012). Consumer environmental awareness and competition in two-stage supply chains. *European Journal of Operational Research, 218*(3), 602–613.

Long, N. V. (2010). *A survey of dynamic games in economics*. World Scientific.

Majumder, P., & Groenevelt, H. (2001). Competition in remanufacturing. *Production and Operations Management, 10*(2), 125–141.

Michel, P., & Rotillon, G. (1995). Disutility of pollution and endogenous growth. *Environmental and Resource Economics, 6*(3), 279–300.

Miliman, S. R., & Prince, R. (1989). Firm incentives to promote technological change in pollution control. *Journal of Environmental Economics and Management, 17*(3), 247–265.

Mishra, B. K., Raghunathan, S., & Yue, X. (2009). Demand forecast sharing in supply chains. *Production and Operations Management, 18*(2), 152–166.

Ni, D., Li, K. W., & Tang, X. (2010). Social responsibility allocation in two-echelon supply chains: Insights from wholesale price contracts. *European Journal of Operational Research, 207*(3), 1269–1279.

Piao, S. L., Ciais, P., Friedlingstein, P., Peylin, P., Reichstein, M., Luyssaert, S., Margolis, H., Fang, J. Y., Barr, A., Chen, A. P., Grelle, A., Hollinger, D. Y., Laurila, T., Lindroth, A., Richardson, A. D., & Vesala, T. (2008). Net carbon dioxide losses of northern ecosystems in response to autumn warming. *Nature, 451*, 49–52.

Pierce, R. J. (1998). Small is not beautiful: The case against special regulatory treatment of small firms. *Administrative Law Review, 50*(3), 537–578.

Piot-Lepetit, I. (2014). Technological externalities and environmental policy: How to simulate manure management regulation within a DEA framework. *Annals of Operations Research, 214*(1), 31–48.

Simpson, D. (2010). Use of supply relationships to recycle secondary materials. *International Journal of Production Research, 48*(1), 227–249.

Spengler, J. (1950). Vertical integration and antitrust policy. *Journal of Political Economy, 4*(58), 347–352.

Subramanian, R., Gupta, S., & Talbot, B. (2007). Compliance strategies under permits for emissions. *Production and Operations Management, 16*(6), 763–779.
Tang, C. S., & Zhou, S. (2012). Research advances in environmentally and socially sustainable operations. *European Journal of Operational Research, 223*(3), 585–594.
United States Department of State. (2010, June). *U.S. Climate Action Report 2010*. Global Publishing Services.
Vachon, S., & Klassen, R. D. (2008). Environmental management and manufacturing performance: The role of collaboration in the supply chain. *International Journal of Production Economics, 111*(2), 299–315.
Wald, M. L. (2013, June 18). New efforts to quantify 'social cost' of pollution. *New York Times*.
Wong, E. (2014, December 31). Fines total $26 million for polluters in China. *The New York Times*.
Xu, X., & Hopp, W. J. (2006). A monopolistic and oligopolistic stochastic flow revenue management model. *Operations Research, 54*(6), 1098–1109.
Zhang, L., Wang, J., & You, J. (2015). Consumer environmental awareness and channel coordination with two substitutable products. *European Journal of Operational Research, 241*(1), 63–73.
Zhao, M., & Running, S. W. (2010). Drought-induced reduction in global terrestrial net primary production from 2000 through 2009. *Science, 329*, 940–943.

Chapter 6
Pollution Accumulation and Abatement Policies in Two Supply Chains Under Vertical and Horizontal Competition and Strategy Types

Fouad El Ouardighi, Jeong Eun Sim, and Bowon Kim

Abstract This research investigates the impact of both horizontal and vertical competition, on the one hand, and strategy types (commitment-based *versus* contingent-based equilibrium strategies), on the other hand, on the pollution accumulated by two supply chains over time. We consider a two-stage game model where two manufacturers and two retailers are involved in a wholesale price contract, to supply the demand over a finite time horizon. In the first stage of the game, the manufacturers set their respective optimal transfer prices. During the second stage, polluting emissions are created over time in proportion to demand, which is controlled by the retailers' respective consumer prices. In this stage, the manufacturers are involved in emissions abatement. In this setup, we seek to identify the combination of market structure and strategy type that leads the two supply chains to generate the lowest pollution intensity and the highest level of abatement intensity.

Keywords Pollution accumulation · Price competition · Double marginalization · Supply chains

F. El Ouardighi (✉)
ESSEC Business School, Cergy Pontoise Cedex, France
e-mail: elouardighi@essec.fr

J. E. Sim
Sauder School of Business, University of British Columbia, Vancouver, BC, Canada

B. Kim
KAIST Business School, Seoul, Korea

F. El Ouardighi (ed.), *Essays on Pollution Control in Economics and Management Science*, Lecture Notes in Operations Research,
https://doi.org/10.1007/978-3-031-78227-5_6

Introduction

Although much has been written about the importance of environmental performance (e.g., Chava, 2014; Flammer, 2015) and the impact of environmental regulation on firms' competitiveness (Porter, 1991; Porter & van der Linde, 1995; Xepapadeas & de Zeeuw, 1999), the influence of competition on pollution accumulation has been neglected. Competition has two dimensions: horizontal and vertical. Horizontal competition is usually based on product substitutability that leads producers to compete either on price or on quantity (e.g., Singh & Vives, 1984). In contrast, a usual feature of vertical competition between manufacturers and retailers is the well-known double marginalization (Spengler, 1950), which results in higher consumer prices, lower demand, and smaller profits than if decisions were centralized.

The environmental economics literature has dealt with the consequences of a lack of horizontal internalization of pollution externalities in a decentralized setting where noncompeting producers draw utility from polluting emissions as a by-product of production and incur a cost related to the pollution stock [for a survey, see, e.g., Jørgensen et al. (2010), Benchekroun and Long (2011)]. One serious outcome of this setting is that strategy types, i.e., whether producers' strategy is precommitted to emissions reduction or contingent upon the pollution stock do matter. Along with strategy types, both horizontal and vertical competition can affect pollution accumulation.

Consider the apparel industry where polluting emissions have been a serious concern. While many fast fashion companies such as H&M and Gap use outsourcing extensively without keeping in-house capacity, some retailers such as Zara and Uniqlo maintain a vertically integrated chain where the design and the production also take place in-house. In the oil industry, major companies such as Shell, Chevron, and BP manage from upstream exploration to marketing and sales of refined products, whereas nonintegrated super-independents such as Anadarko, Apache, or Talisman Energy consist in more disintegrated, i.e., competing, supply chains (Boscheck, 2006). Mergers and acquisitions also take place actively in this industry, e.g., Chevron's recent acquisition of Anadarko in a $33 billion deal (Casey et al., 2019), implying the tendency toward horizontal integration.

Distinct patterns of competition between firms, vertically and horizontally, influence abatement and pricing decisions and the pollution of firms in industries, and several studies have explored this aspect based on the static approach as a first step. However, an important element that has been largely ignored is the intertemporal dimension of the pollution externalities. For instance, pollutants are generally long-lasting and cumulative in nature (Arrow et al., 1995) and firms typically pay attention to pollution from the long-term perspective e.g., BP planning to keep its emissions around 50 million tons of CO_2 by 2025, and Shell is setting emission targets for 2030. This study directly incorporates the intertemporal feature of the pollution externalities based on dynamic games in investigating competition and pollution accumulation, thus filling the important missing element in the literature and also responding to the call for studies on dynamic settings with regard to environmental problems (e.g., Xu et al., 2017; Galbreth & Ghosh, 2013).

Another important observation that motivates our research relates to strategy types. While contingent strategies are common in practice, a growing number of global producers (e.g., Ikea, Coca-Cola, Dell, General Mills, Sony,…) recently decided to opt for commitment strategies (https://sciencebasedtargets.org/companies-taking-action). Whether to adopt a commitment or contingent strategy is of particular concern for many firms these days along with a global movement to tackle climate change. For instance, Shell, which previously hesitated to employ any pre-committed strategies, recently made an important decision to precommit to their strategies for 3–5 years to reduce emissions, by 20% by 2035 (Raval et al., 2018; Van Beurden, 2018). BP also pledged to invest $100 million over the next 3 years for emission reduction in exploration and production (Herron, 2019). Top apparel companies also recently adopted commitment strategies to emission reductions, including major brands such as Gap, Nike, Levi's, Guess, H&M, Zara, and Puma (WRI, 2017). Similar examples abound across a wide variety of industries, e.g., Siemens precommitted to investing $110 million over 3 years specifying the details on how the investment in emission reduction will be realized (Kaeser, 2015), and Procter and Gamble committed to their set of actions to cut its emissions in half (P&G, 2018).

Which strategy types (commitment versus contingent) firms adopt is not only of practical importance, but also a critical factor in studying competition and pollution. Since firms under competition interact with each other through their strategic choices, the way firms make their decisions can completely alter the competitive landscape between firms. In this regard, whether firms precommit to their course of actions at the beginning, or revise them continuously throughout the competitive interaction has been regarded as an important theoretical distinction in terms of the behaviors of competing firms (Kamien & Schwartz, 2012; Dockner et al., 2000). These two strategies make contrasting assumptions on firms' behaviors, and strategic interactions of competing firms are usually more pronounced under contingent strategy compared to under commitment strategy. Therefore, we examine how the strategy types are affected by competition and integration in an industry to determine the pollution accumulation and abatement policies of firms.

The following research questions are addressed:

- How are pollution accumulation and abatement related to the presence of competition and integration in an industry? That is, how do horizontal and vertical internalization of pollution externalities affect abatement policies and pollution accumulation in a bilateral duopoly?
- How do strategy types modify the relative impact of horizontal and vertical competition on pollution accumulation, if any? More specifically, how does the lack of precommitment by firms influence pollution intensity, abatement intensity, and firms' profits?

We briefly summarize the key results as follows:

- First, horizontal competition leads to more pollution than horizontal integration. The rationale behind this is twofold: a "horizontally-induced volume effect" due

to more competitive sales prices, leading to greater sales hence more pollution, on the one hand, and a "horizontally-induced polluting effect" resulting from lower incentives for abatement efforts due to competition, making the manufacturing process dirtier.
- Second, vertical competition results in lower pollution than vertical integration. This is because double marginalization under vertical competition reduces the gross emissions via sales, on the one hand, and induces relatively greater manufacturers' abatement effort, on the other hand. These two effects are defined as "vertically-induced volume effect" and "vertically-induced polluting effect."
- Third, the vertically-induced volume effect is greater than the horizontally-induced volume effect and the horizontally-induced polluting effect is greater than the vertically-induced polluting effect. That is, the volume effect depends more on the vertical configuration, whereas the polluting effect depends more on the horizontal configuration.
- Finally, the impact of a strategy type depends on market configurations. Under horizontal competition, a commitment strategy leads to mitigating the polluting effect compared to a contingent strategy. The impact of a strategy type on the volume effect depends on the market structure. Under horizontal competition, strategy types influence the abatement and pricing paths of competing firms and a commitment strategy leads to higher profitability for supply chain entities. In the context of horizontal integration, both strategy types coincide.

The paper proceeds as follows. Second section reviews the relevant literature. In third section, we develop a dynamic game model with two manufacturers and two retailers. Fourth section successively derives the optimal policies. Fifth section compares the strategies and their outcomes both analytically and numerically. Sixth section concludes the paper.

Literature Review

This study is related to two streams of literature on pollution accumulation: competition and strategy types. We first review the studies on the impacts of horizontal and vertical competition on pollution, and then review the literature on the strategy type, i.e., pre-commitment and contingent strategies, and its impact on pollution accumulation.

Horizontal and Vertical Competition and Pollution

Horizontal Competition and Pollution

Many studies examine the impact of regulation or consumer awareness under a horizontally competitive market in the environmental context (e.g., Ouchida & Goto,

2016; Subramanian et al., 2007). Sabzevar et al. (2017) investigate the effects of cap-and-trade policy, and analyze whether tighter policies increase the competitive firms' profits. Galbreth and Ghosh (2013) examine the impact of environmental awareness on two horizontally competing firms with different sustainability levels. They find that higher awareness may benefit not only the more sustainable firm, but also the less sustainable one. The impacts of different power structures are also often examined under horizontal competition. Meng et al. (2018) show that firms' optimal product strategies between common and low-carbon products depend on the power structure, i.e., Nash versus Stackelberg structures. Chen et al. (2017) also examine how the market power between two competing manufacturers influences the pricing and emissions reduction strategies.

How horizontal competition influences the strategic interaction between firms and the environment is gaining increasing attention, wherein the advantage of horizontal competition has been largely supported. Yalabik and Fairchild (2011) show that if the degree of environmental competition between firms is high, it encourages more environmental innovation in competing firms. Yenipazarli (2019) finds that a firm under duopoly competition exerts more abatement efforts than a monopolist firm to differentiate itself from its competitor when the abatement requires a relatively large up-front investment. Ouchida and Goto (2016) further show that even when competing firms can coordinate emission-reducing R&D to maximize joint profits, such cooperation can be socially inferior in cases of severe environmental damage and low investment costs. Luo et al. (2016) investigate the competing manufacturers' pricing and abatement strategies under the cap-and-trade policy and argue that while horizontal competition increases emissions, it leads to higher green technology investments. Contrary to the aforementioned papers based on static games, we find that horizontal competition lessens the abatement incentive of competing firms because the intertemporal pollution externalities are not internalized by the competitive manufacturers. We also find that strategy types influence the degree to which horizontal competition reduces the abatement incentive of firms.

Vertical Competition and Pollution

Due to the well-known double marginalization phenomenon, a rich body of literature on vertical competition explores how coordinating contracts in the decentralized chain can achieve system efficiency in the centralized chain (see Tsay et al. (1999) for a review). As a direct extension, many environmental studies on vertical competition explore how diverse contracts may influence environmental performance (e.g., Zhou et al., 2016; Du et al., 2015). Ghosh and Shah (2015) show that while the integrated chain leads to the highest product greening levels, the greening level in the competitive chain becomes also higher with the cost-sharing than without. Yang and Chen (2018) show that under consumer awareness and carbon tax, sharing both revenue and abatement investment costs results in the same equilibrium outcome as sharing revenue only, thus cost-sharing contracts become dispensable. Hong and Guo (2018) examine price-only, green-marketing cost-sharing, and two-part tariff contracts in a supply chain and show that increased cooperation may not be beneficial to all entities in a supply chain, but it can lead to the highest product greenness. Some environmental studies consider bargaining power between manufacturers and retailers in the vertically competitive chain (e.g., Ni et al., 2010; Ghosh & Shah, 2012). Wang et al. (2016) examine the emissions reduction under the retailer dominant and the power-balanced cases, with the vertically integrated chain as a benchmark.

The underlying theme of the aforementioned studies is the inefficiency of vertical competition in the environmental context. However, recent studies suggest that this belief may not always hold and vertical competition may not be detrimental, leading to the same or even better environmental sustainability. Zhu and He (2017) investigate the effects of supply chain structures, product types, and competition types (price or greenness). They suggest that although the lack of vertical coordination results in less sales and less profit, it may not affect the optimal degree of product greenness. Xu et al. (2017) similarly show that the abatement level remains the same regardless of whether a manufacturer and a retailer coordinate their decisions, despite a decrease in production quantities and profits in a decentralized chain. Jaber et al. (2013) find that vertical coordination can reduce the total supply chain cost, but it can increase emissions and penalty costs. Chen et al. (2019) show that the vertically coordinated chain may result in higher emissions than the competitive chain when firms make order replenishment decisions. Toptal and Cetinkaya (2017) also find that, depending on production and inventory parameters, carbon emissions can be larger under a vertically coordinated system.

Our study extends this line of research that suggests the adverse environmental impact of vertical integration. However, there are important differences between our paper and the afore-cited studies in that Jaber et al. (2013), Chen et al. (2019), Toptal and Cetinkaya (2017), and Zhu and He (2017) do not consider pollution abatement and often assume a fixed demand, whereas we explicitly consider firms' investment in pollution abatement and assume an endogenously determined demand. Our study also differs from Xu et al. (2017) in that while they focus on coordinating contracts, we focus on strategy types that govern firms' behaviors over time. Furthermore, none of those studies consider intertemporal pollution externalities or the strategy types of firms.

Overall Contribution to the Literature on Competition and Pollution

Based on the literature review above, we summarize how this study contributes to the studies on the competition (horizontal and vertical) and the environment. First, despite the growing interest in the impacts of competition and integration on pollution, theoretical lenses that provide a systematic understanding of the results are largely lacking. To the best of our knowledge, our study is one of the first to identify and compare drivers of the impacts of competition on pollution accumulation, i.e., "volume effects" and "polluting effects," both vertically- and horizontally-induced.

Second, we find that strategy type may influence the effects of market structures, thus filling the gap in the literature where these interdependences are largely ignored. For instance, strategy type affects the degree of the polluting effect and the volume effect underlying the impact of competition on pollution accumulation and also the abatement and pricing trajectories of firms under distinct market structures.

Third, our study extends the literature in that we jointly investigate two dimensions of competition, i.e., vertical competition or integration, and horizontal competition or integration. This setting allows us to examine how the respective impacts interact in the markets where both types of competition exist, which is common in practice. A few recent studies consider both dimensions of competition along with environmental issues. Yang et al. (2017) suggest that vertical cooperation leads to a higher abatement rate and upstream horizontal coordination may reduce downstream profits, from which the revenue-sharing contract can bring economic and environmental benefits. Creti and Sanin (2017) find that horizontal integration increases profits in the vertically-related market due to the decrease in emission permit price. Madani and Rasti-Barzoki (2017) show that horizontal cooperation under vertically competitive chains, where one chain sells green products and the other sells non-green products, leads to both eco-friendlier products and higher profits. Our study differs from these studies in two critical aspects: (1) Although these studies consider two dimensions of competition, the interactive impacts between vertical and horizontal structures remain mostly unexplored. In contrast, we show that vertical structure can significantly alter the effects of horizontal structure, and vice versa. (2) While these papers adopt the static games model, we employ dynamic games to incorporate the intertemporal feature of pollution externalities and investigate strategy types of competing firms, which determine the competitive landscape of firms.

Strategy Types

While the majority of environmental studies are based on static models, our work is based on a dynamic game model and shows how firms' emissions abatement, production, and pricing behaviors would change over time. This approach is particularly useful in investigating the dynamics of pollution accumulation, instead of one-shot emission (Jørgensen et al., 2010). In this regard, we extend the bilateral monopoly game model in El Ouardighi et al. (2016) to a bilateral duopoly structure to assess the relative effects of price competition and double marginalization. This study differs from El Ouardighi et al. (2016) in that we characterize the main drivers of pollution accumulation, i.e., volume effects and polluting effects, and investigate their relative importance under horizontal and vertical market configurations, which shows that while volume effects are influenced more by vertical configurations, polluting effects are influenced more by horizontal configurations. We also investigate the interaction between competition and strategy types and find that the effects of strategy types can be significant depending on market configurations.

In the dynamic game models for firms' emissions, the impact of strategy types has been investigated, especially for a commitment strategy (i.e., an open-loop Nash equilibrium; OLNE) and for a contingent strategy (i.e., a feedback Nash equilibrium; FBNE). Contingent strategy has been generally considered to generate lower public goods since it often leads to free-riding behaviors between players, which results in higher pollution accumulation than commitment strategy (see Dockner et al., 2000; El Ouardighi et al., 2018b). Hoel (1993) also suggests that contingent strategies result in higher total emissions, however, they reveal that the tax giving a Pareto optimal can be the same for both strategies. El Ouardighi et al. (2016) further show that the two strategies may coincide due to the linear state property involved. Bertinelli et al. (2014) also find that the contingent strategies may even lead to lower long-run carbon emissions if the pollution is initially high. Feenstra et al. (2001) find that the strategy type can alter the effectiveness of tax versus standard: although standards are better under commitment strategy, tax can be better due to the increased strategic interaction if firms adopt contingent strategies. Stimming (1999) shows that under duopolistic competition, stricter policy toward one firm may result in higher total emissions due to the higher investment of its competitor and contingent strategies may strengthen this effect.

We fill the gap in this literature in two major aspects. First, although the aforementioned studies explore different strategy types in the context of pollution accumulation, the impacts of market configurations are largely overlooked. To be specific, most studies consider horizontal competition as a given context, while neglecting vertical competition, which is another important dimension of strategic interaction between firms. Zhang et al. (2016) examine vertical competition for both strategy types and show that the level of energy efficiency is higher under contingent strategies. However, they focus on demand evolution and energy efficiency, without considering any pollution-specific dynamics. Zu et al. (2018) and Zhou and Ye (2018) consider vertical competition with environmental regulation for an infinite horizon, but they do not investigate the effects of different strategy types.

Second, although previous studies reported varying results on the effects of strategy types, the driving forces remain largely unclear. We add to the literature by showing that the underlying effect of strategy types can further depend on market structures. For instance, while contingent strategies lead to more pollution than commitment strategies due to lower abatement under TC, the environmental demerit is strengthened under HCVI due to both lower abatement and greater gross emissions.

Model Formulation

To answer the research questions, we consider a bilateral duopoly with two manufacturers and two retailers (Fig. 6.1). Manufacturer i sells a finished good to retailer i, $i = 1, 2$, which resells it, meeting a consumer demand. The retailers compete on price for the demand. The manufacturers generate polluting emissions and incur a

Fig. 6.1 Price-competing supply chains

cost related to their accumulation. To reduce the pollution stock, the manufacturers have the opportunity to invest in abatement efforts.

In this study, we employ the dynamic games approach that allows a much deeper understanding of the pollution problem than static games. Specifically, dynamic games capture the intertemporal dimension of pollution externalities, which cannot be dealt with in static games. This intertemporal feature implies that firms' strategic decisions and pollution essentially evolve together over time, both impacting each other: on the one hand, the decisions of competing firms influence pollution accumulation, but on the other hand, competing firms' decisions are also shaped by the pollution stock based upon which they can condition their profit-maximizing abatement, pricing and production paths over time. Dynamic games are of considerable value in dealing with these intertemporal dynamics underlying pollution externalities. In our study, another important advantage of the dynamic games approach is that it provides mathematically distinctive equilibrium concepts related to strategy types (commitment, contingent), which govern strategic interaction between firms competing over time. In this regard, dynamic games facilitate a more rigorous analysis of our key research questions.

Most often, transactions between a manufacturer and a retailer are governed by a conventional wholesale price contract (WPC), under which the transfer price for each unit ordered by the retailer is set to maximize the manufacturer's profit. Under a WPC, double marginalization is common. To study the role of double marginalization, we consider a two-stage game model where each manufacturer and one retailer first agree on a WPC, and then supply the demand over a finite time horizon. At the first stage of the game, the manufacturers set their respective optimal transfer prices. During the second stage, polluting emissions are generated over time in proportion to demands, which are in turn controlled by the competing retailers' respective consumer prices. In this stage, the manufacturers are involved in emissions abatement.

We seek to analyze how pollution accumulation and abatement are related to the presence of competition and integration in an industry. Based on competition or integration at vertical and horizontal levels, four market structures are considered:

- A monopoly market, where the manufacturers and retailers are all centralized to maximize their joint profits (total integration);
- A bilateral duopoly market, where the manufacturers and retailers are all decentralized to maximize their individual profits (total competition);
- A bilateral monopoly market, where integrated manufacturers compete with integrated retailers to maximize their horizontal profits (horizontal integration and vertical competition);

– A duopoly market, where integrated supply chains compete with each other to maximize their vertical profits (horizontal competition and vertical integration).

In our model, integration typically corresponds to perfect coordination between firms, while competition reflects a lack of coordination. Except in the total integration case, the firms can select either a commitment strategy (OLNE) or a contingent strategy (FBNE):

– Firms' strategies depend on time only, that is, each firm commits to its entire course of action at the outset, and the information on pollution stock is disregarded by (or unavailable to) firms (OLNE).
– Firms condition their actions both on the current value of this stock and on time (FBNE).

In addition, we assume a make-to-order system where no firm carries an inventory. Each manufacturer controls its transfer price to its retailer, which in turn determines its consumer price. Retailers compete for the demand whose substitution effects are based on the consumer price.

We assume price-dependent demands with imperfect product substitutability. Function $D_i(p_i(t), p_j(t)) \geq 0$, $i = 1, 2$, $i \neq j$, represents retailer i 's demand, which is determined by price competition such that it decreases in retailer i 's price and increases in the rival's price. Therefore, price competition enables a retailer to improve its demand at its rival's expense. Demand $D_i(t)$ is specified as a linear function, that is:

$$D_i(t) = \alpha_i - \beta_i p_i(t) + \gamma_i p_j(t) \tag{6.1}$$

where $\alpha_i > 0$, denotes retailer i's potential market, and $0 \leq \gamma_i < \beta_i$ are parameters, $i = 1, 2$, $i \neq j$, such that β_i denotes the retailer i's demand sensitivity to its own price, and γ_i reflects the degree of product substitutability between retailer i and its rival, $i = 1, 2$ (e.g., Ha et al., 2011; Wu et al., 2009). Note that $\gamma_i = 0$, $i = 1, 2$, corresponds to the case where the retailers are not in direct competition. Accordingly, the demands are nonnegative if $p_i(t) \leq (\alpha_i + \gamma_i p_j(t))/\beta_i$, $i = 1, 2$, $i \neq j$.

Assuming that each unit of a good produced and sold results in ψ_i units of emitted pollutants, the dynamics of the pollution stock, $y(t) \geq 0$, are given by:

$$\dot{y}(t) = \sum_{i \neq j} \psi_i \left(\alpha_i - \beta_i p_i(t) + \gamma_i p_j(t)\right) - \sum_i \varphi_i v_i(t), \quad y(0) = 0 \tag{6.2}$$

where $v_i(t) \geq 0$ is the manufacturer i's abatement effort, and $\varphi_i > 0$ is the marginal effectiveness of the manufacturer i's abatement effort, $i = 1, 2$.

In Eq. 6.2, the total current emissions result from the sum of the demands, and are therefore influenced both by the horizontal interaction between the retailers (consumer price competition) and by the vertical interaction between manufacturers and retailers (double marginalization). Further, the manufacturer i's abatement efforts are independent of the corresponding emissions, $\psi_i(\alpha_i - \beta_i p_i(t) + \gamma_i p_j(t))$, $i = 1, 2$. In this regard, the constraints $v_i(t) \leq \psi_i(\alpha_i - \beta_i p_i(t) + \gamma_i p_j(t))/\varphi_i$, $i = 1, 2$, $i \neq j$, should be satisfied so that the manufacturer i's net current emissions remain nonnegative, which implies that $\dot{y}(t) \geq 0$, $\forall t \geq 0$.

To account for recent empirical observations that terrestrial and oceanic natural pollution sinks are tending to exhaustion and will eventually become pollution sources (Cox et al., 2000; Zhao & Running, 2010), no natural abatement of the pollution stock is considered in Eq. 6.2, which is also consistent with the assumption of a finite time horizon. To account for the accountability of the bilateral duopoly in the pollution accumulation, the initial pollution stock in Eq. 6.2 is normalized to zero, i.e., $y(0) = 0$.

Given that each firm's objective is to maximize its cumulative profits over a finite time horizon $[0, T]$, the discounting of future profits is omitted. Manufacturer i's instantaneous gross profit is given by $(\omega_i - c_i)(\alpha_i - \beta_i p_i(t) + \gamma_i p_j(t))$, where $\omega_i > 0$ is the manufacturer i's transfer price to retailer i, and $c_i \geq 0$ is manufacturer i's unit production cost, $i = 1, 2$, $i \neq j$. Manufacturer i's transfer price to retailer i, ω_i, $i = 1$, 2, is chosen to determine an optimal WPC that maximizes the manufacturer i's individual profit. The retailer can then accept or reject the contract.

In general, most of the polluting emissions are related to the manufacturing process (Krass et al., 2013). In return, the accumulation of polluting emissions causes environmental externalities (or damage), which reduce the flow of useful services generated by the natural system to the manufacturing system. That is, the pollution stock entails a cost that is incurred by the manufacturers only. The manufacturers can thus invest in abatement efforts to reduce the pollution stock and the related costs. The negative externalities of the pollution stock are valued as an increasing convex cost function of pollution, that is, $f_i y(t)^2/2$, where $f_i > 0$, is manufacturer i's pollution cost coefficient, $i = 1, 2$. This quadratic form, which implies that the manufacturer i's cost would increase sharply when the pollution level becomes very high, is standard in the literature on pollution control (e.g., Kossioris et al., 2008; Jørgensen et al., 2010; El Ouardighi et al., 2018a, 2018b). Due to diminishing returns to scale, the manufacturers incur a quadratic cost function for the abatement efforts, that is, $e_i v_i(t)^2/2$ where $e_i > 0$, is manufacturer i's abatement effort cost coefficient, $i = 1, 2$. A similar quadratic form is used in Liu et al. (2012) and Bertinelli et al. (2014).

In contrast, the retailer i's current net profit is the difference between its gross sales revenue and its procurement cost from the manufacturer, that is, $(p_i(t) - \omega_i)(\alpha_i - \beta_i p_i(t) + \gamma_i p_j(t))$, $i = 1, 2$, $i \neq j$.

We assume equivalent parameter values for both retailers, that is, $\alpha_i \equiv \alpha > 0$, $\beta_i \equiv \beta > 0$, $\gamma_i \equiv \gamma \geq 0$, $i = 1, 2$, so that any differences in the way sales are affected by price competition do not stem from arbitrary decisions concerning parameter values. Also, for simplicity, we assume symmetric parameter values for the manufacturer i's abatement effort's marginal effectiveness, $\varphi_i \equiv \varphi > 0$, manufacturer i's pollution cost coefficient, $f_i \equiv f > 0$, and the pollutant per unit production, $\psi_i \equiv \psi > 0$, $i = 1, 2$. Without loss of generality, manufacturer i's unit production cost, c_i, and manufacturer i's abatement cost coefficient, e_i, are respectively normalized as $c_i = 0$ and $e_i = 1$, $i = 1, 2$.

Table 6.1 Main variables and parameters

$y(t) \geq 0$	Pollution stock at t
$v_i(t) \geq 0$	Manufacturer i's pollution abatement effort at t, $i = 1, 2$
$p_i(t) \geq 0$	Retailer i's consumer price at t, $i = 1, 2$
$D_i(t) \geq 0$	Retailer i's demand at t, $i = 1, 2$
$\omega_i \geq 0$	Manufacturer i's transfer price, $i = 1, 2$
J^{m_i}, J^{r_i}	Manufacturer i's cumulative profit, retailer i's cumulative profit, $i = 1, 2$
J^m, J^r	Cumulative profit of the manufacturers, cumulative profit of the retailers
J, J^i	Cumulative profit of the entire system, cumulative profit of the supply chain i, $i = 1, 2$
$\overline{v}_i$	Manufacturer i's average pollution abatement effort, i.e., $\overline{v}_i = \int_0^T v_i(t)dt/T$
ξ_i	Manufacturer i's pollution abatement effort per unit of cumulative demand, i.e., $\xi_i = \int_0^T v_i(t)dt / \int_0^T D_i(t)dt$
$\overline{p}_i$	Retailer i's average consumer price, $i = 1, 2$; i.e., $\overline{p}_i = \int_0^T p_i(t)dt/T$
D_{cum}	Retailer i's cumulative demand, $i = 1, 2$; i.e., $D_{cum} = \sum_i \int_0^T D_i(t)dt$
PI	The terminal pollution stock per unit of cumulative demand, i.e., $PI = y(T)/\sum_i \int_0^T D_i(t)$
AI	The total cumulative abatement effort per unit of terminal pollution stock, i.e., $AI = \sum_i \int_0^T v_i(t)/y(T)$
$\alpha_i > 0$	Retailer i's potential market, $i = 1, 2$;
$\beta_i > 0$	Marginal sensitivity of retailer i's demand to own price, $i = 1, 2$;
$\gamma_i > 0$	Marginal sensitivity of retailer i's demand to retailer j's price, $i, j = 1, 2,\ i \neq j$
$\varphi_i > 0$	Marginal effectiveness of manufacturer i's abatement effort, $i = 1, 2$
$c_i \geq 0$	Manufacturer i's unit production cost, $i = 1, 2$
$e_i > 0$	Manufacturer i's abatement effort cost coefficient, $i = 1, 2$
$f_i > 0$	Manufacturer i's pollution cost coefficient, $i = 1, 2$
$\psi_i > 0$	Gross pollutant (before abatement) per unit production of manufacturer i, $i = 1, 2$
$t \in [0, T]$	Time, $T < \infty$

Table 6.1 summarizes the main variables and parameters of the model. Note that we may use superscripts in notations to indicate the corresponding market configuration: *TI, TC, HIVC,* and *HCVI* denote total integration, total competition, horizontal integration and vertical competition, and horizontal competition and vertical integration, respectively. We may also use subscripts to distinguish the strategy type: *ol* and *fb* denote the OLNE strategy and the FBNE strategy, respectively.

The differential game model is then given as:

$$\underset{\omega_i, v_i \geq 0}{\text{Max}}\ J^{m_i} = \int_0^T \left[\omega_i(\alpha - \beta p_i + \gamma p_j) - \frac{v_i^2}{2} - \frac{fy^2}{2}\right] \mathrm{d}t \tag{6.3}$$

Table 6.2 Main characteristics of the four configurations

	Vertical integration	Vertical competition
Horizontal integration	*Total integration* – Monopoly market structure – No double marginalization – Both vertical and horizontal internalization of pollution externalities – Price coordination between retailers	*Horizontal integration and Vertical competition* – Bilateral monopoly market structure – Double marginalization – Only horizontal internalization of pollution externalities – Price coordination between retailers
Horizontal competition	*Horizontal competition and Vertical integration* – Duopoly market structure – No double marginalization – Only vertical internalization of pollution externalities – Price competition between retailers	*Total competition* – Bilateral duopoly market structure – Double marginalization – Neither vertical nor horizontal internalization of pollution externalities – Price competition between retailers

$$\underset{p_i \geq 0}{\text{Max}} J^{r_i} = \int_0^T (p_i - \omega_i)(\alpha - \beta p_i + \gamma p_j)\mathrm{d}t \tag{6.4}$$

subject to:

$$\dot{y}(t) = \sum_{i \neq j} \psi(\alpha - \beta p_i(t) + \gamma p_j(t)) - \varphi \sum_i v_i(t), \quad y(0) = 0 \tag{6.5}$$
$$p_i(t) \leq (\alpha + \gamma p_j(t))/\beta, \quad v_i(t) \leq \psi(\alpha - \beta p_i(t) + \gamma p_j(t))/\varphi$$

$i = 1, 2, i \neq j$. As will be shown below, the control constraints are never binding. We therefore do not explicitly derive the related optimality conditions. Four market configurations, whose main characteristics are described in Table 6.2, are analyzed.

Analysis

The next four subsections derive the pricing and abatement decisions in the contexts of monopoly, bilateral duopoly, bilateral monopoly, and duopoly. Proofs of the following theorems and lemmas are provided in Appendix. Note also that cumulative profits under optimal strategies for all cases in subsections from 'Total Integration' to 'Horizontal Competition and Vertical Integration' are derived in Table 6.12 in the Appendix.

Total Integration

This subsection analyzes total integration, where the manufacturers and the retailers jointly centralize their decisions to maximize the joint profits, as in a monopoly. In this case, which corresponds to a one-stage game, there is perfect coordination between the firms both at the vertical level, because the marginalization is totally mitigated, and at the horizontal level, because the pollution externalities and the substitution effects are both vertically and horizontally internalized in the manufacturers' and retailers' decisions.

Theorem 1. *The equilibrium consumer prices and abatement efforts are unique and given by:*

$$v_i^{TI}(t) = \frac{\alpha\varphi\psi}{((\beta-\gamma)\psi^2+2\varphi^2)}\left[1-\frac{\mathrm{ch}(\theta t)}{\mathrm{ch}(\theta T)}\right] \tag{6.6}$$

$$p_i^{TI}(t) = \frac{1}{2}\left\{\frac{\alpha}{(\beta-\gamma)}+\frac{\alpha\psi^2}{((\beta-\gamma)\psi^2+2\varphi^2)}\left[1-\frac{\mathrm{ch}(\theta t)}{\mathrm{ch}(\theta T)}\right]\right\} \tag{6.7}$$

$i = 1, 2$, *where the superscript TI refers to "total integration," and* $\theta \equiv \sqrt{2f((\beta-\gamma)\psi^2+2\varphi^2)}$. *Accordingly, the equilibrium pollution stock is unique and given by:*

$$y^{TI}(t) = \frac{\alpha\psi\,\mathrm{sh}(\theta t)}{\theta\,\mathrm{ch}(\theta T)} \tag{6.8}$$

From Eqs. 6.6 and 6.7, we observe that the greater the abatement efforts the higher the consumer prices. An intuitive explanation on the positive relationship is that manufacturers' and retailers' abatement and pricing decisions can be coordinated to reduce the pollution through greater abatement efforts and smaller volume, because pollution externalities are perfectly internalized in this case.

From Eq. 6.7, the retailer i's consumer price is initially high and decreases concavely over time $t \in [0, T]$, which implies convexly increasing demands. From Eq. 6.6, the manufacturer i's abatement effort is also initially high and concavely decreasing over time. These imply that firms under perfect coordination put more emphasis on mitigating the pollution accumulation during the initial periods, whereas they focus more on enlarging the demand volume in the later periods. Thus, the pollution stock is strictly and convexly increasing over time as shown in Eq. 6.8.

Total Competition

This subsection deals with the total competition case, where the manufacturers and the retailers make noncooperative decisions to maximize their individual profits, as in a bilateral duopoly. In this case, there is no coordination between the players at the vertical level because double marginalization prevails between manufacturers and retailers, or at the horizontal level, because of the imperfect internalization of pollution externalities by the manufacturers and the price competition for demand between the retailers. In addition, the pollution stock is payoff-irrelevant to the retailers, which implies that pollution externalities are not at all internalized in the retailers' consumer prices.

Under total competition, Eqs. 6.3–6.4–6.5 form a two-stage game whose sequence of events is as follows:

- In the first stage of the game, $t = 0$, the manufacturer i sets the optimal transfer price for the retailer i;
- During the second stage, assuming that an agreement on the optimal WPC has been reached between the manufacturer i and retailer i, $i = 1, 2$, each firm deploys its strategy over the time interval $[0, T]$.
- In the final period, $t = T$, the profits are allocated among the firms according to the WPC.

As usual in a two-stage game, one solves the game backward.

Theorem 2. *Under commitment strategies (OLNE), the manufacturer i's transfer price and abatement effort and the retailer i's consumer price are unique and given by:*

$$\omega_{iol}^{TC} = \frac{\left\{\beta(\beta-\gamma)\psi^2\left[\left(2-\operatorname{sech}(\rho T)^2\right)\rho T - \operatorname{th}(\rho T)\right] + 2\varphi^2\rho(2\beta-\gamma)T\right\}}{\left\{\beta(\beta-\gamma)\psi^2\left[\left(2-\operatorname{sech}(\rho T)^2\right)\rho T - \operatorname{th}(\rho T)\right] + 4\varphi^2\rho(2\beta-\gamma)T\right\}} \frac{\alpha}{(\beta-\gamma)} \tag{6.9}$$

$$v_{iol}^{TC}(t) = \left[1 - \frac{\operatorname{ch}(\rho t)}{\operatorname{ch}(\rho T)}\right] \frac{2\alpha\beta\varphi\rho T\psi}{\beta(\beta-\gamma)\psi^2\left[\rho T\left(2-\operatorname{sech}(\rho T)^2\right) - \operatorname{th}(\rho T)\right] + 4\varphi^2\rho(2\beta-\gamma)T} \tag{6.10}$$

$$p_{iol}^{TC}(t) = \frac{\alpha + \beta\omega_{iol}^{TC}}{2\beta-\gamma} \tag{6.11}$$

$i = 1, 2$, where the superscript TC refers to "total competition," and $\rho \equiv \varphi\sqrt{2f}$. Accordingly, the pollution stock is unique and given by:

$$y_{ol}^{TC}(t) = \frac{4\alpha\beta\varphi^2 T\psi}{\beta(\beta-\gamma)\psi^2\left[\rho T\left(2-\operatorname{sech}(\rho T)^2\right) - \operatorname{th}(\rho T)\right] + 4\varphi^2\rho(2\beta-\gamma)T} \frac{\operatorname{sh}(\rho t)}{\operatorname{ch}(\rho T)} \tag{6.12}$$

From Eqs. 6.10 and 6.12, abatement efforts are initially high and concavely decrease to zero and the pollution stock is strictly increasing over time, similar to the total integration case. However, from Eqs. 6.9 and 6.11, we observe that the transfer prices and the consumer prices under commitment strategies are both time-independent, which implies constant demands over the whole planning horizon. In particular, constant consumer prices result from the fact that pollution externalities are not internalized in the retailers' decision-making. Thus, retailers simply set their consumer prices to be proportional to the transfer price, considering the trade-off between their margin and demand, but without consideration of the externalities of their pricing decisions on pollution.

Theorem 3. *Under contingent strategies (FBNE), the manufacturer i's transfer price and abatement effort and the retailer i's consumer price are:*

$$\omega_{ifb}^{TC} = \frac{\{[4\beta(\beta-\gamma)\psi^2 + 3\varphi^2(2\beta-\gamma)]\tau T - 4\beta(\beta-\gamma)\psi^2 \text{th}(\tau T)\}}{\{[2\beta(\beta-\gamma)\psi^2 + 3\varphi^2(2\beta-\gamma)]\tau T - 2\beta(\beta-\gamma)\psi^2 \text{th}(\tau T)\}} \frac{\alpha}{2(\beta-\gamma)} \tag{6.13}$$

$$v_{ifb}^{TC}(t) = \frac{\sqrt{f}\,\text{th}[\tau(T-t)]y(t)}{\sqrt{3}} + \frac{2\beta\psi\left[\alpha-(\beta-\gamma)\omega_{ifb}^{TC}\right]\{1-\text{sech}[\tau(T-t)]\}}{3\varphi(2\beta-\gamma)} \tag{6.14}$$

$$p_{ifb}^{TC}(t) = \frac{\alpha+\beta\omega_{ifb}^{TC}}{2\beta-\gamma} \tag{6.15}$$

$i = 1, 2$, $\tau \equiv \varphi\sqrt{3f}$, where $y(t)$ is given by:

$$\dot{y}_{fb}^{TC}(t) = \frac{\beta\psi\left[2\alpha-(\beta-\gamma)\sum_i\omega_{ifb}^{TC}\right]}{2\beta-\gamma} - \varphi\sum_i v_{ifb}^{TC}(t), \quad y(0)=0 \tag{6.16}$$

From Eqs. 6.13–6.16, we see that the transfer prices, the consumer prices and the abatement efforts under contingent strategies have a similar pattern to those under commitment strategies. The consumer prices are constant over time, yet the abatement efforts should increase along with the pollution stock.

A noteworthy observation is that although the pollution stock is payoff-irrelevant to the retailers, retailers' pricing strategies are not equivalent to commitment and contingent strategies. This is due to the strategic interaction between the manufacturers in terms of abatement efforts during the second stage of the game. In fact, the retailers set constant consumer prices regardless of the pollution accumulation and, as shown from Eqs. 6.11 and 6.15, the retailers' optimal responses remain qualitatively the same regardless of the strategy type. However, manufacturers' transfer prices vary depending on whether they revise abatement strategies depending upon the pollution stock, because interaction between competing manufacturers generates different state trajectories. This in turn influences consumer prices of the retailers indirectly.

Horizontal Integration and Vertical Competition

This subsection combines horizontal integration and vertical competition where the manufacturers and the retailers centralize their decisions separately to maximize their horizontal profits, as in a bilateral monopoly. In this case, there is horizontal internalization of pollution externalities by the manufacturers and horizontal coordination of substitution effects on demand by the retailers. However, there is no vertical coordination and double marginalization prevails while pollution externalities are not internalized in retailers' consumer prices. As in subsection 'Total Competition', a two-stage game approach is followed.

Theorem 4. *Under commitment strategies (OLNE), the manufacturer i's transfer price and abatement effort and the retailer i's consumer price are unique and given by:*

$$\omega_{iol}^{HIVC} = \left[\frac{\delta T((\beta-\gamma)\psi^2+2\varphi^2)-(\beta-\gamma)\psi^2\text{th}(\delta T)}{\delta T((\beta-\gamma)\psi^2+4\varphi^2)-(\beta-\gamma)\psi^2\text{th}(\delta T)}\right]\frac{\alpha}{(\beta-\gamma)} \tag{6.17}$$

$$v_{iol}^{HIVC}(t) = \frac{\alpha\delta\varphi T\psi}{[\delta T((\beta-\gamma)\psi^2+4\varphi^2)-(\beta-\gamma)\psi^2\text{th}(\delta T)]}\left[1-\frac{\text{ch}(\delta t)}{\text{ch}(\delta T)}\right] \tag{6.18}$$

$$p_{iol}^{HIVC}(t) = \left[1+\frac{\delta T((\beta-\gamma)\psi^2+2\varphi^2)-(\beta-\gamma)\psi^2\text{th}(\delta T)}{\delta T((\beta-\gamma)\psi^2+4\varphi^2)-(\beta-\gamma)\psi^2\text{th}(\delta T)}\right]\frac{\alpha}{2(\beta-\gamma)} \tag{6.19}$$

$i = 1, 2$, *where the superscript HIVC refers to "horizontal integration and vertical competition," and* $\delta \equiv 2\varphi\sqrt{f}$. *Accordingly, the pollution stock is unique and given by:*

$$y_{ol}^{HIVC}(t) = \frac{2\alpha\varphi^2 T\psi}{[((\beta-\gamma)\psi^2+4\varphi^2)\delta T-(\beta-\gamma)\psi^2\text{th}(\delta T)]}\frac{\text{sh}(\delta t)}{\text{ch}(\delta T)} \tag{6.20}$$

From Eqs. 6.17 and 6.19, the transfer prices and the consumer prices under commitment strategies are both constant as in the total competition case. From Eq. 6.18, the manufacturer i's abatement efforts are also initially high and concavely decreasing until zero, and from Eq. 6.20, the pollution stock increases over time.

Lemma 1. *In horizontal integration and vertical competition cases, commitment strategies are equivalent to contingent strategies for both manufacturers and retailers.*

The equivalence between commitment and contingent strategies results not only from the fact that the pollution stock is payoff-irrelevant to the retailers, but also from the absence of horizontal competition, and hence of strategic interaction between the manufacturers in terms of abatement efforts or between the retailers in terms of consumer prices during the second stage of the game. These imply that when firms coordinate their decisions horizontally within each tier, manufacturers and retailers may choose either strategy, committing strategies with each other or continuously revising them based on the pollution stock depending on the circumstances, e.g., resources required for each strategy and the availability of the information regarding pollution stock.

Horizontal Competition and Vertical Integration

This subsection combines horizontal competition and vertical integration where manufacturer i and retailer i, $i = 1, 2$, centralize their vertical decisions to maximize their vertical profits, as in a duopoly. In this case, there is vertical coordination because double marginalization is totally mitigated, but no horizontal coordination due to improper internalization of pollution externalities by the manufacturers and price competition for demand by the retailers. The pollution externalities are only vertically internalized in the supply chain decisions.

Theorem 5. *Under commitment strategies (OLNE), the manufacturer i's abatement effort and the retailer i's consumer price are unique and given by:*

$$v_{iol}^{HCVI}(t) = \frac{\alpha\beta\varphi\psi}{(\beta-\gamma)^2\psi^2 + \varphi^2(2\beta-\gamma)}\left[1 - \frac{\mathrm{ch}(\vartheta t)}{\mathrm{ch}(\vartheta T)}\right] \tag{6.21}$$

$$p_{iol}^{HCVI}(t) = \left(\frac{\beta-\gamma}{2\beta-\gamma}\right)\left\{\frac{\alpha}{\beta-\gamma} + \frac{\alpha\beta\psi^2}{(\beta-\gamma)^2\psi^2 + \varphi^2(2\beta-\gamma)}\left[1 - \frac{\mathrm{ch}(\vartheta t)}{\mathrm{ch}(\vartheta T)}\right]\right\} \tag{6.22}$$

$i = 1, 2$, *where the superscript HCVI refers to "horizontal competition and vertical integration" and* $\vartheta \equiv \sqrt{\frac{2f\left[(\beta-\gamma)^2\psi^2 + \varphi^2(2\beta-\gamma)\right]}{2\beta-\gamma}}$. *Accordingly, the pollution stock is unique and given by*:

$$y_{ol}^{HCVI}(t) = \frac{\sqrt{2}\alpha\beta\psi}{\sqrt{f(2\beta-\gamma)\left[(\beta-\gamma)^2\psi^2 + \varphi^2(2\beta-\gamma)\right]}}\frac{\mathrm{sh}(\vartheta t)}{\mathrm{ch}(\vartheta T)} \tag{6.23}$$

In Eq. 6.22, the retailers' consumer prices are concavely decreasing over time, which implies convexly increasing demands. From Eq. 6.21, abatement efforts follow the same pattern as the retailers' consumer prices, i.e., concavely decreasing until zero. From Eq. 6.23, the pollution stock is strictly increasing over time. Similar to the total integration case, the manufacturer's abatement efforts and the retailer's consumer prices are positively associated. These results imply that patterns of pricing and abatement decisions are heavily influenced by vertical market configurations compared to horizontal market configurations.

Theorem 6. *Under contingent strategies (FBNE), the manufacturer i's abatement effort and the retailer i's consumer price are:*

$$v_{ifb}^{HCVI}(t)=\frac{\varphi(2\beta-\gamma)f}{\sigma}\text{th}\left(\frac{\sigma(T-t)}{2\beta-\gamma}\right)y(t)+\frac{\alpha\varphi f\psi\left(4\beta^2-3\beta\gamma+\gamma^2\right)}{\sigma^2}\left[1-\text{sech}\left(\frac{\sigma(T-t)}{2\beta-\gamma}\right)\right] \tag{6.24}$$

$$p_{ifb}^{HCVI}(t)=\frac{\alpha}{2\beta-\gamma} +f\psi(\beta-\gamma)\left\{\frac{1}{\sigma}\text{th}\left(\frac{\sigma(T-t)}{2\beta-\gamma}\right)y(t)+\frac{\alpha\psi\left(4\beta^2-3\beta\gamma+\gamma^2\right)}{\sigma^2(2\beta-\gamma)}\left[1-\text{sech}\left(\frac{\sigma(T-t)}{2\beta-\gamma}\right)\right]\right\} \tag{6.25}$$

$i = 1, 2$, $\sigma\equiv\sqrt{f\left[6\beta^3\psi^2+2\beta\gamma(5\gamma\psi^2-6\varphi^2)-2\beta^2(7\gamma\psi^2-6\varphi^2)-\gamma^2(2\gamma\psi^2-3\varphi^2)\right]}$, where $y(t)$ is given by:

$$\dot{y}_{fb}^{HCVI}(t)=\frac{2\alpha\beta\psi}{2\beta-\gamma}-\left[\frac{(\beta-\gamma)^2\psi^2}{(2\beta-\gamma)\varphi}+\varphi\right]\sum\nolimits_i v_{ifb}^{HCVI}(\text{t}),\quad y(0)=0 \tag{6.26}$$

In this configuration, strategic interaction involves the two vertically integrated chains that compete on prices and make noncooperative abatement efforts. As a consequence, the resulting pollution stock follows different trajectories depending on the strategy type, leading to distinct pricing and abatement efforts between commitment and contingent strategies.

Comparative Analysis

Based on our previous results, we now assess both analytically and numerically the relative impact of horizontal and vertical competition, on the one hand, and strategy types, on the other hand, on the pricing and abatement strategies, the pollution stock, and the profits generated by the two chains.

Analytical Comparisons

Theorem 7. *In terms of the terminal pollution stock,*

(i) *In vertically integrative supply chains, horizontal integration results in a lower terminal pollution stock than horizontal competition under commitment strategy, i.e.,* $y_{ol}^{HCVI}(T)>y^{TI}(T)$.

(ii) *In horizontally integrative supply chains, vertical integration results in a larger terminal pollution stock than vertical competition, i.e.,* $y^{TI}(T) > y^{HIVC}(T)$.

In the TI case and the HCVI case, the two supply chains are vertically integrated, and the only difference is whether they are horizontally competing or integrated. By comparing these cases, the effect of horizontal structure in the case of vertical integration can be isolated. Theorem 7(i) implies that horizontal competition leads to more accumulated pollution than horizontal integration, assuming that the two supply chains are vertically integrated. The rationale behind this is that price competition decreases sales prices, as shown in the subsequent numerical illustrations, and increases the sales volume, which raises polluting emissions. Furthermore, pollution externalities are only partially internalized, which lessens the incentive for each manufacturer to invest in abatement efforts.

On the other hand, in the TI case and the HIVC case, the two supply chains are horizontally integrated, and the only difference is whether they are vertically competing or integrated. By comparing these cases, the influence of vertical structure in the horizontal integration case can be isolated. Theorem 7(ii) states that vertical competition results in lower pollution stock than vertical integration, assuming that industries are horizontally integrated. A possible explanation is that the double marginalization reduces the sales and then the emissions and it also promotes manufacturers' incentives for abatement efforts, which will be confirmed in the subsequent numerical results.

Note that the underlying drivers of the environmental effects in Theorem 7 are in fact related to the two factors: output volume and incentives for abatement. We term these drivers as "volume effect" and "polluting effect" respectively, and investigate them in detail in the subsequent section, showing that each effect can be more pronounced under distinct configurations.

Theorem 8. *In vertically competitive supply chains, there exists* $\overline{\gamma} > 0$ *such that for any* $\gamma > \overline{\gamma}$, *horizontal integration results in a lower terminal pollution stock than horizontal competition under commitment strategy, i.e.,* $y_{ol}^{TC}(T) > y^{HIVC}(T)$.

In TC and HIVC, the two supply chains are vertically competing, and the only difference is whether they are horizontally competing or integrated. By comparing these cases, the influence of horizontal structure in the case of vertical competition can be isolated. Theorem 8 implies that horizontal competition will lead to more accumulated pollution than horizontal integration, assuming that the supply chains are vertically competing and that price competition prevails. Though Theorem 7 assumes vertical integration and Theorem 8 assumes vertical competition, both theorems isolate the effect of horizontal competition and horizontal integration and therefore the underlying mechanisms are fundamentally very similar. Note that Theorem 8 states a sufficient condition, that is, $\gamma > \overline{\gamma}$, for $y_{ol}^{TC}(T) > y^{HIVC}(T)$ to hold. For $0 \leq \gamma \leq \overline{\gamma}$, however, $y_{ol}^{TC}(T)/y^{HIVC}(T) > 1$ still holds in our subsequent numerical analysis. For reference, the best estimated minimum of $y_{ol}^{TC}(T)/y^{HIVC}(T)$ from Mathematica with constraints $\beta, \psi, \varphi, f, T > 0, 0 \leq \gamma \leq \overline{\gamma}$ is larger than 1 (i.e., 1).

Theorem 9. *In horizontally competitive supply chains, there exists $\overline{\varphi} > 0$ such that for any $\varphi > \overline{\varphi}$, vertical integration results in the larger terminal pollution stock than vertical competition under commitment strategy, i.e., $y_{ol}^{HCVI}(T) > y_{ol}^{TC}(T)$.*

In the TC and the HCVI cases, the two supply chains are horizontally competing, and the only difference is whether they are vertically competing or integrated. Therefore, by comparing these cases, the effect of vertical structure can be isolated in the case of horizontal competition. Assuming that industries are horizontally competing and have sufficiently effective abatement efforts, Theorem 9 implies that vertical competition leads to less pollution than vertical integration. Though Theorem 7 assumes horizontal integration and Theorem 9 assumes horizontal competition, both theorems suggest that the underlying mechanisms are fundamentally similar. Theorem 9 states a sufficient condition for $y_{ol}^{HCVI}(T) > y_{ol}^{TC}(T)$. For $0 < \varphi \leq \overline{\varphi}$, however, $y_{ol}^{HCVI}(T)/y_{ol}^{TC}(T) > 1$ still holds in our numerical analysis. For reference, the minimum of $y_{ol}^{HCVI}(T)/y_{ol}^{TC}(T)$ from Mathematica with constraints $\beta, \gamma, \psi, f, T > 0$ and $0 < \varphi \leq \overline{\varphi}$ is larger than 1 (i.e., 2).

Theorem 10. *There exist $\tilde{\beta} > 0$ and $\tilde{\varphi} > 0$ such that for any $\beta > \tilde{\beta}$ and $\varphi > \tilde{\varphi}$, perfect integration horizontally and vertically results in the larger terminal pollution stock than no integration under commitment strategy, i.e., $y^{TI}(T) > y_{ol}^{TC}(T)$.*

In contrast to Theorem 7 and to Theorem 9, Theorem 10 compares the relative effects of horizontal structure and vertical structure. The TI configuration combines vertical integration and horizontal integration, while the TC configuration relates vertical competition and horizontal competition. For sufficiently large values of marginal sensitivity of retailer i's demand to own price and marginal effectiveness of abatement efforts, the pollution stock is greater in the TI than the TC, implying that the relative impact of vertical integration on pollution accumulation is stronger than that of horizontal integration. Although internalization of pollution externalities between manufacturers and horizontal coordination of substitution effects between retailers drive lower pollution stock, the elimination of double marginalization may increase the pollution stock more rapidly, making the perfect coordination more polluting than no coordination.

Based on the previous theorems, Table 6.3 summarizes the combinative effect of horizontal and vertical competition on pollution accumulation as follows. Overall, when price competition and effectiveness of abatement efforts are both relatively high ($\gamma > \overline{\gamma}, \varphi > \overline{\varphi}$), horizontally competing and vertically integrated supply chains generate the highest pollution stock while horizontally integrated and vertically competing supply chains result in the lowest pollution stock. The pollution stock of the total integration and total competition cases are in between. These relationships may hold in other conditions as well. Furthermore, when the sensitivity of demand to the firm's own price is relatively high ($\beta > \tilde{\beta}$) and the effectiveness of abatement efforts is sufficiently high ($\varphi > \tilde{\varphi}$), the relative effect of vertical structure (i.e., vertical competition *vs.* vertical integration) is more likely to outweigh the relative effect of horizontal structure (i.e., horizontal competition *vs.* horizontal integration).

Table 6.3 Influence of horizontal and vertical competition on pollution accumulation under commitment strategies

Pollution accumulation ($y(T)$)

	$\gamma < \bar{\gamma}$	$\gamma > \bar{\gamma}$
$\vartheta > \bar{\vartheta}$	HIVC < TI TI, TC < HCVI	HIVC < TI, TC < HCVI
$\vartheta < \bar{\vartheta}$	HIVC < TI < HCVI	HIVC < TI, TC TI < HCVI

(Row boundary: $\bar{\vartheta}$; column boundary: $\bar{\gamma}$)

Numerical Comparisons

We conduct a numerical study to further understand the underlying mechanisms regarding the impact of competition on pollution accumulation. The parameter values for the numerical study are chosen based on the apparel industry. The apparel industry has been known as a highly polluting industry (Conca, 2015). In this industry, curbing manufacturing emissions is important because most of the polluting emissions are generated in the manufacturing stage. It has been reported that manufacturing one T-shirt generates 4.8 kg CO_2e (Muthu (2014); p.166), therefore we set $\psi = 5$. We specify the market potential (α) as 100 and the price sensitivity (β) as 5, in accordance with the studies for the apparel industry (e.g., Marti et al., 2015; Ghosh & Shah, 2012). The cross-price sensitivity (γ) is specified as 1 to satisfy $\gamma < \beta$. We assume that $\varphi = 15$ and $f = 0.1$ to ensure that firms have nonnegative profit. Note that, though $\varphi > \psi$, since the abatement cost increases quadratically in v_i for each manufacturer while the revenue increases in demand linearly, increasing output can be considerably polluting due to the abatement cost burden. The planning horizon is 12 months ($T = 12$) due to the short finite horizon. We consider a broad set of parameter values for α, β, γ, φ, f, ψ to ensure the robustness of the numerical results. The base case values and the ranges of parameter values used in the numerical study are summarized in Table 6.4. In particular, we examine diverse situations with regard to the level of abatement difficulty as shown in the wide range of the parameter values of φ and ψ in the table. Similarly, we also consider various situations regarding pollution burden as shown in the broad range of the parameter

Table 6.4 Base case and range values

Parameter	α	β	γ	φ	f	ψ
Base case	100	5	1	15	0.1	5
Range	[10,1000]	[1.4,27]	[0,3.3]	[8,30]	[0.01,0.38]	[0.1,10]

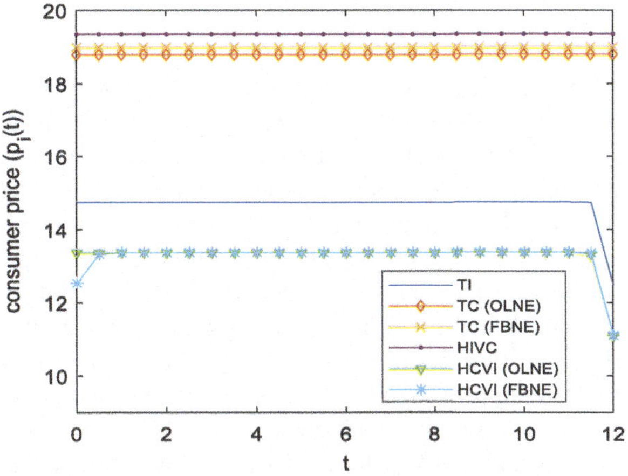

Fig. 6.2 Retailer i's consumer price ($pi(t)$)

values of f. For brevity, we summarize the major results of the numerical analysis, focusing on the robustness of the insights under the wide sets of parameters.

Comparisons involving the cooperative, commitment, and contingent strategies on a numerical basis are provided hereafter. We first illustrate the pricing strategies and demand (section 'Pricing Strategies') and the abatement strategies (section 'Abatement Strategies') in each market structure. We then discuss their impacts on pollution accumulation (section 'The Environmental Performance') and the implications on the supply chain profits (section 'Supply Chain Profits').

Pricing Strategies

In this subsection, we examine the effects of competition and the strategy type on prices. We then elaborate on the implications on demand, which also relates to the "volume effect" on the pollution stock.

Figure 6.2 illustrates the evolution of consumer prices over time in the base case. Throughout the planning horizon, the case of horizontal integration and vertical competition (HIVC) results in the highest consumer price, followed by total competition (TC), total integration (TI), and horizontal competition and vertical integration (HCVI) regardless of the strategy types. To compare the impacts of competition on consumer prices more precisely, we also examine the average consumer price over the planning period (i.e., $\overline{p}_i = \frac{1}{T}\int_0^T p_i(t)dt$). In line with Fig. 6.2, the average consumer price ($\overline{p}_i$) has the following order, i.e., $\overline{p}_i^{HIVC} > \overline{p}_i^{TC} > \overline{p}_i^{TI} > \overline{p}_i^{HCVI}$. The

Table 6.5 Pricing strategies in the base case

	TI	TC (OLNE)	TC (FBNE)	HIVC	HCVI (OLNE)	HCVI (FBNE)
$\overline{p}_i$	14.75	18.81	19.03	19.37	13.38	13.37
ω_i	n/a	13.87	14.25	13.74	n/a	n/a
$D_{cum} = \sum_i \int_0^T D_i(t)\mathrm{d}t$	983.55	593.81	573.26	540.47	1115.9	1116.4

optimal pricing strategies and the consequent demand in the base case are reported in Table 6.5, from which we make several observations as follows.

First, the impacts of strategy types on the pricing strategy may vary significantly depending on the market configurations. Whether firms must set higher or lower prices under commitment strategy compared to under contingent strategy is determined by market configurations. Although a contingent strategy results in higher consumer prices and smaller demand than a commitment strategy in the TC case, it leads to lower prices and greater demand in the HCVI case. The reason is that, different from the HCVI case, the TC case exhibits the double marginalization effect, thus strategic interactions between manufacturers result in higher transfer prices, leading to higher consumer prices. The effects of competition on consumer prices do not depend heavily on the strategy type. We observe that the same order of consumer prices holds for all parameter ranges across commitment and contingent strategies, except for $\gamma < 0.5$ or $\psi > 8.2$ where TC yields higher $\overline{p}_i$ than HIVC due to relatively low competitive intensity and high transfer price under the contingent strategy. The cumulative demand D_{cum} is thus generally the largest in the case of HCVI, followed by TI, TC, and HIVC.

Second, strategy types may also change the equilibrium path of price. In HCVI configuration, retailers' prices under commitment strategies are concavely decreasing in line with Theorem 5. In contrast, prices under contingent strategies first increase and then decrease over time. That is, while sales prices are at their highest at the beginning under commitment strategies, they are at their highest during the intermediate periods under contingent strategies. These impacts can be more apparent under lower f, where the lessened pollution burden causes more dynamic price adjustments of firms.

Third, the transfer prices ω_i are lower in the HIVC case than in the TC case under all ranges of parameters. Given that the retailers in HIVC would charge relatively higher sales prices due to the internalization of substitution effects, which leads to lower demand, the manufacturers have less room to further raise the transfer prices. This implies that the retailers in the HIVC case get higher profit margins than in the TC case. In other words, horizontal integration strengthens double marginalization for retailers' benefit.

The numerical results also confirm discussions on Theorems 7–9 by showing that horizontal competition generally leads to lower consumer prices than horizontal integration, when the market is either vertically integrative or competitive, and that vertical competition increases consumer prices regardless of the horizontal structure. Under horizontal competition, retailers do not fully internalize the substitution effects between demands, leading to more competitive consumer prices. In contrast, the overpricing strategies under vertical competition are mainly due to the double marginalization.

Therefore, when the market structure changes either from horizontal integration to horizontal competition, or from vertical competition to vertical integration, the sales volume increases due to lower prices, and the gross volume of emissions enlarges as a consequence, which implies a "*volume effect*" in both cases, i.e., "horizontally-induced volume effect" and "vertically-induced volume effect."

Note that the vertically-induced volume effect is generally greater than the horizontally-induced volume effect. That is, the volume effect is influenced more significantly by vertical market configurations. Table 6.6 shows that while both horizontal competition and vertical integration generally lead to relatively higher D_{cum} compared to horizontal integration and vertical competition respectively, the impacts of vertical configuration are greater than the impacts of horizontal configuration. For instance, when α varies from 10 to 1000, horizontal competition leads to D_{cum} which is 106~114% of that of horizontal integration, while vertical integration leads to D_{cum} which is 182~195% of that of vertical competition. Both horizontally-induced and vertically-induced volume effects become greater under smaller β and greater γ, implying that the consumers' buying behaviors influence the degree of the volume effects significantly. The effects of strategy types on the volume effects also depend on the configurations. While the horizontally-induced volume effect can be more prominent under commitment strategies under vertical competition, the vertically-induced volume effect is intensified under contingent strategies under horizontal competition.

Abatement Strategies

In this subsection, we investigate the effects of competition and the strategy type on abatement efforts. We also discuss the manufacturer's incentives for abatement, which relates to the "polluting effects."

Figure 6.3 depicts the evolution of abatement efforts over time in the base case. Among the configurations, HCVI generally yields the largest abatement efforts, followed by TI, TC, and HIVC. In terms of the average abatement efforts over the planning horizon (i.e., $\overline{v}_i = \frac{1}{T}\int\limits_0^T v_i(t)\mathrm{d}t$), it also holds that $\overline{v}_i^{HCVI} > \overline{v}_i^{TI} > \overline{v}_i^{TC} > \overline{v}_i^{HIVC}$ (Table 6.7). The low level of manufacturers' abatement efforts in HIVC is adequately reflected in the corresponding low level of manufacturers' transfer prices. The same rank also holds for most of the parameter ranges in both commitment and contingent strategies.

Different from the effects of strategy types on prices, the commitment strategy yields higher abatement efforts and abatement rates compared to the contingent strategy in either TC or HCVI case, confirming that the manufacturing process becomes cleaner under the commitment strategy whenever the manufacturers are

Table 6.6 Impacts of competition on the cumulative demand

Impact		Strategy	α	β	γ	φ	f	ψ
Horizontal competition vs. horizontal integration	Under VI: $D_{cum}^{HCVI}/D_{cum}^{TI}$	OLNE	[1.13,1.13]	[1.57,1.03]	[1.00,1.56]	[1.16,1.12]	[1.14,1.13]	[1.11,1.17]
		FBNE	[1.14,1.14]	[1.57,1.04]	[1.00,1.56]	[1.16,1.12]	[1.14,1.13]	[1.11,1.17]
	Under VC: $D_{cum}^{TC}/D_{cum}^{HIVC}$	OLNE	[1.10,1.10]	[1.55,1.01]	[1.00,1.46]	[1.09,1.11]	[1.10,1.10]	[1.11,1.08]
		FBNE	[1.06,1.06]	[1.54,0.89]	[0.96,1.43]	[1.02,1.10]	[1.06,1.06]	[1.11,0.98]
Vertical integration vs. vertical competition	Under HI: $D_{cum}^{TI}/D_{cum}^{HIVC}$		[1.82,1.82]	[1.98,1.42]	[1.78,1.91]	[1.67,1.95]	[1.82,1.82]	[2.00,1.55]
	Under HC: $D_{cum}^{HCVI}/D_{cum}^{TC}$	OLNE	[1.90, 1.90]	[2.01,1.45]	[1.79,2.04]	[1.78,1.97]	[1.89,1.88]	[2.00,1.69]
		FBNE	[1.95,1.95]	[2.03,1.66]	[1.86,2.09]	[1.91,1.98]	[1.95,1.95]	[2.00,1.87]

VI vertical integration, *VC* vertical integration, *HI* horizontal integration, *HC* horizontal competition

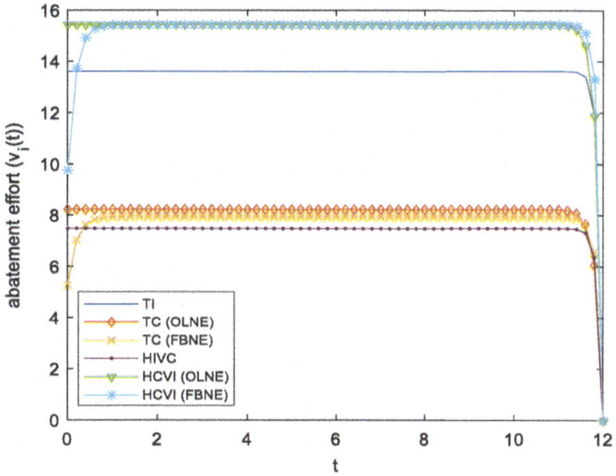

Fig. 6.3 Manufacturer i's abatement efforts ($vi(t)$)

Table 6.7 Abatement strategies in the base case

	TI	TC (OLNE)	TC (FBNE)	HIVC	HCVI (OLNE)	HCVI (FBNE)
$\overline{v}_i$	13.53	8.14	7.85	7.44	15.29	15.25
$\xi_i = \int_0^T v_i(t)\mathrm{d}t / \int_0^T D_i(t)\mathrm{d}t$	0.3301	0.3292	0.3284	0.3304	0.3288	0.3279

competitive. An intuitive explanation can be that contingent strategy where abatement efforts of competing manufacturers are continuously revised intensifies the effects of imperfect internalization of pollution externalities, resulting in lower pollution abatement. Strategy types may also influence the equilibrium path of abatement efforts under horizontal competition. Similar to pricing paths, abatement efforts follow concavely decreasing trajectories under commitment strategies in line with Theorems 2 and 5, being at their highest levels during the initial periods under commitment strategies, whereas they become the highest during the intermediate periods under contingent strategies. These impacts can be more prominent under lower f, where firms have relatively low incentives to reduce emissions.

The numerical results show that the market configuration with lower sales prices generally yields larger abatement. For instance, HCVI which exhibits the lowest sales price yields the largest abatement efforts. This relationship is intuitive in that lower sales prices enlarge the sales demand and thus gross emissions volume, which necessitates more abatement efforts. However, whether the largest abatement efforts in HCVI are driven by the manufacturer's incentive for abatement for a given demand level, or whether they are merely driven by the largest volume is not straightforward.

We thus introduce the abatement rate (ξ), which measures the average abatement efforts per demand, i.e., $\xi_i = \int_0^T v_i(t)dt / \int_0^T D_i(t)dt$, to examine the impacts of competition on the manufacturer's incentives for abatement efforts more precisely. Table 6.7 shows that $\xi_i^{HIVC} > \xi_i^{TI} > \xi_i^{TC} > \xi_i^{HCVI}$ holds under either strategy type in the base case. In all ranges of parameter values, it also holds that ξ_i^{HIVC} is the largest and ξ_i^{HCVI} is the smallest. This implies that when the demand level is taken into account, the manufacturers have the highest abatement incentive in HIVC and the lowest incentive in HCVI.

More specifically, horizontal competition leads to lower abatement rates for a given vertical structure. Since the pollution externalities are not internalized by competitive manufacturers, the manufacturers' abatement incentives decrease accordingly. On the other hand, vertical integration leads to lower abatement rates than vertical competition for a given horizontal configuration. A possible explanation is that vertically integrated chains can control pricing and abatement decisions simultaneously, thus having opportunities to partially replace costly abatement efforts with adjustment of demand.

As horizontal competition and vertical integration lead to relatively lower abatement rates respectively, the degree of net emissions for a given demand increases, implying a "horizontally-induced polluting effect" and "vertically-induced polluting effect." Table 6.8 shows that contrary to volume effects, the horizontally-induced polluting effect is generally greater than the vertically-induced polluting effect. That is, the polluting effect depends more on horizontal market configurations. For instance, while horizontal competition reduces ξ_i by 0.4~0.7% compared to horizontal integration over the range of parameter α, vertical integration reduces ξ_i by 0.1~0.2% compared to vertical competition. The polluting effects become more evident under lower φ, lower f and higher ψ, implying that difficulties in abatement and the government pressures influence the degree of the polluting effects. Different from volume effects under which the effects of strategy types depend on market configuration, polluting effects, either horizontally-induced or vertically-induced, are generally intensified under contingent strategies, which result in relatively lower abatement rates.

The Environmental Performance

Then, what are the comprehensive impacts of competition on the final pollution stock considering the pricing and abatement strategies? Table 6.9 shows that $y^{HIVC}(T) < y^{TC}(T) < y^{TI}(T) < y^{HCVI}(T)$ holds in the base case. The numerical analysis also confirms that the final pollution stock is the largest in HCVI, the smallest in HIVC, and intermediate in TI and TC for all ranges of parameter values regardless of strategy types, which also supports Theorems 7–10. While the TI case generally results in the larger $y(T)$ than the TC, the opposite may also hold under small β or large γ where the competitive intensity is relatively large and thus the impact of horizontal internalization becomes influential. These results imply that horizontal competition and vertical integration result in the larger terminal pollution stock respectively as a result of the volume effects and the polluting effects.

One important question remaining here is whether the environmentally detrimental effect of horizontal competition or vertical integration remain even after its higher

Table 6.8 Impacts of competition on the abatement rates

Impact		Strategy	α	β	γ	φ	f	ψ
Horizontal competition vs. horizontal integration	Under VI: ξ_i^{HCVI}/ξ_i^{TI}	OLNE	[0.996,0.996]	[0.996,0.995]	[0.996,0.996]	[0.994,0.998]	[0.988,0.998]	[0.996,0.996]
		FBNE	[0.993,0.993]	[0.994,0.991]	[0.993,0.994]	[0.989,0.997]	[0.978,0.997]	[0.994,0.992]
	Under VC: ξ_i^{TC}/ξ_i^{HIVC}	OLNE	[0.996,0.996]	[0.996,0.996]	[0.996,0.996]	[0.995,0.998]	[0.988,0.998]	[0.996,0.996]
		FBNE	[0.994,0.994]	[0.994,0.994]	[0.994,0.994]	[0.991,0.997]	[0.981,0.997]	[0.994,0.994]
Vertical integration vs. vertical competition	Under HI: ξ_i^{TI}/ξ_i^{HIVC}		[0.999,0.999]	[0.999,0.995]	[0.999,1.000]	[0.997,1.000]	[0.997,1.000]	[1.000,0.997]
	Under HC: ξ_i^{HCVI}/ξ_i^{TC}	OLNE	[0.999,0.999]	[0.999,0.993]	[0.998,1.000]	[0.996,1.000]	[0.997,0.999]	[1.000,0.996]
		FBNE	[0.998,0.998]	[0.999,0.992]	[0.998,0.999]	[0.995,1.000]	[0.995,0.999]	[1.000,0.995]

VI vertical integration, *VC* vertical integration, *HI* horizontal integration, *HC* horizontal competition

Table 6.9 The environmental performance in the base case

	TI	TC (OLNE)	TC (FBNE)	HIVC	HCVI (OLNE)	HCVI (FBNE)
$y(T)$	47.67	36.88	42.10	23.74	75.68	91.43
PI	0.049	0.062	0.073	0.044	0.068	0.082
AI	6.81	5.30	4.47	7.52	4.85	4.00

demand is taken into account. We thus examine the pollution intensity (*PI*), which is measured as the terminal pollution stock per unit of cumulative demand, i.e., $PI = y(T)/\sum_i \int_0^T D_i(t)$, between the four configurations. Table 6.9 shows that $PI^{HIVC} < PI^{TI} < PI^{TC} < PI^{HCVI}$ holds for a given strategy type in the base case. The numerical analysis for all ranges of parameter values also confirms that for a given strategy type, the pollution intensity is the highest in HCVI and the lowest in HIVC. In terms of the intermediate ranks, TC leads to larger pollution intensity than TI in most cases, except for $\beta \geq 19.5$. These results thus confirm the environmental demerits of horizontal competition and vertical integration.

When horizontal competition prevails, the contingent strategies result in the larger terminal pollution stock and the larger pollution intensity compared to the commitment strategies, because manufacturers behave more opportunistically when they are able to update their abatement decisions. Compared to TC, the polluting effect of contingent strategies is stronger under HCVI because of lower abatement, and also lower price, hence more gross emissions. Thus, contingent strategies differ from commitment strategies in that they lead both to higher emissions and greater savings on abatement efforts, but also to higher pollution costs.

Table 6.9 shows that the abatement intensity, denoted by *AI* and measured as cumulative abatement efforts per unit of terminal pollution stock, i.e., $AI = \sum_i \int_0^T v_i(t)/y(T)$, has the following order, i.e., $AI^{HIVC} > AI^{TI} > AI^{TC} > AI^{HCVI}$, for a given strategy type. As expected, it is the same as the one for abatement rates and the opposite to the one for pollution intensity, supporting the horizontally-induced and vertically-induced polluting effects. The sensitivity analysis shows that the abatement intensity is the largest in HIVC and the smallest in HCVI under FBNE strategies for all parameter ranges. In between, TI generally has higher abatement intensity than TC except for $\beta \geq 19.5$.

Supply Chain Profits

In this subsection, we discuss the effects of competition and strategy types on supply chain profits. Let J, J^m, J^r denote the supply chains', the manufacturers', and the retailers' total profits respectively. In the base case, $J^{TI} > J^{HCVI} > J^{TC} > J^{HIVC}$ holds for both strategy types (Table 6.10).

It is straightforward that the TI case, where all firms centralize their decisions, yields the largest supply chains' profits. Apart from TI, the sensitivity analysis shows that the orders between HCVI, TC, and HIVC depend on the parameter values. For instance, $J^{TI} > J^{HCVI} > J^{TC} > J^{HIVC}$ holds for $0 \leq \gamma < 2.2$, $J^{TI} > J^{TC} > J^{HCVI} > J^{HIVC}$ for $2.2 \leq \gamma < 3$, and $J^{TI} > J^{TC} > J^{HIVC} > J^{HCVI}$ for $\gamma \geq 3$.

Table 6.10 Supply chain profits in the base case

	TI	TC (OLNE)	TC (FBNE)	HIVC	HCVI (OLNE)	HCVI (FBNE)
J	12,294	10,361	9904.3	9798.8	12,064	10,695
J^m	8389.3	7422.6	7165.8	6755.9	8274.0	7561.1
J^r	3905.1	2938.4	2738.6	3042.8	3789.8	3133.9

It is noteworthy that the impacts of vertical configurations on supply chains' profits significantly depend on the horizontal configurations, and vice versa. To be specific, horizontal competition results in lower supply chains' profits than horizontal integration due to intense price competition when the firms are vertically integrative. Similarly, when the firms are horizontally integrative, vertical competition results in lower supply chains' profits than vertical integration due to double marginalization. However, such economic impacts of horizontal (vertical) competition do not necessarily hold under vertical (horizontal) competition. For instance, the TC case mostly leads to larger supply chains' profits than the HIVC case because the demand-enhancing effect of lower prices under horizontal competition becomes significant when double marginalization exists. The TC case may also yield higher profits than the HCVI case e.g., under high γ, implying that lower emissions under vertical competition can enhance the economic performance when price competition is intense.

Another observation is that for a given configuration, the contingent strategy results in lower supply chains' profits than the commitment strategy for all ranges of parameter values. That is, commitment strategy has not only environmental virtue as discussed in the preceding section, but also economic merit to the entire supply chains, especially under HCVI. Note also that contingent strategies are less beneficial to both manufacturers and retailers than commitment strategies in the HCVI and TC cases. This suggests that whenever contingent strategies are possible options, their related tradeoff, which is characterized by more emissions with less abatement and greater pollution costs than in commitment strategy, is detrimental to all firms.

The manufacturers' profits (J^m) and the retailers' profits (J^r) are the highest under the TI case where the Nash bargaining scheme (NBS) (see Dockner et al., 2000) is used to compute J^m and J^r based on those obtained from the TC case under the commitment strategy. While the manufacturers' profits are at their lowest in the HIVC case, the retailers' profits may be at their lowest in the TC case (computed with an NBS based on the TC case). That is, the HIVC case, which is the least polluting, is generally less beneficial to manufacturers than to retailers.

Conclusions

This research has shown that horizontal and vertical market structure has a critical influence on pollution accumulation by analyzing four market configurations using differential game models.

Our results provide several insights with regard to the environmental impacts of market structures. On the one hand, vertical competition is less pollution-intensive and more abatement-intensive than vertical integration. On the other hand, horizontal competition is more pollution-intensive and less abatement-intensive than horizontal integration. The major mechanisms underlying these impacts of competition are the volume effects and the polluting effects, both are vertically as well as horizontally-induced. The four configurations in Table 6.11 can be summarized as follows:

- Both vertical and horizontal internalization of the pollution externalities (TI) solely involve vertically-induced "volume effect" and "polluting effect";
- Neither vertical nor horizontal internalization of the pollution externalities (TC) exclusively involve horizontally-induced "volume effect" and "polluting effect";
- Horizontal internalization of the pollution externalities (HIVC) involves neither "volume effect" nor "polluting effect";
- Vertical internalization of the pollution externalities (HCVI) involves both vertically and horizontally induced "volume effect" and "polluting effect."

The vertically-induced volume effect is relatively larger than the horizontally-induced volume effect. That is, an increase in gross emissions in the market is

Table 6.11 Main differences between configurations

Bilateral duopoly	Vertical integration	Vertical competition
Horizontal integration	– Intermediate consumer prices, demands, abatement efforts, and abatement rates – Intermediate terminal pollution stock – Intermediate pollution and abatement intensive – Highest supply chain profit	– Highest consumer prices, lowest demands, and abatement efforts, highest abatement rates – Lowest terminal pollution stock – Least pollution-intensive, most abatement-intensive – Lowest (intermediate) supply chain profit
Horizontal competition	– Lowest consumer prices, highest demands, and abatement efforts, lowest abatement rates – Greatest terminal pollution stock – Most pollution-intensive, least abatement-intensive – Intermediate (lowest) supply chain profit – Greater terminal pollution stock and lower profits in contingent than in commitment strategy	– Intermediate consumer prices, demands, abatement efforts, and abatement rates – Intermediate terminal pollution stock – Intermediate pollution and abatement intensive – Intermediate supply chain profit – Greater terminal pollution stock and lower profits in contingent than in commitment strategy

derived more from the elimination of double marginalization than the introduction of competition in the market. On the contrary, the vertically-induced polluting effect is generally smaller than the horizontally-induced polluting effect, implying that the competitive pressures in the market weaken the manufacturers' abatement incentives more significantly than the vertical internalization of pollution externalities.

Strategy types also play a significant role in the context of horizontal competition. Contingent strategies result in less environment-friendly configurations under horizontal competition. The environmental demerit of contingent strategies arises mainly due to the reduced abatement incentives. The impact of contingent strategies on the gross emissions volume is modified by vertical configuration: while contingent strategies lead to smaller emissions volume under vertical competition, they result in larger emissions volume under vertical integration.

Our analysis provides valuable guidance for managers as follows.

First, we offer guidance on the profit-maximizing pricing and abatement policies which depend on the market structures and strategy types. For instance, managers need to invest relatively heavily in abatement efforts for a given output and to set a relatively higher price, thus reducing the output level, if they operate under horizontally integrated or vertically competitive markets. At the same time, managers need to be aware that abatement and pricing policies must be revised accordingly depending on strategy types. If commitment strategies are adopted, it is advisable to increase abatement intensity, but whether to raise sales price compared to under contingent strategies depends further on market configurations. Additionally, while managers need to invest in abatement efforts most heavily during intermediate periods under contingent strategies, managers must invest actively from the beginning under commitment strategies.

Second, the growing adoption of commitment strategies in practice (e.g., Shell, BP, Gap, Nike, etc.) can be not only environmentally but also economically advantageous in the horizontally competing industries. In this context, commitment strategies serve as a measure to alleviate negative effects due to the imperfect internalization of intertemporal pollution externalities. However, in horizontally integrative industries, firms are indifferent between commitment and contingent strategies because horizontal internalization of pollution externalities and substitution effects suffice to eliminate incentives of firms to deviate from the optimal paths under commitment strategies.

Third, managers should not overlook the environmental dimension when formulating strategies that influence structural configurations in the industry. For instance, while vertical integration with supply chain partners can eliminate double marginalization, it may result in lower profits due to the increased pollution burden. Our analysis also suggests that horizontal integration, e.g., Chevron's takeover of Anadarko, in industries, can lead to not only lower gross emissions but also stronger abatement incentives for firms, contradicting the argument that horizontal integration may be "just a temporary solution" for curbing emissions and results in lower greening investments (Luo et al., 2016).

Our study also offers policy advice for regulators to curb pollution emissions in industries.

First, regulators may differentiate their focus in employing policies to curb firms' emissions depending on market structures. Specifically, as polluting effects are more induced by horizontal configurations, policies to incentivize firms' abatement efforts can be preferentially implemented in horizontally competitive markets. On the other hand, as volume effects are more driven by vertical configurations, regulators may monitor whether vertically integrated chains offer excessively low prices, which cause large gross emissions as a byproduct.

Second, regulators may promote commitment strategies in horizontally competitive markets, which lead to lower pollution emissions. One indirect way is to limit the information on pollution. Although regulators typically pursue higher visibility regarding this matter, e.g., Canada's Greenhouse Gas Reporting Program publicizing information on the pollution of each facility, our results suggest that regulators may strategically restrict information visibility on the pollution based upon which firms can employ contingent strategies and revise their actions continuously.

Third, the least polluting configuration, horizontal integration and vertical competition, is more unfavorable to manufacturers than retailers. Thus, regulators can focus more on providing sufficient incentives for upstream firms to coordinate their decisions, to reduce environmental damage in industries.

A possible avenue for future research consists of building an empirical test to assess the relative importance of vertically and horizontally induced volume and polluting effects in the industry.

Appendix

Table 6.12 Profits under different cases and strategy types. Plugging the derived abatement and pricing strategies gives profits under commitment strategies and the TI case. The payoff functions are obtained from the value functions directly under contingent strategies. The detailed proofs are omitted for brevity

Case	Strategy	Profits
TI		$J=\frac{\alpha^2}{((\beta-\gamma)\psi^2+2\varphi^2)}\left[\frac{\varphi^2 T}{\beta-\gamma}+\frac{\psi^2\mathrm{th}(\theta T)}{2\theta}\right]$
TC	Commitment	$J_{ol}^{m_i}=\frac{\alpha^2\beta T^2}{2\left\{\beta(\beta-\gamma)\psi^2\left[\rho T\left(2-\mathrm{sech}(\rho T)^2\right)-\mathrm{th}(\rho T)\right]+4\varphi^2\rho(2\beta-\gamma)T\right\}^2}$ $\left\{\frac{4\varphi^3\sqrt{2f}\left\{\beta(\beta-\gamma)\psi^2\left[\rho T\left(2-\mathrm{sech}(\rho T)^2\right)-\mathrm{th}(\rho T)\right]+2\varphi^2\rho(2\beta-\gamma)T\right\}}{(\beta-\gamma)}+\beta\varphi^2\psi^2\mathrm{sech}(\rho T)^2[\rho\mathrm{sh}(2\rho T)-2\rho^2 T\mathrm{ch}(2\rho T)]\right\}$ $J_{ol}^{r_i}=\frac{4\alpha^2\beta\varphi^4\rho^2T^3}{\left\{\beta(\beta-\gamma)\psi^2\left[\rho T\left(2-\mathrm{sech}(\rho T)^2\right)-\mathrm{th}(\rho T)\right]+4\varphi^2\rho(2\beta-\gamma)T\right\}^2}$
	Contingent	$J_{fb}^{m_i}=\frac{\beta\left[\alpha-(\beta-\gamma)\omega_{ifb}^{TC}\right]}{3\varphi^2\tau\ (2\beta-\gamma)^2}\left\{\left[\left(2\beta^2\psi^2-3\gamma\varphi^2-2\beta\gamma\psi^2+6\beta\varphi^2\right)\omega_{ifb}^{TC}-2\alpha\beta\psi^2\right]\tau T+2\beta\left[\alpha-(\beta-\gamma)\omega_{ifb}^{TC}\right]\ \mathrm{th}(\tau T)\psi^2\right\}$ $J_{fb}^{r_i}=\frac{\beta\left[\alpha-(\beta-\gamma)\omega_{ifb}^{TC}\right]^2 T}{(2\beta-\gamma)^2}$
HIVC	Commitment & contingent	$J_{ol}^{m}=\frac{\delta T^2\alpha^2\varphi^2}{(\beta-\gamma)(\delta T(4\varphi^2+(\beta-\gamma)\psi^2)-(\beta-\gamma)\psi^2\mathrm{th}(\delta T))}$ $J_{ol}^{r}=\frac{2\alpha^2\delta^2\varphi^4T^3}{(\beta-\gamma)[((\beta-\gamma)\psi^2+4\varphi^2)\delta T-(\beta-\gamma)\psi^2\mathrm{th}(\delta T)]^2}$
HCVI	Commitment	$J_{ol}^{i}=\frac{1}{8\vartheta}\left\{\varphi^2\kappa^2\mathrm{sech}(\vartheta T)^2[3\mathrm{sh}(2\vartheta T)-2\vartheta(2+\mathrm{ch}(2\vartheta T))T]+\frac{\kappa^2\vartheta^2\mathrm{sech}(\vartheta T)^2(2\vartheta T-\mathrm{sh}(2\vartheta T))}{f}\right.$ $\left.+\frac{8e^{2\vartheta T}\left[2\vartheta T\left(\alpha^2\beta-2(\beta-\gamma)^3\kappa^2\psi^2-\alpha\gamma(\beta-\gamma)\kappa\psi\right)+2\vartheta T\left(\alpha\beta+(\beta-\gamma)^2\kappa\psi\right)(\alpha-(\beta-\gamma)\kappa\psi)\mathrm{ch}(2\vartheta T)+(\beta-\gamma)\psi\left(3(\beta-\gamma)^2\kappa\psi+2\alpha\gamma\right)\kappa\mathrm{sh}(2\vartheta T)\right]}{(1+e^{2\vartheta T})^2(2\beta-\gamma)^2}\right\}$ where $\kappa=-\frac{\alpha\beta\psi}{(\beta-\gamma)^2\psi^2+\varphi^2(2\beta-\gamma)}$
	Contingent	$J_{fb}^{i}=\frac{\alpha^2 f}{2\sigma^3(2\beta-\gamma)}\left\{\sigma T(2\beta-\gamma)\left[6\beta\varphi^2-(\beta+\gamma)^2\psi^2\right]+\left(4\beta^2-3\beta\gamma+\gamma^2\right)^2\psi^2\ \mathrm{th}\left(\frac{\sigma T}{2\beta-\gamma}\right)\right\}$

Proof of Theorem 1.
Skipping the time index for convenience, the corresponding joint objective criterion is:

$$\underset{p_1,p_2,v_1,v_2\geq 0}{\text{Max}} J=\sum_i (J^{m_i}+J^{r_i})=\int_0^T \left[\sum_{i\neq j} p_i(\alpha-\beta p_i+\gamma p_j)-\frac{\sum_i v_i^2}{2}-fy^2\right]dt$$

subject to Eq. 6.5. The Hamiltonian in the total integration case is written as:

$$H=\sum_{i\neq j} p_i(\alpha-\beta p_i+\gamma p_j)-\frac{\sum_i v_i^2}{2}-fy^2+\eta\left[\psi\sum_{i\neq j}(\alpha-\beta p_i+\gamma p_j)-\varphi\sum_i v_i\right]$$

from which necessary conditions for optimality with symmetric controls are,

$$\dot{\eta}=2fy, H_{v_i}=-v_i-\varphi\eta=0\Rightarrow v_i=-\varphi\eta, H_{p_i}=\alpha-2\beta p_i+2\gamma p_j-(\beta-\gamma)\eta\psi=0$$
$$\Rightarrow p_i=\frac{\alpha-(\beta-\gamma)\eta\psi}{2(\beta-\gamma)}, i=1,2, i\neq j.$$

The Hamiltonian is concave in (v_1, v_2, p_1, p_2) because the Hessian is definitely negative and therefore the Legendre-Clebsch condition is satisfied (Grass et al., 2008). However, it can be shown that the necessary conditions are also sufficient for optimality, which implies that the optimal solution is unique. The canonical system then writes $\dot{\eta}=2fy$, $\dot{y}=\alpha\psi+((\beta-\gamma)\psi^2+2\varphi^2)\eta$, which gives a general solution $\eta(t)=A_{11}e^{m_{11}t}+A_{12}e^{m_{12}t}+K_{11}$, where $m_{11}=\sqrt{2f((\beta-\gamma)\psi^2+2\varphi^2)}$, $m_{12}=-m_{11}$, $K_{11}=-\frac{\alpha\psi}{((\beta-\gamma)\psi^2+2\varphi^2)}$. Also, from $y(0)=0$ and $\eta(T)=0$, we have $A_{11}=\frac{-m_{11}K_{11}+2fy_0e^{-m_{11}T}}{m_{11}(e^{m_{11}T}+e^{-m_{11}T})}$ and $A_{12}=-\frac{m_{11}K_{11}+2fy_0e^{m_{11}T}}{m_{11}(e^{m_{11}T}+e^{-m_{11}T})}$. Plugging the derived $\eta(t)$ completes the proof. Furthermore, from Eq. 6.7, it can be shown that $p_i(t)\leq\frac{\alpha}{\beta-\gamma}$, $i=1, 2$, which guarantees that the control constraints $p_i(t)\leq(\alpha+\gamma p_j(t))/\beta$ are never binding and therefore the retailers' demands are nonnegative. Strictly and convexly increasing pollution stock ensures that the control constraint of $v_i^{TI}(t)$ is never binding and therefore that the manufacturers' net current emissions (i.e., current emissions minus abatement efforts) are nonnegative. □

Proof of Theorem 2. In the total competition case, the respective Hamiltonians write:

$$H^{m_i}=\omega_i(\alpha-\beta p_i+\gamma p_j)-\frac{v_i}{2}-\frac{fy^2}{2}+\lambda_i\left[\psi\sum_{i\neq j}(\alpha-\beta p_i+\gamma p_j)-\varphi\sum_i v_i{}'\right]$$
$$H^{r_i}=(p_i-\omega_i)(\alpha-\beta p_i+\gamma p_j)+\mu_i\left[\psi\sum_{i\neq j}(\alpha-\beta p_i+\gamma p_j)-\varphi\sum_i v_i\right]$$

$i = 1, 2$, $i \neq j$, from which the equations of the costate variables are given by $\dot{\lambda}_i = -H_y^{m_i} = fy$, $\dot{\mu}_i = -H_y^{r_i} = 0$. Given $\dot{\mu}_i = 0$ and the transversality condition $\mu_i(T) = 0$, we get $\mu_i(t) = 0$, $i = 1, 2$. Manufacturer i's and retailer i's necessary conditions for optimality yield respectively $H_{v_i}^{m_i} = 0 \rightarrow v_i = -\varphi\lambda_i$, $H_{p_i}^{r_i} = 0 \rightarrow p_i = \frac{\alpha + \gamma p_j + \beta\omega_i}{2\beta}$. It can be shown that the necessary conditions are also sufficient for optimality, which implies that the equilibrium is unique. Since $\lambda_i = \lambda_j = \lambda$, the manufacturers' controls are symmetric, that is $v_i = v_j = -\varphi\lambda$. Also, $\omega_i = \omega_j = \omega$, for such symmetric manufacturers and therefore we get $p_i = \frac{\alpha + \beta\omega}{2\beta - \gamma}$, which is constant over time. Therefore, the canonical system writes $\dot{\lambda} = fy$, $\dot{y} = 2\left[\psi\left(\alpha - \frac{(\beta - \gamma)(\alpha + \beta\omega)}{2\beta - \gamma}\right) + \varphi^2\lambda\right]$, which has a general solution $\lambda(t) = A_{31}e^{m_{31}t} + A_{32}e^{m_{32}t} + K_{31}$ with $m_{31} = \varphi\sqrt{2f}$, $m_{32} = -m_{31}$ and $K_{31} = -\frac{\beta\psi[\alpha - (\beta - \gamma)\omega]}{\varphi^2(2\beta - \gamma)}$. $A_{31} = A_{32} = \frac{-K_{31}}{(e^{m_{31}T} + e^{-m_{31}T})}$ are derived from $y(0) = 0$ and $\lambda(T) = 0$. Plugging the expressions and simplifying gives

$$J^{m_i} = \frac{\beta[\alpha - (\beta - \gamma)\omega]}{2\beta - \gamma}\left\{\omega T + \frac{\beta[\alpha - (\beta - \gamma)\omega]\psi^2}{4m_{31}(2\beta - \gamma)\varphi^2}\left[-2m_{31}T + m_{31}T\,\mathrm{sech}(m_{31}T)^2 + \mathrm{th}(m_{31}T)\right]\right\}$$

$i = 1, 2$. We derive the right-hand side of J^{m_i} with respect to ω and solve to obtain, using $\rho \equiv \varphi\sqrt{2f}$:

$$\omega = \frac{\left\{\beta(\beta - \gamma)\psi^2\left[\left(2 - \mathrm{sech}(\rho T)^2\right)\rho T - \mathrm{th}(\rho T)\right] + 2\varphi^2\rho(2\beta - \gamma)T\right\}}{\left\{\beta(\beta - \gamma)\psi^2\left[\left(2 - \mathrm{sech}(\rho T)^2\right)\rho T - \mathrm{th}(\rho T)\right] + 4\varphi^2\rho(2\beta - \gamma)T\right\}}\frac{\alpha}{(\beta - \gamma)}$$

Plugging the expression of ω and simplifying completes the proof. Furthermore, from Eq. 6.9, it is obvious that the manufacturer i's OLNE transfer price, ω_{iol}^{TC}, is strictly positive and lower than $\frac{\alpha}{\beta - \gamma}$, $i = 1, 2$, which implies, from Eq. 6.11, that the retailers' demands are nonnegative. Strictly increasing pollution stock ensures that the manufacturers' net current emissions are nonnegative. □

Proof of Theorem 3. Since the game is symmetric, we assume that the optimal value functions are the same for the manufacturers and for the retailers, that is, $V^{m_i} = V^{m_j} = V^m$ and $V^{r_i} = V^{r_j} = V^r$. In this regard, $v_i = v_j$, we assume symmetric transfer prices such that $\omega_i = \omega_j = \omega$. We first construct the Hamilton-Jacobi-Bellman (HJB) equations:

$$-\dot{V}^m = \mathrm{Max}_{v_i}\left\{\omega(\alpha - \beta p_i + \gamma p_j) - \frac{v^{i2}}{2} - \frac{fy^2}{2} + V_y^m\left[\psi\sum\nolimits_i(\alpha - \beta p_i + \gamma p_j) - \varphi\sum\nolimits_i v_i\right]\right\}$$

$$-\dot{V}^r = \mathrm{Max}_{p_i}\left\{(p_i - \omega)(\alpha - \beta p_i + \gamma p_j) + V_y^r\left[\psi\sum\nolimits_i(\alpha - \beta p_i + \gamma p_j) - \varphi\sum\nolimits_i v_i\right]\right\}$$

$i = 1, 2, i \neq j$. The maximization on the right-hand side of the HJB equations with respect to v_i and p_i gives $v_i = -\varphi V_y^m$ and $p_i = \frac{\alpha + \gamma p_j + \beta\omega - (\beta - \gamma)\psi V_y^r}{2\beta}$. Since $V^{m_i} = V^{m_j} = V^m$, $v_i = v_j$, and then $p_i = p_j = \frac{\alpha + \beta\omega - \psi(\beta - \gamma)V_y^r}{2\beta - \gamma}$. We make the conjectures for the value functions: $V^m(y, t) = A(t)y^2 + B(t)y + C(t)$, $V^r(y, t) = D(t)y + E(t)$. To ensure that the HJB equations and the boundary conditions hold for all $y \geq 0$ and all t, after substituting the conjectured value functions into the HJB equations, the following set of equations should be met:

$$\dot{A} = \frac{f}{2} - 6\varphi^2 A^2, \quad \dot{B} = -\frac{2A\left(3B(2\beta - \gamma)\varphi^2 + 2\psi\left(\beta(\alpha - \omega(\beta - \gamma)) + D(\beta - \gamma)^2\psi\right)\right)}{2\beta - \gamma}$$

$$\dot{C} = -\frac{2\omega\alpha\beta - 2\omega^2\beta(\beta - \gamma) - 2\omega(\beta - \gamma)(2B\beta - D(\beta - \gamma))\psi + B\left(3B(2\beta - \gamma)\varphi^2 + 4\psi\left(\alpha\beta + D(\beta - \gamma)^2\psi\right)\right)}{2(2\beta - \gamma)}$$

$$\dot{D} = -4\varphi^2 AD, \dot{E} = -\left(2BD\varphi^2 + \frac{\left(\beta(\alpha - \omega(\beta - \gamma)) + D(\beta - \gamma)^2\psi\right)(\alpha - \omega(\beta - \gamma) + D(3\beta - \gamma)\psi)}{(2\beta - \gamma)^2}\right)$$

with $A(T) = B(T) = C(T) = 0$ and $D(T) = E(T) = 0$. Solving the differential equations and plugging the solutions yields Eqs. 6.14–6.16. Given that $y_0 = 0$, the payoff functions rewrite:

$$J^{m_i} = V^m(y_0, 0) = \beta[\alpha - (\beta - \gamma)\omega]\{2\beta\sqrt{3}$$

$$\frac{[\alpha - (\beta - \gamma)\omega] \operatorname{th}\left(\varphi T\sqrt{3f}\right)\psi^2 - 3\varphi T\sqrt{f}\left[2\alpha\beta\psi^2 - \left(2\beta^2\psi^2 - 3\gamma\varphi^2 - 2\beta\gamma\psi^2 + 6\beta\varphi^2\right)\omega\right]\}}{9\varphi^3(2\beta - \gamma)^2\sqrt{f}},$$ and

$J^{r_i} = V^r(y_0, 0) = \frac{\beta[\alpha - (\beta - \gamma)\omega]^2 T}{(2\beta - \gamma)^2}$. Deriving J^{m_i} with respect to ω gives Eq. 6.13, which completes the proof. Furthermore, from Eq. 6.13, the manufacturer i's transfer price, ω_{ifb}^{TC}, appears to be strictly positive and lower than $\frac{\alpha}{\beta - \gamma}$, $i = 1, 2$, which implies, from Eq. 6.15, retailers' nonnegative demands. Using Eq. 6.14 in Eq. 6.16, the pollution stock is strictly increasing over time because the condition $y_{fb}^{TC}(t) < \frac{\beta\psi\left[\alpha - (\beta - \gamma)\omega_{ifb}^{TC}\right]\{2 + \operatorname{ch}[\tau(T - t)]\}}{\tau(2\beta - \gamma)\operatorname{sh}[\tau(T - t)]}$ holds $\forall t \in [0, T]$. This ensures that manufacturers' net current emissions are nonnegative. □

Proof of Theorem 4.

In the horizontal integration and vertical competition case, the corresponding objective criteria are:

$$\underset{\omega_1, \omega_2, v_1, v_2 \geq 0}{\text{Max}} J^m = \int_0^T \left[\sum\nolimits_{i \neq j} \omega_i(\alpha - \beta p_i + \gamma p_j) - \frac{\sum_i v_i^2}{2} - fy^2\right] dt$$

$$\underset{p_1,p_2\geq 0}{\text{Max}}\ J^r = \int_0^T \sum\nolimits_{i\neq j}(p_i-\omega_i)\left(\alpha-\beta p_i+\gamma p_j\right)\mathrm{d}t$$

subject to Eq. 6.5. The respective Hamiltonians write:

$$H^m = \sum\nolimits_{i\neq j}\omega_i\left(\alpha-\beta p_i+\gamma p_j\right)-\frac{\sum_i v_i^2}{2}-fy^2+\lambda\left[\psi\sum\nolimits_i\left(\alpha-\beta p_i+\gamma p_j\right)-\varphi\sum\nolimits_i v_i\right]$$

$$H^r = \sum\nolimits_{i\neq j}(p_i-\omega_i)\left(\alpha-\beta p_i+\gamma p_j\right)+\mu\left[\psi\sum\nolimits_i\left(\alpha-\beta p_i+\gamma p_j\right)-\varphi\sum\nolimits_i v_i\right]$$

$i = 1, 2, i \neq j$, from which the equations of the costate variables are given by $\dot{\lambda} = -H_y^m = 2fy$, $\dot{\mu} = -H_y^r = 0$, which yields $\mu(t) = 0$ given the transversality condition $\mu(T) = 0$. Manufacturer's and retailer's necessary conditions for optimality are respectively $H_{v_i}^m = 0 \rightarrow v_i = -\varphi\lambda$, $H_{p_i}^r = 0 \rightarrow p_i = \frac{\alpha+2\gamma p_j+\beta\omega_i-\gamma\omega_j}{2\beta}$. The Hamiltonians H^m and H^r are concave in (v_i, v_j) and (p_i, p_j) respectively, because the Hessians are definite negative and therefore the Legendre-Clebsch is satisfied (Grass et al., 2008). It can be shown that the necessary conditions are also sufficient for optimality, which implies that the equilibrium is unique. Manufacturers' controls are symmetric, that is, $v_i = v_j = v$. Also, $\omega_i = \omega_j = \omega$, for such symmetric manufacturers and therefore we get $p_i = \frac{\alpha+(\beta-\gamma)\omega}{2(\beta-\gamma)}$, which is constant over time. The canonical system writes $\dot{\lambda} = 2fy$, $\dot{y} = \psi(\alpha-(\beta-\gamma)\omega)+2\varphi^2\lambda$, which has a general solution $\lambda(t) = A_{41}e^{m_{41}t}+A_{42}e^{m_{42}t}+K_{41}$ with $m_{41} = 2\varphi\sqrt{f}$, $m_{42} = -m_{41}$ and $K_{41} = -\frac{[\alpha-(\beta-\gamma)\omega]\psi}{2\varphi^2}$. Also, $y(0) = y_0$ and $\lambda(T) = 0$ gives $A_{41} = \frac{-m_{41}K_{41}+2fy_0e^{-m_{41}T}}{m_{41}\left(e^{m_{41}T}+e^{-m_{41}T}\right)}$ and $A_{42} = -\frac{m_{41}K_{41}+2fy_0e^{m_{41}T}}{m_{41}\left(e^{m_{41}T}+e^{-m_{41}T}\right)}$. Plugging the expressions and simplifying gives $J^m = [\alpha-(\beta-\gamma)\omega]\left\{\omega T-\frac{[\alpha-(\beta-\gamma)\omega]\psi^2}{4m_{41}\varphi^2}[m_{41}T-\mathrm{th}(m_{41}T)]\right\}$. Deriving J^m with respect to ω and solving for ω gives Eq. 6.17. Plugging $\lambda(t)$ and ω completes the proof. Furthermore, from Eq. 6.17, the manufacturer i's OLNE transfer price, ω_{iol}^{HIVC}, is strictly positive and lower than $\alpha/(\beta-\gamma)$, $i = 1, 2$, which, using Eq. 6.19, implies retailers' nonnegative demands. The pollution stock only increases over time, which ensures manufacturers' nonnegative net current emissions. □

Proof of Lemma 1. We first derive the FBNE equilibrium. The Hamilton-Jacobi-Bellman (HJB) equations are respectively:

$$-\dot{V}^m = \text{Max}_{v_i,v_j}\left\{\sum\omega_i\left(\alpha-\beta p_i+\gamma p_j\right)-\frac{\sum v_i^2}{2}-fy^2+V_y^m\left[\psi\sum\nolimits_i\left(\alpha-\beta p_i+\gamma p_j\right)-\varphi\sum\nolimits_i v_i\right]\right\}$$

$$-\dot{V}^r = \text{Max}_{p_i}\left\{\sum(p_i-\omega_i)\left(\alpha-\beta p_i+\gamma p_j\right)+V_y^r\left[\psi\sum\nolimits_i\left(\alpha-\beta p_i+\gamma p_j\right)-\varphi\sum\nolimits_i v_i\right]\right\}$$

$i = 1, 2, i \neq j$. The maximization on the right-hand side of the HJB equations with respect to v_i and p_i gives $v_i = v_j = -\varphi V_y^m$ and $p_i = \frac{\alpha+(\beta-\gamma)\omega_i-(\beta-\gamma)V_y^r\psi}{2(\beta-\gamma)}$. Since each manufacturer has a symmetric profit structure and $v_i = v_j$, we assume symmetric manufacturers such that $\omega_i = \omega_j = \omega$. We make the following conjectures for the value functions $V^m(y, t) = A(t)y^2 + B(t)y + C(t)$, $V^r(y, t) = D(t)y + E(t)$. To ensure that the HJB equations and the boundary conditions hold for all $y \geq 0$ and all t, after substituting the conjectured value functions into the HJB equations, the following set of equations should be met:

$$\dot{A} = f - 4A^2\varphi^2, \;\; \dot{B} = -4AB\varphi^2 + 2A\omega_i(\beta-\gamma)\psi - 2A\psi(\alpha + D(\beta-\gamma)\psi)$$

$$\dot{C} = -\alpha\omega_i + {\omega_i}^2(\beta-\gamma) - B^2\varphi^2 + B\omega_i(\beta-\gamma)\psi - D\omega_i(\beta-\gamma)\psi - B\psi(\alpha + D(\beta-\gamma)\psi)$$

$$\dot{D} = -4\varphi^2 AD, \dot{E} = -\frac{\alpha^2 + {\omega_i}^2(\beta-\gamma)^2 + 4BD(\beta-\gamma)\varphi^2 + 2D\omega_i(\beta-\gamma)(-\beta+\gamma)\psi + D^2(\beta-\gamma)^2\psi^2 - 2\alpha(\beta-\gamma)(\omega_i - D\psi)}{2(\beta-\gamma)}$$

with $A(T) = B(T) = C(T) = 0$ and $D(T) = E(T) = 0$. Solving the differential equations and using $J^{m_i} = V^m(y_0, 0)$ given that $y_0 = 0$, we obtain the equilibriums satisfying $\omega_{iol}^{HIVC}(t) = \omega_{ifb}^{HIVC}(t)$, $p_{iol}^{HIVC}(t) = p_{ifb}^{HIVC}(t)$, $v_{iol}^{HIVC}(t) = v_{ifb}^{HIVC}(t)$, which completes the proof. □

Proof of Theorem 5. The objective criteria in the horizontal competition and vertical integration case are:

$$\underset{p_i, v_i \geq 0}{\text{Max}} J^i = \int_0^T \left[p_i(\alpha - \beta p_i + \gamma p_j) - \frac{v_i^2}{2} - \frac{fy^2}{2}\right] dt$$

subject to Eq. 6.5, $i = 1, 2$. In an OLNE, the Hamiltonians write:

$$H^i = p_i(\alpha - \beta p_i + \gamma p_j) - v_i^2/2 - fy^2/2 + \pi_i\left[\psi \sum_i (\alpha - \beta p_i + \gamma p_j) - \varphi \sum_i v_i\right]$$

$i = 1, 2, i \neq j$, from which the equations of the costate variables are given by $\dot{\pi}_i = fy$ and the transversality conditions $\pi_i(T) = 0$. Because $\pi_i = \pi_j = \pi$, we can assume symmetric players. Necessary conditions for optimality are respectively $H^i_{v_i} = 0 \Rightarrow v_i = -\varphi\pi$, $H^i_{p_i} = 0 \Rightarrow p_i = [\alpha - (\beta-\gamma)\pi\psi]/(2\beta-\gamma)$. The Hamiltonians are concave in (v_i, p_i) since the Legendre-Clebsch conditions are satisfied. On the other hand, it can be shown that the necessary conditions are also sufficient for optimality, which implies that the optimal solution is unique. The canonical system then writes $\dot{\pi} = fy$, $\dot{y} = 2\left\{\beta\alpha\psi + \left[(\beta-\gamma)^2\psi^2 + \varphi^2(2\beta-\gamma)\right]\pi\right\}/(2\beta-\gamma)$, which yields a general solution $\pi(t) = A_{21}e^{m_{21}t} + A_{22}e^{m_{22}t} + K_{21}$, with $m_{21} = \sqrt{\frac{2f\left[(\beta-\gamma)^2\psi^2 + \varphi^2(2\beta-\gamma)\right]}{2\beta-\gamma}}$, $m_{22} = -m_{21}$ and $K_{21} = -\frac{\alpha\beta\psi}{\left[(\beta-\gamma)^2\psi^2 + \varphi^2(2\beta-\gamma)\right]}$. We obtain $A_{21} = A_{22} = \frac{-K_{21}}{(e^{m_{21}T} + e^{-m_{21}T})}$ from $y(0) = 0$ and $\pi(T) = 0$. Plugging the expressions and simplifying completes the proof. Furthermore, from Eq. 6.21, retailer i's OLNE consumer price, p_{iol}^{TC}, $i = 1, 2$, is strictly positive and lower than $\frac{\alpha}{\beta - \gamma}$, which ensures nonnegative retailers' demands. The pollution stock is strictly increasing over time, which guarantees nonnegative manufacturers' net emissions. □

Proof of Theorem 6. Since the game is symmetric, we assume that the optimal value functions are the same for each supply chain, that is, $V^i = V^j = V$. We first construct the Hamilton-Jacobi-Bellman (HJB) equations:

$$-\dot{V} = \underset{v_i, p_i}{\mathrm{Max}} \left\{ p_i(\alpha - \beta p_i + \gamma p_j) - \frac{v_i^2}{2} - \frac{fy^2}{2} + V_y \left[\psi \sum\nolimits_i (\alpha - \beta p_i + \gamma p_j) - \varphi \sum\nolimits_i v_i \right] \right\}$$

$i = 1, 2$, $i \neq j$. The maximization on the right-hand side of the HJB equations with respect to v_i and p_i gives $v_i = -\varphi V_y$ and $p_i = \{\alpha - (\beta - \gamma)V_y\psi\}/(2\beta - \gamma)$. We make the conjectures for the value functions: $V(y, t) = A(t)y^2 + B(t)y + C(t)$. To ensure that the HJB equations and the boundary conditions hold for all $y \geq 0$ and all t, after substituting the conjectured value functions into the HJB equations, the following set of equations should be met:

$$\dot{A} = -\frac{2Q_1 A^2}{(2\beta - \gamma)^2} + \frac{f}{2}, \quad \dot{B} = -\frac{4A}{2(2\beta - \gamma)^2} \left[\alpha\psi(4\beta^2 - 3\beta\gamma + \gamma^2) + Q_1 B \right]$$

$$\dot{C} = -\frac{1}{2(2\beta - \gamma)^2} \left[2\alpha(4\beta^2 - 3\beta\gamma + \gamma^2) B\psi + Q_1 B^2 + 2\alpha^2\beta \right]$$

with $A(T) = B(T) = C(T) = 0$, where $Q_1 \equiv [6\beta^3\psi^2 + 2\beta\gamma(5\gamma\psi^2 - 6\varphi^2) - 2\beta^2(7\gamma\psi^2 - 6\varphi^2) + \gamma^2(-2\gamma\psi^2 + 3\varphi^2)]$. Solving the differential equations and plugging the solutions completes the proof. Furthermore, from Eq. 6.25, retailer i's FBNE consumer price, p_{ifb}^{TC}, is strictly positive. Using Eq. 6.24 in Eq. 6.26, it can be shown that the pollution stock is strictly increasing over time whenever $\beta[6\beta^3\psi^2 + 2\beta\gamma(5\gamma\psi^2 - 6\varphi^2) - 2\beta^2(7\gamma\psi^2 - 6\varphi^2) - \gamma^2(2\gamma\psi^2 - 3\varphi^2)] > (4\beta^2 - 3\beta\gamma + \gamma^2)[(\beta - \gamma)^2\psi^2 + (2\beta - \gamma)\varphi^2]$ is fulfilled. This ensures that manufacturers' net current emissions are nonnegative. From Eqs. 6.24–6.25, we observe that the abatement efforts and the consumer prices in an FBNE are both increasing with the pollution stock at a decreasing rate. □

Proof of Theorem 7. After rearranging, we get: $\frac{y_{ol}^{HCVI}(T)}{y^{TI}(T)} = \frac{2\beta}{(2\beta - \gamma)} \cdot \frac{\frac{\mathrm{th}(\vartheta T)}{\vartheta T}}{\frac{\mathrm{th}(\theta T)}{\theta T}}$. Note that $\frac{\mathrm{th}(x)}{x}$ is a decreasing function for $x > 0$ and $\vartheta < \theta$. These lead to $\frac{\mathrm{th}(\vartheta T)}{\vartheta T} > \frac{\mathrm{th}(\theta T)}{\theta T}$. Therefore, $y_{ol}^{HCVI}(T) > y^{TI}(T)$ holds. Let $\beta - \gamma \equiv 2k\varphi^2$ $(k > 0)$. Then, $\theta = \delta\sqrt{(k\psi^2 + 1)} > \delta$, and $\frac{y^{TI}(T)}{y^{HIVC}(T)} = \left\{ \frac{[\delta T - \mathrm{th}(\delta T)]}{\delta T} \frac{k\psi^2}{\sqrt{k\psi^2 + 1}} + \frac{2}{\sqrt{k\psi^2 + 1}} \right\} \cdot \frac{\mathrm{th}(\theta T)}{\mathrm{th}(\delta T)}$. Let $\bar{k} \equiv k\psi^2$. Then, for $\bar{k} < 3$, it holds that $\frac{2}{\sqrt{\bar{k} + 1}} > 1$. Therefore, for any $0 < \bar{k} < 3$,

$\frac{y^{TI}(T)}{y^{HIVC}(T)} > 1$ holds. For $\bar{k} \geq 3$, a sufficient condition for $\frac{y^{TI}(T)}{y^{HIVC}(T)} > 1$ is that $\frac{\delta T - \mathrm{th}(\delta T)}{\delta T} > \frac{\sqrt{\bar{k}+1}-2}{\bar{k}}$. Let $\bar{k}^* = \mathrm{argmax}_{\bar{k} \in [3,\infty)}\left(\frac{\sqrt{\bar{k}+1}-2}{\bar{k}}\right)$, then $\frac{\sqrt{\bar{k}+1}-2}{\bar{k}}_{|\bar{k}^*} = 0.134$ and $\lim_{\bar{k} \to \infty} \frac{\sqrt{\bar{k}+1}-2}{\bar{k}} = 0$. Also, let $\tau = \delta T$ and, for >1, $\tau^* = \mathrm{argmin}_{\tau \in [1,\infty)}\left(\frac{\delta T - \mathrm{th}(\delta T)}{\delta T}\right)$. Then, $\frac{\delta T - \mathrm{th}(\delta T)}{\delta T}_{|\tau^*} = 0.238$ and $\lim_{\tau \to \infty} \frac{\delta T - \mathrm{th}(\delta T)}{\delta T} = 1$. Therefore, the sufficient condition $\frac{\delta T - \mathrm{th}(\delta T)}{\delta T} > \frac{\sqrt{\bar{k}+1}-2}{\bar{k}}$ holds for $\bar{k} \geq 3$ and $\tau > 1$, which leads to $\frac{y^{TI}(T)}{y^{HIVC}(T)} > 1$. Lastly, when $\bar{k} \geq 3$ and $0 < \tau \leq 1$, $\frac{y^{TI}(T)}{y^{HIVC}(T)} > 1$ hold if $\frac{[\tau - \mathrm{th}(\tau)]}{\tau} \frac{\bar{k}}{\sqrt{\bar{k}+1}} + \frac{2}{\sqrt{\bar{k}+1}} > \frac{\mathrm{th}(\tau)}{\mathrm{th}\left(\tau\sqrt{\bar{k}+1}\right)}$. In the case of $\tau = 1$, the latter inequality holds for any $\bar{k} \geq 3$. When $\tau \to 0$, it holds that $\lim_{\tau \to 0}\left\{\frac{[\tau - \mathrm{th}(\tau)]}{\tau} \frac{\bar{k}}{\sqrt{\bar{k}+1}} + \frac{2}{\sqrt{\bar{k}+1}}\right\} = \frac{2}{\sqrt{\bar{k}+1}}$ and $\lim_{\tau \to 0}\left\{\frac{\mathrm{th}(\tau)}{\mathrm{th}\left(\tau\sqrt{\bar{k}+1}\right)}\right\} = \frac{1}{\sqrt{\bar{k}+1}}$. Thus, $\frac{y^{TI}(T)}{y^{HIVC}(T)} > 1$ holds when $\tau < 1$ approaches zero. Also, for all (k, τ) where $\bar{k} \geq 3$ and $0 < \tau \leq 1$, the minimum value of $\left\{\frac{\tau - \mathrm{th}(\tau)}{\tau} \frac{\bar{k}}{\sqrt{\bar{k}+1}} + \frac{2}{\sqrt{\bar{k}+1}}\right\} \cdot \frac{\mathrm{th}\left(\tau\sqrt{\bar{k}+1}\right)}{\mathrm{th}(\tau)}$ is strictly larger than 1, when numerically computed, confirming that $\frac{y^{TI}(T)}{y^{HIVC}(T)} > 1$. Furthermore, when $\bar{k} \to \infty$, $\frac{[\tau - \mathrm{th}(\tau)]}{\tau} \frac{\bar{k}}{\sqrt{\bar{k}+1}} + \frac{2}{\sqrt{\bar{k}+1}} > \frac{\mathrm{th}(\tau)}{\mathrm{th}\left(\tau\sqrt{\bar{k}+1}\right)}$ is satisfied, thereby $\frac{y^{TI}(T)}{y^{HIVC}(T)} > 1$ holds. □

Proof of Theorem 8. $\frac{y^{TC}_{ol}(T)}{y^{HIVC}(T)} = \frac{\sqrt{2}\beta\left[\sqrt{2}\left((\beta-\gamma)\psi^2+4\varphi^2\right)\rho T - (\beta-\gamma)\psi^2 \mathrm{th}\left(\sqrt{2}\rho T\right)\right]}{\beta(\beta-\gamma)\psi^2\left[\rho T\left(2-\mathrm{sech}(\rho T)^2\right) - \mathrm{th}(\rho T)\right] + 4\varphi^2\rho T(2\beta-\gamma)} \frac{\frac{\mathrm{th}(\rho T)}{\rho T}}{\frac{\mathrm{th}\left(\sqrt{2}\rho T\right)}{\sqrt{2}\rho T}}$.

Since $\frac{\frac{\mathrm{th}(\rho T)}{\rho T}}{\frac{\mathrm{th}\left(\sqrt{2}\rho T\right)}{\sqrt{2}\rho T}} > 1$, a sufficient condition (though not necessary) for $\frac{y^{TC}_{ol}(T)}{y^{HIVC}(T)} > 1$ is $\frac{\sqrt{2}\beta\left[\sqrt{2}\left((\beta-\gamma)\psi^2+4\varphi^2\right)\rho T - (\beta-\gamma)\psi^2 \mathrm{th}\left(\sqrt{2}\rho T\right)\right]}{\beta(\beta-\gamma)\psi^2\left[\rho T\left(2-\mathrm{sech}(\rho T)^2\right) - \mathrm{th}(\rho T)\right] + 4\varphi^2\rho T(2\beta-\gamma)} > 1 \leftrightarrow$ $\beta^2\psi^2\left[\rho T \mathrm{sech}(\rho T)^2 + \mathrm{th}(\rho T) - \sqrt{2}\mathrm{th}\left(\sqrt{2}\rho T\right)\right] > \gamma\{\beta\psi^2 \left[\rho T \mathrm{sech}(\rho T)^2 + \mathrm{th}(\rho T) - \sqrt{2}\mathrm{th}\left(\sqrt{2}\rho T\right)\right] - 4\varphi^2\rho T\}$. Since $\rho T \mathrm{sech}(\rho T)^2 + \mathrm{th}(\rho T) - \sqrt{2}\mathrm{th}\left(\sqrt{2}\rho T\right) < 0$, the sufficient condition for $\frac{y^{TC}_{ol}(T)}{y^{HIVC}(T)} > 1$ is $\gamma > \bar{\gamma} = \frac{\beta^2\psi^2\left[\sqrt{2}\mathrm{th}\left(\sqrt{2}\rho T\right) - \rho T \mathrm{sech}(\rho T)^2 - \mathrm{th}(\rho T)\right]}{\beta\psi^2\left[\sqrt{2}\mathrm{th}\left(\sqrt{2}\rho T\right) - \rho T \mathrm{sech}(\rho T)^2 - \mathrm{th}(\rho T)\right] + 4\varphi^2\rho T} > 0$, where $\rho = \varphi\sqrt{2f}$. □

Proof of Theorem 9.

$$\frac{y^{HCVI}(T)}{y^{TC}(T)} = \frac{\beta(\beta-\gamma)\psi^2\left[\varphi T\sqrt{2f}\left(2-\mathrm{sech}\left(\varphi T\sqrt{2f}\right)^2\right) - \mathrm{th}\left(\varphi T\sqrt{2f}\right)\right] + 4\varphi^3 T\sqrt{2f}(2\beta-\gamma)}{2\sqrt{2}\varphi^2 T\sqrt{f(2\beta-\gamma)\left[(\beta-\gamma)^2\psi^2+\varphi^2(2\beta-\gamma)\right]}} \frac{\mathrm{th}(\vartheta T)}{\mathrm{th}\left(\varphi\sqrt{2f}T\right)}.$$

A sufficient condition (though not necessary) for $\frac{y^{TC}(T)}{y^{HIVC}(T)} > 1$ is

$$\frac{\beta(\beta-\gamma)\psi^2\left[\varphi T\sqrt{2f}\left(2-\mathrm{sech}\left(\varphi T\sqrt{2f}\right)^2\right)-\mathrm{th}\left(\varphi T\sqrt{2f}\right)\right]+4\varphi^3T\sqrt{2f}(2\beta-\gamma)}{2\sqrt{2}\varphi^2T\sqrt{f(2\beta-\gamma)\left[(\beta-\gamma)^2\psi^2+\varphi^2(2\beta-\gamma)\right]}}>1\leftrightarrow\beta(\beta-\gamma)\psi^2\left[\varphi T\sqrt{2f}\left(2-\mathrm{sech}(\varphi T\sqrt{2f})^2\right)-\mathrm{th}(\varphi T\sqrt{2f})\right]$$

$$+4\varphi^3T\sqrt{2f}(2\beta-\gamma)-2\sqrt{2}\varphi^2T\sqrt{f(2\beta-\gamma)\left[(\beta-\gamma)^2\psi^2+\varphi^2(2\beta-\gamma)\right]}>0$$

as $\frac{\mathrm{th}(\vartheta T)}{\mathrm{th}\left(\varphi\sqrt{2f}T\right)}>1$,. Because $\left[\varphi T\sqrt{2f}\left(2-\mathrm{sech}(\varphi T\sqrt{2f})^2\right)-\mathrm{th}(\varphi T\sqrt{2f})\right]>0$, the above condition is satisfied if $\varphi>\overline{\varphi}=\frac{(\beta-\gamma)\psi}{\sqrt{3(2\beta-\gamma)}}$. □

Proof of Theorem 10. $\frac{y^{TI}(T)}{y^{TC}_{ol}(T)}=\frac{\beta(\beta-\gamma)\psi^2\left[\rho T\left(2-\mathrm{sech}(\rho T)^2\right)-\mathrm{th}(\rho T)\right]+4\varphi^2\rho(2\beta-\gamma)T}{4\beta\theta\varphi^2T}\frac{\mathrm{th}(\theta T)}{\mathrm{th}(\rho T)}$ and $\frac{\mathrm{th}(\theta T)}{\mathrm{th}(\rho T)}>1$. A sufficient condition (though not necessary) for $\frac{y^{TI}(T)}{y^{TC}_{ol}(T)}>1$ is $\beta(\beta-\gamma)\psi^2[\rho T(2-\mathrm{sech}\,(\rho T)^2)-\mathrm{th}(\rho T)]+4\varphi^2\rho(2\beta-\gamma)T>4\beta\theta\varphi^2T$. As $[\rho T(2-\mathrm{sech}\,(\rho T)^2)-\mathrm{th}(\rho T)]>0$, which holds if $\varphi>\tilde{\varphi}=\beta\psi\sqrt{\frac{\beta-\gamma}{(2\beta-\gamma)^2-2\beta^2}}$ for $\beta>\tilde{\beta}=\frac{\gamma}{(2-\sqrt{2})}$. □

References

Arrow, K., Bolin, B., Costanza, R., Dasgupta, P., Folke, C., Holling, C. S., Jansson, B.-O., Levin, S., Mäller, K.-G., Perrings, C., & Pimentel, D. (1995). Economic growth, carrying capacity, and the environment. *Science, 268*, 520–521.

Benchekroun, H., & Long, N. V. (2011). Game theory: Static and dynamic games. In A. A. Batabyal & P. Nijkamp (Eds.), *Research tools in natural resource and environmental economics*. World Scientific Publishing.

Bertinelli, L., Camacho, C., & Zou, B. (2014). Carbon capture and storage and transboundary pollution: a differential game approach. *European Journal of Operational Research, 237*(2), 721–728.

Boscheck, R. (2006). *Assessing 'new' upstream business model*. IMD Working Paper 2006-06, 21 pp.

Casey, S., Blas, J., & Gilbolm, K. (2019, April 12). Chevron leaps to 'Ultramajor' oil status with Anadarko deal. *Bloomberg*.

Chava, S. (2014). Environmental externalities and cost of capital. *Management Science, 60*(9), 2223–2247.

Chen, X., Luo, Z., & Wang, X. (2017). Impact of efficiency, investment, and competition on low carbon manufacturing. *Journal of Cleaner Production, 143*, 388–400.

Chen, X., Benjaafar, S., & Elomri, A. (2019). On the effectiveness of emission penalties in decentralized supply chains. *European Journal of Operational Research, 274*(3), 1155–1167.

Conca, J. (2015, December 3). Making climate change fashionable – the garment industry takes on global warming. *Forbes*.

Cox, P. M., Betts, R. A., Jones, C., Spall, S. A., & Totterdell, I. (2000). Acceleration of global warming due to carbon-cycle feedbacks in a coupled climate model. *Nature, 408*, 184–187.

Creti, A., & Sanin, M.-E. (2017). Does environmental regulation create merger incentives? *Energy Policy, 105*(C), 618–630.

Dockner, E. J., Jørgensen, S., Van Long, N., & Sorger, G. (2000). *Differential games in economics and management science*. Cambridge University Press.

Du, S., Zhu, J., Jiao, H., & Ye, W. (2015). Game-theoretical analysis for supply chain with consumer preference to low carbon. *International Journal of Production Research, 53*(12), 3753–3768.
El Ouardighi, F., Sim, J. E., & Kim, B. (2016). Pollution accumulation and abatement policy in a supply chain. *European Journal of Operational Research, 248*(3), 982–996.
El Ouardighi, F., Kogan, K., Sanguinetti, M., & Gnecco, G. (2018a). Commitment-based equilibrium environmental strategies under time-dependent absorption efficiency. *Group Decision and Negotiation, 27*(2), 235–249.
El Ouardighi, F., Kogan, K., Sanguinetti, M., & Gnecco, G. (2018b). Transboundary pollution control and environmental absorption efficiency management. *Annals of Operations Research.* https://doi.org/10.1007/s10479-018-2927-7
Feenstra, T., Kort, P. M., & de Zeeuw, A. (2001). Environmental policy instruments in an international duopoly with feedback investment strategies. *Journal of Economic Dynamics & Control, 25*(10), 1665–1687.
Flammer, C. (2015). Does corporate social responsibility lead to superior financial performance? A regression discontinuity approach. *Management Science*, forthcoming.
Galbreth, M., & Ghosh, B. (2013). Competition and sustainability: The impact of consumer awareness. *Decision Sciences, 44*(1), 127–159.
Ghosh, D., & Shah, J. (2012). A comparative analysis of greening policies across supply chain structures. *International Journal of Production Economics, 135*(2), 568–583.
Ghosh, D., & Shah, J. (2015). Supply chain analysis under green sensitive consumer demand and cost sharing contract. *International Journal of Production Economics, 164*, 319–329.
Grass, D., Caulkins, J. P., Feichtinger, G., Tragler, G., & Behrens, D. A. (2008). *Optimal control of nonlinear processes with applications in drugs, corruption, and terror*. Springer.
Ha, A. Y., Tong, S., & Zhang, H. (2011). Sharing demand information in competing supply chains with production diseconomies. *Management Science, 57*(3), 566–581.
Herron, J. (2019, March 26). BP pledges $100 million to cut emissions from oil operations. *Bloomberg*.
Hoel, M. (1993). Intertemporal properties of an international carbon tax. *Resource and Energy Economics, 15*(1), 51–70.
Hong, Z., & Guo, X. (2018). Green product supply chain contracts considering environmental responsibilities. *Omega*. https://doi.org/10.1016/j.omega.2018.02.010
Jaber, M., Glock, C., & El Saadany, A. (2013). Supply chain coordination with emissions reduction incentives. *International Journal of Production Research, 51*(1), 69–82.
Jørgensen, S., Martín-Herrán, G., & Zaccour, G. (2010). Dynamic games in the economics and management of pollution. *Environmental Modeling and Assessment, 15*(6), 433–467.
Kaeser, J. (2015, September 22). Industry can lead on climate change. *The New York Times*.
P&G. (2018, April 16). P&G announces new environmental sustainability goals focused on enabling and inspiring positive impact in the world.
Kamien, M. I., & Schwartz, N. L. (2012). *Dynamic optimization: The calculus of variations and optimal control in economics and management*. Courier Corporation.
Kossioris, G., Plexousakis, M., Xepapadeas, A., de Zeeuw, A. J., & Mäler, K.-G. (2008). Feedback Nash equilibria for non-linear differential games in pollution control. *Journal of Economic Dynamics and Control, 32*(4), 1312–1331.
Krass, D., Nedorezov, T., & Ovchinnikov, A. (2013). Environmental taxes and the choice of green technology. *Production and Operations Management, 22*(5), 1035–1055.
Liu, Z., Anderson, T. D., & Cruz, J. M. (2012). Consumer environmental awareness and competition in two-stage supply chains. *European Journal of Operational Research, 218*(3), 602–613.
Luo, Z., Chen, X., & Wang, X. (2016). The role of co-opetition in low carbon manufacturing. *European Journal of Operational Research, 253*(2), 392–403.
Madani, S. R., & Rasti-Barzoki, M. (2017). Sustainable supply chain management with pricing, greening and governmental tariffs determining strategies: A game-theoretic approach. *Computers & Industrial Engineering, 105*, 287–298.

Martí, J. M. C., Tancrez, J.-S., & Seifert, R. W. (2015). Carbon footprint and responsiveness trade-off in supply chain network design. *International Journal of Production Economics, 166*, 129–142.

Meng, X., Yao, Z., Nie, J., Zhao, Y., & Li, Z. (2018). Low-carbon product selection with carbon tax and competition: Effects of the power structure. *International Journal of Production Economics, 200*, 224–230.

Muthu, S. S. (2014). *Assessing the environmental impact of textiles and the clothing supply chain.* Elsevier.

Ni, D., Li, K. W., & Tang, X. (2010). Social responsibility allocation in two-echelon supply chain: Insights from wholesale price contracts. *European Journal of Operational Research, 207*(3), 1269–1279.

Ouchida, Y., & Goto, D. (2016). Cournot duopoly and environmental R&D under regulator's precommitment to an emissions tax. *Applied Economics Letters, 23*(5), 324–331.

Porter, M. (1991). America's green strategy. *Scientific American, 264*(4), 168.

Porter, M., & van der Linde, C. (1995). Toward a new conception of the environment competitiveness relationship. *Journal of Economic Perspective, 9*(4), 97–118.

Raval, A., Hook, L., & Mooney, A. (2018, December 4). Shell yields to investors by setting target on carbon footprint. *Financial Times.*

Sabzevar, N., Enns, S. T., Bergerson, J., & Kettunen, J. (2017). Modeling competitive firms' performance under price-sensitive demand and cap-and-trade emissions constraints. *International Journal of Production Economics, 184*, 193–209.

Singh, N., & Vives, X. (1984). Price and quantity competition in a differentiated duopoly. *Rand Journal of Economics, 15*(4), 546–554.

Spengler, J. (1950). Vertical integration and antitrust policy. *Journal of Political Economy, 4*, 347–352.

Stimming, M. (1999). Capital accumulation subject to pollution control: Open-loop versus feedback investment strategies. *Annals of Operations Research, 88*, 309–336.

Subramanian, R., Gupta, S., & Talbot, B. (2007). Compliance strategies under permits for emissions. *Production and Operations Management, 16*(6), 763–779.

Toptal, A., & Cetinkaya, B. (2017). How supply chain coordination affects the environment: A carbon footprint perspective. *Annals of Operations Research, 250*(2), 487–519.

Tsay, A. A., Nahmias, S., & Agrawal, N. (1999). Modeling supply chain contracts: A review. In S. Tayur, R. Ganeshan, & M. Magazine (Eds.), *Quantitative models for supply chain management* (pp. 299–336). Springer.

Van Beurden, B. (2018, July 5). Shell says firm carbon emissions targets are 'superfluous'. *Financial Times.*

Wang, Q., Zhao, D., & He, L. (2016). Contracting emission reduction for supply chains considering market low-carbon preference. *Journal of Cleaner Production, 120*, 72–84.

WRI. (2017, September 18). Gap Inc., NIKE, Inc., Levi's among retailers joining initiative to tackle climate change.

Wu, D., Baron, O., & Berman, O. (2009). Bargaining in competing supply chains with uncertainty. *European Journal of Operational Research, 197*(2), 548–556.

Xepapadeas, A., & de Zeeuw, A. (1999). Environmental policy and competitiveness: The Porter hypothesis and the composition of capital. *Journal of Environmental Economics and Management, 37*(2), 165–182.

Xu, X., He, P., Xu, H., & Zhang, Q. (2017). Supply chain coordination with green technology under cap-and-trade regulation. *International Journal of Production Economics, 183*(B), 433–442.

Yalabik, B., & Fairchild, R. J. (2011). Customer, regulatory, and competitive pressure as drivers of environmental innovation. *International Journal of Production Economics, 131*, 519–527.

Yang, H., & Chen, W. (2018). Retailer-driven carbon emission abatement with consumer environmental awareness and carbon tax: Revenue-sharing versus Cost-sharing. *Omega, 78*(C), 179–191.

Yang, L., Zhang, Q., & Ji, J. (2017). Pricing and carbon emission reduction decisions in supply chains with vertical and horizontal cooperation. *International Journal of Production Economics, 191*, 286–297.

Yenipazarl, A. (2019). Incentives for environmental research and development: Consumer preferences, competitive pressure and emissions taxation. *European Journal of Operational Research, 276*(2), 757–769.

Zhang, Q., Tang, W., & Zhang, J. (2016). Green supply chain performance with cost learning and operational inefficiency effects. *Journal of Cleaner Production, 112*(4), 3267–3284.

Zhao, M., & Running, S. W. (2010). Drought-induced reduction in global terrestrial net primary production from 2000 through 2009. *Science, 329*, 940–943.

Zhou, Y., Bao, M., Chen, X., & Xu, X. (2016). Co-op advertising and emission reduction cost sharing contracts and coordination in low-carbon supply chain based on fairness concerns. *Journal of Cleaner Production, 133*, 402–413.

Zhou, Y., & Ye, X. (2018). Differential game model of joint emission reduction strategies and contract design in a dual-channel supply chain. *Journal of Cleaner Production, 190*, 592–607.

Zhu, W., & He, Y. (2017). Green product design in supply chains under competition. *European Journal of Operational Research, 258*(1), 165–180.

Zu, Y., Chen, L., & Fan, Y. (2018). Research on low-carbon strategies in supply chain with environmental regulations based on differential game. *Journal of Cleaner Production, 177*, 527–546.

Chapter 7
Voluntary Environmental Effort Under (s, S) Inventory Policy

Alain Bensoussan and Fouad El Ouardighi

This work is dedicated to the memory of Jean Marie Proth. One of the greatest scholars in the domain of production management, Jean Marie became very concerned by environmental considerations. He was convinced that environment could not be absent in operations management modelling. He remains, in this domain, an inspirational figure.

Alain Bensoussan—Also with the School of Data Science, City University Hong Kong. Research supported by the National Science Foundation under grant NSF-DMS 2204795.

Abstract Prior research on inventory control has been wide-ranging, yet the environmental implications of an (s, S) inventory policy remain uninvestigated. This paper seeks to bridge the gap by characterizing a firm's voluntary environmental policy in the setup of an (s, S) inventory control policy. We suggest a mixed model structure wherein, due to the presence of fixed production costs, the inventory is determined *continuously* by sales and *impulsively* with ordering decisions obeying an optimal stopping process, while the uncertain sales process is controlled by continuous-time environmental goodwill-related decisions. We show that a firm should successively use voluntary environmental efforts to stimulate its sales when there is inventory and to increase backlogging to improve its production efficiency. Given the recurrent pattern of this policy, we conclude that voluntary environmental

A. Bensoussan
International Center for Decision and Risk Analysis, Jindal School of Management, University of Texas at Dallas, Richardson, TX, USA

F. El Ouardighi (✉)
ESSEC Business School, Cergy Pontoise Cedex, France
e-mail: elouardighi@essec.fr

F. El Ouardighi (ed.), *Essays on Pollution Control in Economics and Management Science*, Lecture Notes in Operations Research,
https://doi.org/10.1007/978-3-031-78227-5_7

efforts under an (s, S) inventory control is not compatible with using these efforts to generate ephemeral reputation insurance.

Keywords Inventory control · Environmental protection management · Stochastic demand · Voluntary programs

Introduction

It has been shown that a voluntary approach is a valuable environmental policy instrument (e.g., Arimura et al., 2008). Examples of such efforts are compensating for carbon emissions by planting trees or purchasing offsets, investing in clean energy, cleaning rivers, reducing the quantity of water or energy used for production, increasing the proportion of recycled waste used, and improving water quality. A voluntary approach is rewarded by relaxed regulatory scrutiny (e.g., Innes & Sam, 2008), therefore firms facing higher regulatory pressure are more likely to participate in voluntary environmental programs (e.g., Potoski & Prakash, 2005; Wu, 2009).

The question of how to design a voluntary environmental policy to be as effective as possible is crucial (Koehler, 2007; Borck & Coglianese, 2009). This paper seeks to contribute to the environmental management literature by characterizing a firm's optimal voluntary environmental policy. In our setup, environmentally protective initiatives include voluntary efforts that provide goodwill due to customers' environmental awareness (Heydari et al., 2021; Hong et al., 2020) and therefore promote a firm's sales. In contrast with regulatory environmental initiatives that seek to avoid a penalty associated with the environmental impact of production, these efforts are disconnected from production cost considerations. Examples of prominent profit-oriented firms that voluntarily invest in environmental processes are Shell,[1] Land Rover,[2] Unilever,[3] and Cemex.[4]

A main novelty in this paper lies in the fact that such a policy is characterized by the setup of an (s, S) inventory control policy. Though the concept of (s, S) policy is well-established both in theory and in practice [see the comprehensive presentation proposed by Dolgui and Proth (2010)], and was extended in several directions [see the recent exhaustive survey by Perera and Sethi (2022)], little is known about its implications in terms of environmentally protective management. This paper seeks to bridge the gap by developing an (s, S) inventory control model that extends towards environmental protection. A primary issue here is thus to induce an optimal voluntary environmental policy that is contingent on an (s, S) inventory policy.

---

[1] https://www.shell.com/shellenergy/othersolutions/welcome-to-shell-environmental-products/working-with-customers-to-compensate-for-their-emissions.html

[2] https://www.reliableplant.com/view/22448/land-rover-offsets-co2

[3] https://climatechampions.unfccc.int/unilever-and-the-race-to-halve-emissions-by-2030/

[4] https://cen.acs.org/materials/Chemex-goes-global-carbon-neutral/98/i42

Clearly, the response to this issue involves both operational production decisions and a marketing approach to satisfy market demand.[5] The tradeoff thus requires the design of an operations-marketing interface that delivers both efficiency in production and environmental goodwill for sales promotion. In this regard, we consider a monopolist firm that faces a market demand that is sensitive to the firm's efforts to protect the ecological environment. In this setup, the firm can leverage a decision variable, i.e., a voluntary environmental effort, to stimulate demand. That is, a greener (i.e., more environmentally protective) production will boost demand, though at additional cost. This assumption is consistent with a recent conclusion from a Mc Kinsey & Co. report that suggests that 66% of all respondents of a US cohort survey (and 75% of millennial respondents) say that they consider sustainability when they make a purchase (Mc Kinsey, 2019).

To focus on the interactions between inventory control and environmental protection, we assume that the price is set exogenously once and for all at the beginning of the planning horizon. This means there is no pricing management, which simplifies the mathematical analysis. A diffusion model is suggested that accounts for the marketing impact of voluntary environmental efforts on an uncertain market demand.

The novelty of our approach lies in the fact that our model relies on an interface between operational and marketing instruments combining production and environmental goodwill decisions. More precisely, we suggest a mixed model structure wherein, due to the presence of fixed production costs, the inventory is determined *continuously* by sales and *impulsively* with ordering decisions obeying an optimal stopping process, while the sales process is controlled by continuous-time environmental goodwill-related decisions. The methodological resolution of the model proceeds in two steps: the inventory control is first characterized in terms of an s, S policy, and the environmental goodwill policy is then determined as optimal feedback on the inventory.

Our results show that, in the context of an (s, S) inventory policy, a firm should leverage voluntary environmental efforts on a quasi-cyclical basis as a marketing tool when there is inventory and as a production efficiency tool in the case of backlogging. That is, voluntary environmental management should stimulate demand when the stock is replenished to maximize current profit, and subsequently generate backlogs when the stock is depleted to ensure that the fixed costs of future production can be covered. These results differ from Barcos et al. (2013) in that we find a U-shaped rather than an inverted U-shaped relationship between firms' voluntary environmental initiatives and their inventory levels under an (s, S) inventory policy.

The paper is organized as follows. We review the relevant literature in the next section, and develop our model in third section. In the subsequent section, we analyze the model and derive our results. Fifth section provides numerical illustrations. Sixth section concludes the paper.

---

[5] A similar approach can be found in, e.g., Khmelnitsky and Singer (2015) and Herbon (2021).

Literature Review

Our research lies at the intersection of two kinds of literature, those concerning (s, S) inventory policy and analysis of related production decisions, and corporate environmental management and performance.

Regarding the (s, S) periodic-review inventory model and analysis of related production decisions, an important stream of literature has emerged since the model's introduction by Arrow et al. (1951) and its consolidation by Scarf (1960). Consistent with many real-life scenarios, this model assumes fixed costs of ordering items or setting up a process. The (s, S) policy, where s is the ordering point and S is the order-up-to level, can be described as follows: if, at a certain period, the initial inventory level, x, is lower than the ordering point, s, then an order equal to the difference between the order-up-to level, S, and the initial inventory level, x, should be placed, and otherwise no order should be placed. The optimal expressions of s and S are derived by minimizing a loss function representing the present value of the total expected lost sales incurred over a given time horizon for any initial inventory level.

Perera and Sethi (2022) provide an exhaustive survey of (s, S) policies in various settings, that is, discrete- and continuous-time reviews, finite- and infinite-time horizons, discounted- and average-cost objectives, backlogging and lost-sales settings, standard and generalized demand and cost structures, deterministic and stochastic delivery lead times, single- and multi-product settings, and coordinated pricing-inventory control.[6] Among the most representative extensions are those relating to Markovian demand (Sethi & Cheng, 1997), demand Bayesian learning (Larson et al., 2001), generalized loss functions (Benkherouf & Sethi, 2010), lost sales (Bensoussan et al., 1983; Zipkin, 2008), multiple items (Silver, 1974), cutoff transaction size (Hollier et al., 1995), information delay (Bensoussan et al., 2009), deteriorating items (Ravichandran, 1995), supply chain management (Kelle & Milne, 1999), and competing suppliers (Fox et al., 2006). As in our study, some important papers assume a Wiener process demand where the Bellman equation of dynamic programming for the considered inventory problem reduces to a quasi-variational inequality (QVI) (Bensoussan & Lions, 1987) that is solved to obtain an optimal impulse control policy and hence an optimal inventory policy. Among these papers are those of Bensoussan and Tapiero (1982), Bensoussan et al. (2005, 2009). Although the problem of joint pricing and inventory policy (s, S, p), where p stands for price, has been extensively investigated in a dynamic setting (Chen & Simchi-Levy, 2004a, 2004b; Chen et al., 2006; Song et al., 2009), the problem of joint goodwill advertising and inventory policy remains largely unexamined. Huh and Janakiraman (2008) consider a broad class of decisions, such as pricing and advertising, that influence demand. In their study, a stationary, single-stage inventory system is assumed, so the dynamic impact of these decisions is not considered. To characterize the most protective voluntary environmental effort pattern associated with an (s, S) inventory policy, we follow the approach that consists of reducing the

[6] Earlier surveys are notably Aggarval (1974) and Porteus (1990).

(s, S) inventory problem to a QVI (Bensoussan & Tapiero, 1982; Bensoussan et al., 2005; Benkherouf & Bensoussan, 2009), with the innovation that a continuous control of voluntary environmental efforts determines the state of the system. The optimal voluntary environmental efforts are then determined as a feedback policy on the inventory level which depends on impulse control, given as an increasing sequence of stopping times at which a certain quantity of products is produced.

As for corporate environmental management and performance, King and Lenox (2002) suggest that waste prevention often provides unanticipated innovation offsets while onsite waste treatment often engenders unexpected costs. They find evidence that the benefits of waste prevention alone are responsible for the observed association between lower emissions and profitability. Jacobs et al. (2010) analyze the effects of environmental performance on shareholder value by quantifying the stock market response associated with declarations of environmental performance. They find that, while the market does not react significantly to self-reported corporate efforts to avoid, mitigate, or offset the environmental impacts of the firm's products, services, or processes, announcements of philanthropic gifts for environmental causes are associated with significant positive market reactions. Kroes et al. (2012) show a negative relationship between environmental performance and firm market performance over at least a 3-year period. Barrage et al. (2020) suggest that firms may have incentives to engage in green advertising without investments in environmental stewardship to get reputation insurance (Minor & Morgan, 2011).

Barcos et al. (2013) assume that there is an inverted U-shaped relationship between firms' corporate social responsibility and their inventory levels. The reason is that while customers put pressure on firms to increase inventories, environmental activists force firms to reduce inventories. Their empirical findings support their contention that for low levels of corporate social responsibility (CSR), customers are more relevant; and for higher levels of CSR, the natural environment gains importance.

An important stream of the corporate environmental management literature relates to process improvement with the objective of compliance with regulatory norms [see the reviews in Kleindorfer et al. (2005), Corbett and Klassen (2006), Sarkis and Zhu (2018)]. Recent contributions to the literature on operational compliance with regulatory norms include Drake et al. (2016), Jaber et al. (2013), Krass et al. (2013), Porteous et al. (2015), and Jabbour et al. (2016). More recently, Xiao et al. (2019) used a dynamic, determinist setting to analyze a firm's investment in environmental process improvement to reduce the environmental impact of its manufacturing processes by taking into account various internal firm characteristics and different external regulatory drivers. An article by Chen and Monahan (2010) is the only paper that investigates the relationship between environmental management and production planning and inventory control policies. Based on a static model with both stochastic demand and environmental uncertainties, they analyze the optimal policies of production planning and inventory control under both regulatory and voluntary pollution control approaches, and investigate their operational and environmental effects. They show that a regulatory environmental standard that limits the total amount of waste may induce the firm to raise its planned stock level, which would lead to a higher expected amount of environmental waste before the standard is enforced. However, the additional planned stock level, termed "environmental safety stock," can be reversed by using the voluntary control approach that provides the firm with the flexibility to occasionally exceed the environmental standard. Our paper adopts a different perspective from Barcos et al. (2013) and Chen and Monahan (2010) in that we investigate the effect of a firm's inventory level on its environmental management policy rather than the converse. Our approach also differs from Chen and Monahan (2010) and Xiao et al. (2019) in that we consider a product's uncertain diffusion process over time, which depends on voluntary environmental efforts with no impact on the unit production cost and no required compliance to an environmental regulation.

A growing stream of the corporate environmental management literature, to which we contribute, seeks to maximize the effectiveness of voluntary initiatives out of any regulatory pressure. Recent publications on voluntary environmental initiatives include topics such as polluting emissions abatement in a supply chain's manufacturing and sales processes. For instance, El Ouardighi et al. (2016), Sim et al. (2019), and El Ouardighi et al. (2021) investigate how polluting emissions and abatement are affected by competition and integration in an industry, and eventually how firms' strategy types modify the relative impact of horizontal and vertical competition on pollution. In this paper, we consider a single firm whose voluntary initiatives seek to carry out a broad spectrum of external environmental actions, such as the restoration of natural carbon sinks, which have the effect of promoting sales.

Model Formulation

Background and General Comments

We develop a novel decision-aid model for inventory and production control. The standard inventory control theory particularly leading to the famous (s, S) policy introduced by Arrow et al. (1951) and Scarf (1960) considers the demand as an external stochastic process with independent demands over time. The costs are inventory holding and shortage costs, namely, costs were purely limited to physical aspects of production and inventories. The popularity of (s, S) policy lies in the fact that it is very intuitive and convenient to practitioners. After this remarkable progress in management science and operations research, a lot of effort has been devoted to meaningful extensions (see Perera & Sethi, 2022). The natural extensions have notably dealt with modeling the demand process. A very natural question is: how sensitive is the (s, S) policy to factors influencing the demand process? In this setup, the assumption of independence of successive demands over time is indeed not anymore relevant. As an alternative, the demand process can be modeled as a Markov chain (Kalymon, 1971; Sethi & Cheng, 1997). Another type of extension is not to consider the demand process as external, but linked to the inventory, for instance, a mean reverting connection (Cadenillas et al., 2010). But there are also economic factors that influence the demand, many of them are not fully external but may depend on corporation decisions, for instance, marketing decisions. Generally, the marketing and production decisions emanate from different departments of a company, but the issues are very connected. A famous case, that has generated huge research efforts, is pricing management. In pricing management, the issue is to decide the selling price of the product in connection with the production and inventory decisions. This problem is quite complicated because the price will affect the demand and therefore the production and inventory policy. In this regard, the price should not be decided independently from the inventory policy, as is often the case.

In this paper, we introduce a different connection, related to environmental considerations. While the concern for the environment was not so present when inventory control was developed, it has now become unthinkable to consider manufacturing and retailing activities without taking account of their environmental impact. Indeed, this problem has primarily to do with engineering issues, e.g., developing less polluting production processes, using alternative energy resources, etc. However, beyond compliance with restrictive regulations, the search for environmental improvements also involves an economic tradeoff because it entails additional expenses, on the one hand, and stimulates demand, on the other hand. This comes from the fact that customers decide more and more on their expenses with environmental issues in mind. Our objective is to analyze with a stylized model the cost-benefit problem of how much to spend on environmental aspects, in relation to the production and inventory control issues.

Description of the Model

We build a model which gets inspiration from the way pricing management has been studied. Particularly, we rely on the work of the first author on pricing management (Bensoussan et al., 2018) but with the difference that we take price as a fixed parameter and we introduce a different decision variable that is the effort of the manufacturer toward the environment. As in pricing management, a major issue here is how an economic factor such as environmental effort affects the demand and thus the inventory policy? It is obvious that this problem is not identical to the pricing management problem because the respective impacts of price and environmental effort are different. Here, the decision variable considered, i.e., the manufacturer's effort toward the environment affects the demand in an opposite way to the price in pricing management. Also, environmental effort has a cost, which needs to be taken into account as part of the economic tradeoff. Regarding the inventory control part, the analysis is similar to the standard theory, and we shall obtain an (s, S) policy. The interconnection of the effort towards the environment and the (s, S) policy is the key mathematical challenge. This interaction is different from what occurs in pricing management.

A Short Presentation of the Mathematical Apparatus

We develop a model in continuous time (though discrete time is also possible) with an infinite time horizon to get stationary problems. There are advantages to continuous time. We use stochastic control instead of Markov decision processes and get closed-form solutions. However, we must accept that inventory control is an impulse control and not a continuous control. In discrete time, there is no difference. A continuous control is a rate, and the state (here, the inventory level) evolves continuously. In contrast, impulse control is a jump and the inventory is discontinuous (it changes instantaneously with the size of the jump). The reason we have to consider impulses is because there are fixed costs. Each time a production is decided, whatever its amount, there is a fixed cost to be paid. So deciding continuously will entail an infinite cost, which is impossible. Therefore, impulse control is a sequence of decision times at which we make an order or produce some quantity. Associated with these times are the quantities produced or ordered, which are also decision variables. They correspond to the jumps in the inventory at the decision times. We can neglect the production times in the sense that what accounts is when the production modifies the inventory. Note that these decision times and quantities produced are random. They cannot be decided *ex ante*, since we must take into account the information that will be available in the future. Random times in probability theory are called stopping times. Information is characterized by σ-algebras. A time t, there is a σ-algebra W^t which is the collection of events observable at time t. The effort towards the environment is a function of time, which affects the rate of demand. An (s, S) policy is an example of impulse control (which turns out to be optimal). When the inventory, denoted hereafter by x, attains the threshold s, then the manufacturer replenishes its inventory, and the amount ordered is $S - x$.

Model

Given the need for the operations function to deliver efficiency and the necessity for environmental marketing to impact sales effectively, the trade-off between inventory level and sales-promoting environmental efforts deserves a thorough examination. In this regard, we assume a company that manufactures a product and sells it at a fixed price. While the manufacturing process has a negative impact on the environment, we can assume that the firm complies with the applicable regulatory environmental norms. However, to gain a better reputation, the company is willing to make further environmental efforts that are voluntary. This may come in the form of planting trees and more generally restoring natural carbon sinks, purchasing offsets, reducing the quantity of water or energy used for production, increasing the proportion of recycled waste used, improving water quality, etc. These voluntary environmental efforts increase costs for the manufacturer but they also boost demand due to the marketing impact on customers who exhibit environmental consciousness (Heydari et al., 2021; Hong et al., 2020). We thus introduce a control variable called voluntary environmental efforts, denoted by $Q(t)$. The uncertainties on the demand are captured by a standard Wiener process $w(t)$, which is external. This process is built on a probability space Ω, $\mathcal{A}$, P, and we denote by W^t the filtration generated by the Wiener process (i.e., a sequence of σ-algebras of information W^t). The control $Q(t)$ depends on W^t. This leads to the following control model for the cumulated demand $D(t)$ on $(0, t)$, that is:

$$dD(t) = \nu(Q(t))dt + \sigma dw(t) \tag{7.1}$$

where ν is the rate of demand by a unit of time. This rate depends on the voluntary environmental effort.

The volatility σ is a constant. It may depend on $Q(t)$. The function $Q \to \nu(Q(t))$ is positive and monotone increasing. For mathematical convenience, we assume that it is a linear function. The ordering (or production) policy is not defined by a rate, which means that the inventory is not replenished continuously, but by impulse control. This is because of fixed costs. In this case, we neglect the time of production of each impulse. So, a production by impulses is an increasing sequence of stopping times θ_i and positive random variables ξ_i, such that θ_i is a stopping time with respect to W^t and ξ_i is W^{θ_i} measurable. The variable ξ_i is the amount of product produced at θ_i (the time to produce ξ_i is neglected[7]). If $x(t)$ represents the inventory at time t, it is expressed by the formula:

$$x(t) = x - \int_0^t \nu(Q(s))ds + \sum_{\{i|\theta_i \le t\}} \xi_i - \sigma w(t) \tag{7.2}$$

The impulse control $(\theta_i, \xi_i, i = 1, \ldots)$ is denoted by V. So the state of the system $x(.)$ depends on two controls: a continuous control $Q(.)$ and an impulse control V. The initial value of the inventory is x, which is a parameter. We then associate to the pair $(Q(.), V)$ a payoff functional given by the formula:

$$J_x(Q(.), V) = E\left\{ \int_0^{+\infty} e^{-\alpha t}\left[\varpi\nu(Q(t)) - hx^+(t) - px^-(t) - \beta Q^2(t)\right]dt - \sum_{i=1}^{+\infty} e^{-\alpha\theta_i}(k + c\xi_i) \right\} \tag{7.3}$$

The scalar ϖ is the selling price of the product. Our model accepts backlogs, so the inventory $x(t)$ can be positive or negative. When the inventory is positive, the firm pays a holding cost $hx^+(t)$ per unit of time. When the inventory is negative, the firm pays a shortage cost $px^-(t)$ per unit of time. The payment $\beta Q^2(t)$ per unit of time

---

[7] Note that the impulse times refer to the times when the inventory is modified. So if production takes some time, for example, a fixed amount of time, the stopping time will be the time of delivery.

represents the cost of the effort towards the environment at time t. Finally, at each time θ_i, the firm pays a fixed cost k to produce and a variable cost $c\xi_i$ proportional to the quantity produced, ξ_i. The parameter α is the discount factor.

The model (Eqs. 7.1–7.3) differs from models based on regulatory environmental effort (e.g., Xiao et al., 2019) in two ways: first, in our model, there is no penalty associated with the environmental impact of production, and second, the production cost is not positively affected by the cumulative regulatory environmental efforts. However, one point in common with such models is the positive impact of either effort on the revenue drawn from production.

The objective is to maximize $J_x(Q(.), V)$ on the pair $Q(.)$, V. We set:

$$u(x) = \sup_{Q(.),\, V} J_x(Q(.), V) \tag{7.4}$$

which is the value function. Using Dynamic Programming, we will define an analytic problem of which $u(x)$ is a solution.

Analysis

To simplify, we will take the following function of $\nu(Q(t))$:

$$\nu(Q(t)) = \nu_0 + \nu_1 Q(t) \tag{7.5}$$

We see from (7.3) that:

$$\begin{aligned} J_x(Q(.), V) &\leq E \int_0^{+\infty} e^{-\alpha t}\left[\varpi\nu(Q(t)) - \beta Q^2(t)\right]dt \\ &\leq \frac{sup_Q\left[\varpi\nu(Q(t)) - \beta Q^2(t)\right]}{\alpha} \end{aligned} \tag{7.6}$$

Therefore, we get:

$$u(x) \leq \frac{1}{\alpha}\left(\varpi\nu_0 + \frac{\varpi^2{\nu_1}^2}{4\beta}\right) \tag{7.7}$$

If we take $Q(.) = 0$ and $\theta_i = +\infty$, the state is reduced to:

$$x_0(t) = x - \nu_0 t - \sigma w(t) \tag{7.8}$$

Therefore, we can state:

$$u(x) \geq -E \int_0^{+\infty} e^{-\alpha t}[hx^+(t) + px^-(t)]dt \tag{7.9}$$

which provides a lower bound for the value function (Eq. 7.4). It is possible to compute this lower bound explicitly.

Remark 1 The result (Eq. 7.7) justifies the choice of Q^2 in the cost of effort. In this way, we obtain a finite upper bound for the value function. The lower bound, which is negative tends to $-\infty$ as x tends to $-\infty$. This is normal: if the initial inventory is $-\infty$, it is impossible to get rid of it and the holding cost will lead to an infinite penalty. This means also that the optimization does not guarantee that the business is profitable. We intend to perform a sensitivity analysis in further work.

Dynamic Programming

Quasi-Variational Inequality (QVI)

We first introduce the function $\lambda \rightarrow \Phi(\lambda)$ defined by:

$$\Phi(\lambda) = \inf_{Q \geq 0} \left(\beta Q^2 + (\lambda - \varpi)\nu(Q)\right) \tag{7.10}$$

and with the specific function $\nu(Q)$ in Eq. 7.5, we obtain:

$$\Phi(\lambda) = \nu_0(\lambda - \varpi) - \frac{{\nu_1}^2}{4\beta}\left((\varpi - \lambda)^+\right)^2 \tag{7.11}$$

and:

$$\Phi'(\lambda) = \nu_0 + \frac{{\nu_1}^2}{2\beta}(\varpi - \lambda)^+ \tag{7.12}$$

The function $\Phi(\lambda)$ is monotone increasing and concave. The optimum $\widehat{Q}(\lambda)$ in Eq. 7.10 is given by:

$$\widehat{Q}(\lambda) = \frac{\nu_1}{2\beta}(\varpi - \lambda)^+ \tag{7.13}$$

In the case of impulse control, the Bellman equation of dynamic programming is replaced with a QVI, which can be written as follows:

$$\min\left[-\frac{1}{2}\sigma^2u''(x)+\Phi(u'(x))+\alpha u(x)+hx^+ +px^-, u(x)-\sup_{\xi\geq 0}(u(x+\xi)-c\xi)+k\right]=0,\text{a.e.}x \tag{7.14}$$

Note that the formulation in Eq. 7.14 requires some smoothness of the function $u(x)$, namely $u(x)$ is C^1, with the second derivative $u''(x)$ defined a.e.

Heuristic derivation of Eq. 7.14
Equation 7.9 can be written as a set of inequalities and complementarity slackness conditions. Indeed, Eq. 7.9 is equivalent to:

$$u(x)\geq \sup_{\xi\geq 0}(u(x+\xi)-c\xi)-k \tag{7.14$'$}$$

$$-\frac{1}{2}\sigma^2u''(x)+\Phi(u'(x))+\alpha u(x)+hx^+ +px^- \geq 0 \tag{7.14$''$}$$

and for any x one of the two inequalities is an equality. The reason the value function $u(x)$ satisfies the two inequalities can be seen intuitively as follows. At time 0, the manufacturer may decide to order the quantity ξ, then the inventory jumps from x to $x+\xi$. There is an immediate cost $c\xi+k$. From the optimality principle, the best profit starting with an inventory $x+\xi$ is $u(x+\xi)$. The right-hand side of the first inequality (Eq. 7.14′) is the value function after the decision of ordering immediately at 0. The sup in ξ captures the fact that one can choose ξ. Since it may not be optimal to order immediately, the value function $u(x)$ must be larger than the right-hand side. The second inequality (Eq. 7.14″) can be obtained by considering the alternative of putting an immediate order, namely not ordering for a small interval of time ϵ. At time ϵ, the inventory is approximately (since ϵ is small) $x-\epsilon v(Q)-\sigma w(\epsilon)$. Using the optimality principle, we can write:

$$u(x)\geq \epsilon\left(\varpi v(Q)-hx^+ -px^- -\beta Q^2\right)+(1-\alpha\epsilon)Eu(x-\epsilon v(Q)-\sigma\omega(\epsilon))$$

Expressing the mathematical expectation with Ito's formula, rearranging, and optimizing in Q, we obtain the second inequality (Eq. 7.14″), using the definition of the function Φ. The fact that at time 0, only two decisions are possible, putting an order immediately or postponing for at least a small amount of time implies that one of the inequalities must be an equality. So this is valid for any value of x.

Preliminaries

We begin with the simple transformation:

$$G(x)=u(x)-cx \tag{7.15}$$

We also define the nonlinear operator:

$$M(G)(x) = -k + \sup_{y \geq x} G(y) \tag{7.16}$$

M is an operator. It associates to the function $G \equiv G(x)$ a function $M(G)(x)$. Then problem (Eq. 7.10) becomes:

$$\min\left[-\frac{1}{2}\sigma^2 G''(x) + \Phi(G'(x)+c) + \alpha G(x) + (h+\alpha c)x^+ + (p-\alpha c)x^-, G(x) \ - M(G)(x)\right] = 0, \text{a.e.} x \tag{7.17}$$

Conjectured Solution:

We look for a solution of Eq. 7.17 as follows: Find s and a function $G_s(x)$, $x \geq s$, satisfying:

$$-\frac{1}{2}\sigma^2 G_s''(x) + \Phi\left(G_s'(x) + c\right) + \alpha G(s) + (h + \alpha c)x^+ + (p - \alpha c)x^- = 0, x > s \tag{7.18}$$

$$G_s'(s) = 0, \quad G_s'(x) \text{ bounded as } x \to +\infty$$

$$G_s(s) = M(G_s)(s) \tag{7.19}$$

The logic is as follows: for s fixed, we solve the boundary value differential Eqs. 7.18 and 7.19 is an algebraic equation to obtain the number s.

We shall need the fundamental assumption:

$$\frac{\sigma}{\sqrt{2\alpha}}(p - \alpha c) + \Phi(c) - \Phi\left(\frac{p}{\alpha}\right) > 0 \tag{7.20}$$

This condition means that $p - \alpha c$ must be sufficiently large. Since p is the unit cost of shortage, the higher p the more we produce to avoid shortage. On the other hand, the more we stimulate demand by efforts towards the environment the more we increase the risk of shortage, if production cannot follow.

The important understanding is that in Eq. 7.18 we identify an (s, S). The s is explicit, the S comes from the definition of the operator M (see Eq. 7.16). S attains the maximum of $G_s(y)$ for $y > s$. The optimization of efforts will be driven by another number, Σ, with $s < \Sigma <$ S. The result is described in the next section, with details in the Appendix.

Optimal Control

Main Result

The main result is the following:

Theorem 1. *Assume Eq. 7.20. Set* $\gamma = \frac{\sigma}{\sqrt{2\alpha}}$. *There exists a unique* $s^* < 0$ *such that:*

$$\frac{1}{\gamma}\left[p - \alpha c - (p+h)e^{\gamma s^*}\right] + \Phi(c) - \Phi\left(\frac{p}{c}\right) = 0$$

Then, there exists a set $(s, \Sigma, S, H(s))$ *satisfying:*

$$s < \Sigma < 0, \quad S > \max(\Sigma, s^*), \quad H(s) \text{ is Lipschitz continuous,} \tag{7.21}$$

$$-\frac{1}{2}\sigma^2 H''(x) + \alpha H(x) + \frac{d}{dx}\Phi(H(x)+c) + (h+\alpha c)\mathbb{1}_{x>0} - (p-\alpha c)\mathbb{1}_{x<0} = 0,$$
$$x > s \tag{7.22}$$

$$H(s) = 0, H(+\infty) = -c - \frac{h}{\alpha}, H'(+\infty) = 0,$$

$$H(S) = 0, k = \int_s^S H(x)dx,$$

$$H(x) > 0, \forall x \in (s, S), H(x) < 0, \forall x > S,$$

$$H'^{(x)} > 0, s \leq x < \Sigma < S, H'^{(x)} > 0, x > \Sigma,$$

and also:

$$-k + \int_x^y H(\xi)d\xi \leq 0, \forall s \leq x \leq y, \tag{7.23}$$

with the estimates:

$$-c - \frac{h}{\alpha} \leq H(x) \leq -c + \frac{p}{\alpha}$$

$$|H'(x)| \leq \frac{\sqrt{2}}{\sigma\sqrt{\alpha}} \max\left(\sqrt{2}(p - \alpha c), p + h\right). \blacksquare$$

The meaning of $s < 0$ is intuitive: a certain number of customers' orders are to be ensured before the launching of a production series. That is, $s < 0$ is the maximum backlogging rate required for production.

We next define:

$$G(x) = G(s) + \int_s^x H(\xi)d\xi, x > s$$

$$G(x) = G(s), \text{if } x < s \tag{7.24}$$

$$G(s) = \frac{1}{\alpha}\left[\frac{1}{2}\sigma^2 H'(s) - \Phi(c) + s(p - \alpha c)\right]$$

We have also:

$$G(s) = -k + G(S) \tag{7.25}$$

$$\alpha G(s) + \Phi(c) - s(p - \alpha c) \geq 0 \tag{7.26}$$

and the function $G(x)$ is the solution of Eq. 7.17.

Remark Since $G(x)$ Bellman Eq. 7.17, the s, S policy obtained by Theorem 1 is an optimal impulse control.

Optimal Feedback

The inventory control is governed by an s, S policy, where the pair (s, S) is that defined in Theorem 2. The values of the pair (s, S) are different from those without effort towards environmental considerations. The optimal voluntary environmental effort is given by a feedback expression of the inventory level, defined by the formula:

$$\widehat{Q}(x) = \widehat{Q}(H(x) + c) \tag{7.27}$$

$$= \frac{\nu_1}{2\beta}(\varpi - c), \forall x \leq s \tag{7.28}$$

Assume $\varpi > c$, we get:

$$\widehat{Q}(x) = \frac{\nu_1}{2\beta}(\varpi - c), \forall x \leq s \tag{7.29}$$

and:

$$\widehat{Q}'(x) = \frac{\nu_1}{2\beta}H'(x)\mathbb{1}_{H(x)+c<\varpi} \tag{7.30}$$

Therefore, from Theorem 2, we have:

$$\widehat{Q}'(x) \leq 0, \text{if } x \leq \Sigma \tag{7.31}$$

$$\widehat{Q}'(x) > 0, \text{if } x > \Sigma$$

Study of Condition (Eq. 7.20)

Condition (Eq. 7.20) reads:

$$\Phi\left(\frac{p}{\alpha}\right) - \Phi(c) = \int_c^{\frac{p}{\alpha}} \Phi'(\lambda)d\lambda \leq \frac{1}{\beta}(p - \alpha c)$$

Therefore, from Eq. 7.12 it follows:

$$\nu_0\left(\frac{p}{\alpha} - c\right) + \frac{\nu_1^2}{2\beta}\int_c^{\frac{p}{\alpha}} (\varpi + \lambda)^+ d\lambda \leq \frac{1}{\beta}(p - \alpha c) \tag{7.32}$$

This condition reads:

$$\frac{\nu_1^2}{4\beta}(\varpi - c)^2 \leq \left(\frac{\alpha}{\beta} - \nu_0\right)\left(\frac{p}{\alpha} - c\right) + \frac{\nu_1^2}{4\beta}\left(\left(\varpi - \frac{p}{\alpha}\right)^+\right)^2 \tag{7.33}$$

Discussion and Managerial Implications

From Eqs. 7.29–7.31, it follows that for $+\infty < x < \Sigma$, $\widehat{Q}'(x)$ decreases from $\frac{\nu_1}{2\beta} \times \left(\varpi + \frac{h}{\alpha}\right)$ to $\frac{\nu_1}{2\beta}(\varpi - H(\Sigma) - c)^+$, while from $\Sigma \leq x \leq s$, it increases from $\frac{\nu_1}{2\beta}(\varpi - H(\Sigma) - c)^+$ to $\frac{\nu_1}{2\beta}(\varpi - c)$. These results suggest the existence of a non-monotonic relationship between the voluntary environmental effort and the inventory level. From Theorem 2, the minimum inventory level s is negative, which means that the reaching of a maximum backlogging boundary should precede production. This is justified by the fact that it is less costly to incur a cumulative shortage cost over $s \leq x \leq 0$ rather than a fixed cost of production. On the other hand, the threshold $\Sigma > s$, for which $\widehat{Q}(x)$ reaches a minimum—though positive—value, is also negative from Theorem 2. Regarding the order-up-to level S, we assume that it is positive, which requires that the fixed cost of production is sufficiently large, i.e., $k > \int_s^0 H(x)dx$, to justify a large enough production scale, ξ_i. Below the backlogging threshold Σ, i.e., $S \geq x \geq \Sigma$, the voluntary environmental effort should decrease as the inventory decreases and then as the backlogging increases. Between the backlogging threshold Σ and the maximum backlogging boundary s, the voluntary environmental effort should have an increasing pattern. Along the decreasing portion of the curve, there exists $\chi > 0$ such that $x = \Sigma + \chi \gtreqless 0$, for which the voluntary environmental effort should have the same value as for $x = s$, that is, $\widehat{Q}(x = \Sigma + \chi) = \widehat{Q}(x = s) = \frac{\nu_1}{2\beta}(\varpi - c)$. Assuming that the fixed cost of production is sufficiently large, i.e., $k > \int_s^{\Sigma+\chi} H(x)dx$, so that $S > \Sigma + \chi$, the voluntary environmental effort should exhibit a U-shaped curve with respect to the inventory level as determined by a (s, S) policy over the interval $x \in [\Sigma + \chi, s]$. For $S > x > \Sigma + \chi$, the

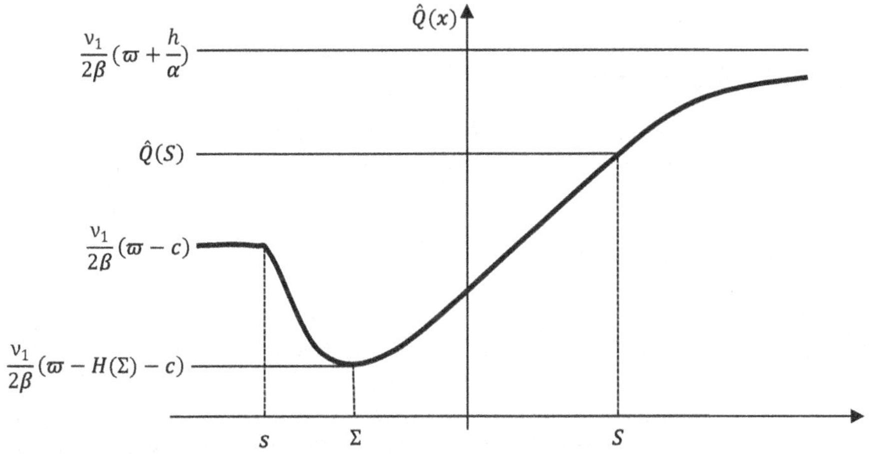

Fig. 7.1 Optimal voluntary environmental effort policy

environmental effort should then decrease in a concave way, due to the convex cost function of voluntary environmental efforts. In any case, the environmental effort associated with a zero-inventory level should be located on the decreasing portion of the curve. That is, even with zero inventory, the voluntary environmental effort remains worthwhile. Along the decreasing portion of the U-curve, the voluntary environmental effort is set at a higher level when the inventory is greater to boost the demand, which can then be instantaneously satisfied. Note that from Eq. 7.5, the instantaneous sales also have a U-shaped relationship pattern with respect to the inventory level. Consequently, the cumulative sales curve is concave-convex while the inventory level has a convex-concave pattern.

Figure 7.1 illustrates the optimal pattern of a voluntary environmental effort policy depending on the evolution from backlogging to inventory, that is, from $x = s$ to $x = S$, where the volatility has been neglected for convenience. Clearly, the voluntary environmental effort stimulates the sales not only to decrease the inventory at hand and thus the related holding cost, if any, but also to some extent to increase the backlogging when the inventory is exhausted. Below the backlogging threshold, the objective is clearly to generate economies of scale to improve production efficiency. Therefore, the backlogging threshold Σ is the turning point for the use of voluntary environmental efforts as an effective sales policy to a cost-efficient production policy. To turn the voluntary environmental efforts toward an increasing pattern, it is therefore essential to identify the backlogging threshold Σ, which corresponds to when the cumulative sales reach a plateau with respect to inventory.

The managerial implications of the above results are summarized as follows: in the context of an (s, S) inventory policy, a firm should leverage voluntary environmental efforts on a quasi-cyclical basis as a marketing tool when there is inventory, and as a production efficiency tool in the case of backlogging. This means that voluntary environmental efforts should be used to stimulate the demand when the

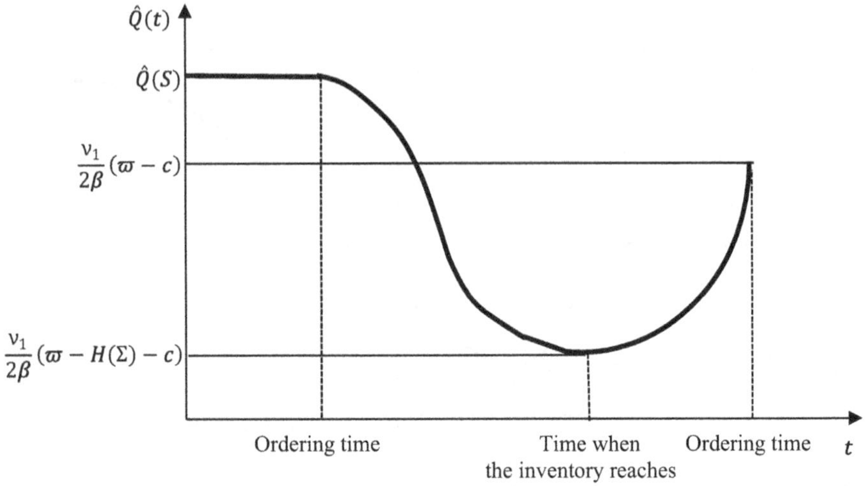

Fig. 7.2 Evolution of voluntary environmental effort in a cycle inventory between two ordering times

stock is replenished to maximize the sales revenue and minimize the holding costs, and generate backlogs when stock is depleted to ensure that the fixed costs of future production are covered (see Fig. 7.2). Relatedly, the implementation of an (s, S) inventory policy should prevent a firm from using voluntary environmental policy as a simple means to get ephemeral reputation insurance. This conclusion is in sharp contrast with Minor and Morgan (2011) and Barrage et al. (2020). The reason lies in the intricacy of production and inventory operations and voluntary environmental policy. That is, the efficiency of production and inventory operations is contingent upon the meticulous deployment of the voluntary environmental policy while the effectiveness of the voluntary environmental policy depends upon the consistent execution of the operational planning.

Finally, a zero or negative inventory should not result in an absence of voluntary environmental efforts. Actually, current voluntary environmental efforts are necessary to prepare for the next sales cycle. In this regard, the fact that increasing or decreasing the voluntary environmental effort is triggered by a threshold of the inventory, is a result of interest to practitioners.

Concluding Remarks

In sharp contrast with Barcos et al. (2013), who assume an inverted U-shaped relationship between firms' environmental initiatives and their inventory levels, we show that a U-shaped relationship actually prevails when an (s, S) inventory policy is implemented. The justification for our result is clear: voluntary environmental efforts are successively deployed for marketing reasons when an inventory exists, and then after for production efficiency reasons when the inventory is exhausted. That is, a firm's voluntary environmental policy should not result from the search for balance between two opposing external pressures, but rather it should serve to reconcile marketing and production objectives.

A numerical study involving a sensitivity analysis of the model with respect to the way the demand is affected by the environmental effort and the parameters that impact the profitability of the firm are needed to complement this study. An important research extension could be to look at price as another important driver of instantaneous sales and to characterize the nature of its interaction with voluntary environmental efforts under an (s, S) inventory policy. Another direction for future research would be to combine both voluntary and regulatory environmental efforts to determine whether both efforts act as mutual complements or substitutes in the context of an (s, S) inventory policy.

Appendix

Solution of Eq. 7.18

We first solve the following problem, obtained by looking at the derivative $H_s(x) = G_s'(x)$, where s is fixed, that is:

$$-\frac{1}{2}\sigma^2 H_s''(x) + \Phi'(H_s(x) + c)H_s'(x) + \alpha H_s(x) + (h + \alpha c)\mathbb{1}_{x>0} + (p - \alpha c)\mathbb{1}_{x<0} = 0,$$

$$x > s \tag{7.34}$$

$$H_s(s) = 0, H_s(+\infty) = -c - \frac{h}{\alpha}$$

We look for a solution $H_s(x)$ which is C^1, with bounded second derivative, satisfying the inequalities:

$$-c - \frac{h}{\alpha} \le H_s(x) \le -c + \frac{p}{\alpha} \tag{7.35}$$

We note that, in view of the boundary conditions, if a solution exists, then from the left inequality (Eq. 7.35):

$$0 \le \Phi'(H_s(x) + c) \le \nu_0 + \frac{\nu_1{}^2}{2\beta}\left(\varpi + \frac{h}{\alpha}\right) \tag{7.36}$$

Theorem 2. *We take $s < 0$. Then there exists one and only one solution of Eq. 7.34, which is $C^1(s, +\infty)$ with bounded second derivative. The second derivative has a discontinuity at 0. The inequalities (Eq. 7.36) are satisfied.* ∎

The existence and uniqueness of $H_s(x)$ follows from standard results on second-order two-point boundary value problems with bounded coefficients.

We can immediately state that the problem:

$$-\frac{1}{2}\sigma^2 G_s''(x) + \Phi(G_s'(x)+c) + \alpha G_s(x) + (h+\alpha c)x^+ + (p-\alpha c)x^- = 0,$$

$$x > s \tag{7.37}$$

$$G_s'(s) = 0, G_s'(+\infty) = -c - \frac{h}{\alpha}$$

has one and only one solution given by:

$$G_s(s) = \frac{1}{\alpha}\left[\frac{1}{2}\sigma^2 H_s'(s) - \Phi(c) + s(p-\alpha c)\right] \tag{7.38}$$

$$G_s(x) = G_s(s) + \int_s^x H_s(\xi)d\xi$$

The fact that $s < 0$ is not necessary to obtain a solution of Eq. 7.34, but we shall need it later on. In fact, we shall restrict further the interval for s. Recalling $\gamma = \frac{\sqrt{2\alpha}}{\sigma}$, problem (Eq. 7.34) is equivalent to the integral equation:

$$H_s(x) = \frac{2}{\sigma^2\gamma}e^{-\gamma(x-s)}\int_s^x e^{\gamma(\xi-s)}\left[p - \alpha c - (p+h)e^{-\gamma\xi^-}\right]d\xi$$
$$+\frac{1}{\sigma^2}\int_s^x e^{-\gamma(x-\xi)}\left[1 + e^{-2\gamma(\xi-s)}\right]\Phi(H_s(\xi)+c)d\xi$$
$$-\frac{1-e^{-2\gamma(x-s)}}{\sigma^2}\int_x^{+\infty} e^{-\gamma(\xi-x)}\Phi(H_s(\xi)+c)d\xi \tag{7.39}$$

We can then compute the derivative:

$$H_s'(x) = \frac{2}{\sigma^2\gamma}$$
$$\times\left\{-\gamma e^{-\gamma(x-s)}\int_s^x e^{\gamma(\xi-s)}\left[p-\alpha c-(p+h)e^{-\gamma\xi^-}\right]d\xi + p - \alpha c - (p+h)e^{-\gamma x^-}\right\}$$
$$+\frac{2}{\sigma^2}\Phi(H_s(x)+c) - \frac{\gamma}{\sigma^2}\int_s^x e^{-\gamma(x-\xi)}\left[1+e^{-2\gamma(\xi-s)}\right]\Phi(H_s(\xi)+c)d\xi$$
$$-\frac{\gamma}{\sigma^2}\left[1+e^{-2\gamma(x-s)}\right]\int_x^{+\infty} e^{-\gamma(\xi-x)}\Phi(H_s(\xi)+c)d\xi$$
$$\tag{7.40}$$

Particularly, for $x = s$, we get, recalling that $s < 0$:

$$H_s'(s) = \frac{2}{\sigma^2\gamma}[p - \alpha c - (p+h)e^{\gamma s}] + \frac{2}{\sigma^2}\left[\Phi(c) - \gamma\int_s^{+\infty} e^{-\gamma(\xi - s)}\Phi(H_s(\xi) + c)d\xi\right] \tag{7.41}$$

Since Φ is monotone increasing, we obtain from Eq. 7.41 the estimate:

$$H_s'(s) \geq \frac{2}{\sigma^2\gamma}[p - \alpha c - (p+h)e^{\gamma s}] + \frac{2}{\sigma^2}\left(\Phi(c) - \Phi\left(\frac{p}{\alpha}\right)\right) \tag{7.42}$$

Recalling assumption (Eq. 7.19) and the definition of s^*, we have:

$$\frac{1}{\gamma}[p - \alpha c - (p+h)e^{\gamma s}] + \Phi(c) - \Phi\left(\frac{p}{\alpha}\right) \geq 0, \forall s \leq s^*, \tag{7.43}$$

therefore $H_s'(s) > 0, \forall s < s^*$.

Proposition 1. *Assume $H_s'(s) > 0$. This assumption is true as soon as $s \leq s^*$. Then the function $H_s(x)$ (a solution of Eq. 7.34) has a unique zero, $S(s) > s$ as well as a unique maximum $\Sigma(s) < 0$. Moreover, $H_s'(x) > 0$, if $s \leq x < \Sigma(s)$ and $H_s'(x) < 0$ if $x > \Sigma(s)$. Also $H_s(x) > 0$, if $s < x < S$ and $H_s(x) < 0$ if $x > S$. If $H_s'(s) \leq 0$ then $H_s'(x) < 0$, $\forall x > s$. Also, $H_s(x) < 0$, $\forall x > s$. In this case, we define $S(s) = s$.* ∎

The proof is technical and relies on maximum principal concepts. The method is almost identical to that of the paper of the first author (Bensoussan et al., 2018).

We next provide estimates on the derivative of $H_s(x)$.

Proposition 2. *Assume $H_s'(s) > 0$. We have the estimates:*

$$0 \leq H_s'(x) \leq \frac{2(p - \alpha c)}{\sigma\sqrt{\alpha}}, \text{if } 0 \leq x \leq \Sigma(s), \tag{7.44}$$

$$-\frac{\sqrt{2}}{\sigma\sqrt{\alpha}}(p+h) \leq H_s'(x) \leq 0, \text{if } \Sigma(s) \leq x \leq +\infty. \blacksquare$$

If $H_s'(s) \leq 0$, then the second estimate holds for $s \leq x \leq +\infty$. So, in all cases:

$$|H_s'(x)| \leq \frac{\sqrt{2}}{\sigma\sqrt{\alpha}}\max\left(\sqrt{2}(p - \alpha c), p + h\right). \tag{7.45}$$

We also state the following:

Proposition 3. *The function $H_s(x)$ is continuous in s, and $H_s'(s)$ is also continuous.* ∎

Consider the next Eq. 7.34 with $s = 0$. It writes:

$$-\frac{1}{2}\sigma^2 H_0''(x) + \frac{d}{dx}\Phi(H_0(x)+c) + \alpha H_0(x) + (h+\alpha c) = 0, \quad x>0 \tag{7.46}$$

$$H_0(0) = 0, H_0(+\infty) = -c - \frac{h}{\alpha}.$$

We can check the properties:

$$H_0'(0) < 0, H_s'(s) > 0, \forall s \leq s^*. \tag{7.47}$$

From the continuity of the function $H_s'(s)$, there exists a point $\overline{s} \in (s^*, 0)$ such that $H_{\overline{s}}'(\overline{s}) = 0$. We can take the smallest one, so that $H_s'(s) > 0$, $\forall s < \overline{s}$. It follows that $S(s) >$ max (s, s^*) for $s < \overline{s}$ and $S(\overline{s}) = \overline{s}$. Moreover, the function $S(s)$ is continuous on $(-\infty, \overline{s})$.

Finding s

We obtain s by solving Eq. 7.19, which amounts to solving:

$$k = \int_s^{S(s)} H_s(x)dx \tag{7.48}$$

where k corresponds to the fixed cost of production (see Eq. 7.3).

We then have:

Proposition 4. *There exists a solution of Eq. 7.48 in the interval* $(-\infty, \overline{s})$*. We take the smallest value, in case there are several ones.* ∎

If we introduce the function:

$$\zeta(s) = \int_s^{S(s)} H_s(x)dx$$

we can check that it is continuous. Moreover, $\zeta(\overline{s}) = 0$ and $\zeta(s) \to +\infty$ as $s \to -\infty$. It follows that Eq. 7.48 has indeed a solution, and the solution is smaller than $\overline{s}$.

Data availability statement Data sharing is not applicable to this article as no new data were created or analyzed in this study.

Disclosure statement The authors ensure that no conflict of interest is directly or indirectly related to the work submitted for publication.

References

Aggarwal, S. C. (1974). A review of current inventory theory and its applications. *International Journal of Production Research, 12*(4), 443–482.

Arimura, T. H., Hibiki, A., & Katayama, H. (2008). Is a voluntary approach an effective environmental policy instrument?: A case for environmental management systems. *Journal of Environmental Economics and Management, 55*(3), 281–295.

Arrow, K., Harris, T., & Marschak, J. (1951). Optimal inventory policy. *Econometrica, 19*(3), 250–272.

Barcos, L., Barroso, A., Surroca, J., & Tribo, J. A. (2013). Corporate social responsibility and inventory policy. *International Journal of Production Economics, 143*, 580–588.

Barrage, L., Chyn, E., & Hastings, J. (2020). Advertising and environmental stewardship: Evidence from the BP oil spill. *American Economic Journal: Economic Policy, 12*(1), 33–61.

Benkherouf, L., & Bensoussan, A. (2009). Optimality of an (*s*,) policy with compound Poisson and diffusion demands: A quasi-variational inequalities approach. *SIAM Journal of Control Optimization, 48*(2), 756–762.

Benkherouf, L., & Sethi, S. P. (2010). Optimality of *(s, S)* policies for a stochastic inventory model with proportional and lump-sum shortage costs. *Operations Research Letters, 38*(4), 252–255.

Bensoussan, A., Cakanyildirim, M., Feng, Q., & Sethi, S. P. (2009). Optimal ordering policies for stochastic inventory problems with observed information delays. *Production and Operations Management, 18*(5), 546–559.

Bensoussan, A., Crouhy, M., & Proth, J.-M. (1983). *Mathematical theory of production planning*. North-Holland.

Bensoussan, A., & Lions, J. L. (1987). *Impulse control and quasi variational inequalities*. Wiley.

Bensoussan, A., Liu, R. H., & Sethi, S. P. (2005). Optimality of an (*s*,) policy with compound Poisson and diffusion demands: A quasi-variational inequalities approach. *SIAM Journal of Control and Optimization, 44*(5), 1650–1676.

Bensoussan, A., Skaaning, S., & Turi, J. (2018). Inventory control with fixed cost and price optimization in continuous time. *Journal of Applied Analysis and Computation, 8*(3), 805–835.

Bensoussan, A., & Tapiero, C. S. (1982). Impulsive control in management: Prospects and applications. *Journal of Optimization Theory and Applications, 37*(4), 419–442.

Borck, J. C., & Coglianese, C. (2009). Voluntary environmental programs: Assessing their effectiveness. *Annual Review of Environment and Resources, 34*, 305–324.

Cadenillas, A., Lakner, P., & Pinedo, M. (2010). Optimal control of a mean-reverting inventory. *Operations Research, 58*(6), 1697–1710.

Chen, C., & Monahan, G. E. (2010). Environmental safety stock: The impacts of regulatory and voluntary control policies on production planning, inventory control, and environmental performance. *European Journal of Operational Research, 207*(3), 1280–1292.

Chen, F., Ray, S., & Song, Y. (2006). Optimal pricing and inventory control policy in periodic-review systems with fixed ordering cost and lost sales. *Naval Research Logistics, 53*(2), 117–136.

Chen, X., & Simchi-Levi, D. (2004a). Coordinating inventory control and pricing strategies with random demand and fixed ordering cost: The finite horizon case. *Operations Research, 52*(6), 887–896.

Chen, X., & Simchi-Levi, D. (2004b). Coordinating inventory control and pricing strategies with random demand and fixed ordering cost: The infinite horizon case. *Mathematics of Operations Research, 29*(3), 698–723.

Corbett, C. J., & Klassen, R. D. (2006). Extending the horizons: Environmental excellence as key to improving operations. *Manufacturing and Service Operations Management, 8*(1), 5–22.

Dolgui, A., & Proth, J. M. (2010). *Supply chain engineering: Useful methods and techniques*. Springer.

Drake, D., Kleindorfer, P. R., & Van Wassenhove, L. N. (2016). Technology choice and capacity portfolios under emissions regulation. *Production and Operations Management, 25*(6), 1006–1025.

El Ouardighi, F., Sim, J. E., & Kim, B. (2021). Pollution accumulation and abatement policies in two supply chains under vertical and horizontal competition and strategy types. *Omega, 98*(102108), 1–35.

El Ouardighi, F., Sim, J. E., & Kim, B. (2016). Pollution accumulation and abatement policy in a supply chain. *European Journal of Operational Research, 248*(3), 982–996.

Fox, E., Metters, R., & Semple, J. (2006). Optimal inventory policy with two suppliers. *Operations Research, 54*(2), 389–393.

Herbon, A. (2021). Managing an expiring product under a market that is heterogeneous in the sensitivity to the retailer's reputation. *International Journal of Production Economics, 232*, 107990.

Heydari, J., Govindan, K., & Basiri, Z. (2021). Balancing price and green quality in presence of consumer environmental awareness: A green supply chain coordination approach. *International Journal of Production Research, 59*(7), 1957–1975.

Hollier, R. H., Makj, K. L., & Lam, C. J. (1995). Continuous review (*s*,) policies for inventory systems incorporating a cutoff transaction size. *International Journal of Production Research, 33*(10), 2855–2865.

Hong, Z., Zhang, Y., Yu, Y., & Chu, C. (2020). Dynamic pricing for remanufacturing within socially environmental incentives. *International Journal of Production Research, 58*(13), 3976–3997.

Huh, W. T., & Janakiraman, G. (2008). (*s*,) optimality in joint inventory-pricing control: An alternate approach. *Operations Research, 56*(3), 783–790.

Innes, R., & Sam, A. G. (2008). Voluntary pollution reductions and the enforcement of environmental law: An empirical study of the 33/50 Program. *The Journal of Law & Economics, 51*(2), 271–296.

Jaber, M. Y., Glock, C. H., & El Saadany, A. M. A. (2013). Supply chain coordination with emissions reduction incentives. *International Journal of Production Research, 51*(1), 69–82.

Jacobs, B. W., Singhal, V. R., & Subramanian, R. (2010). An empirical investigation of environmental performance and the market value of the firm. *Journal of Operations Management, 28*(5), 430–441.

Jabbour, C. J. C., de Sousa Jabbour, A. B. L., Govindan, K., de Freitas, T. P., Soubihia, D. F., Kannan, D., & Latan, H. (2016). Barriers to the adoption of green operational practices at Brazilian companies: Effects on green and operational performance. *International Journal of Production Research, 54*(10), 3042–3058.

Kalymon, B. A. (1971). Stochastic prices in a single-item inventory purchasing model. *Operations Research, 19*(6), 1434–1458.

Kelle, P., & Milne, A. (1999). The effect of (*s*, *S*) ordering policy on the supply chain. *International Journal of Production Economics, 59*(1–3), 113–122.

Khmelnitsky, E., & Singer, G. (2015). An optimal inventory management problem with reputation-dependent demand. *Annals of Operations Research, 231*(1), 305–316.

King, A., & Lenox, M. (2002). Exploring the locus of profitable pollution reduction. *Management Science, 48*(2), 289–299.

Kleindorfer, P. R., Singhal, K., & Van Wassenhove, L. N. (2005). Sustainable operations management. *Production and Operations Management, 14*(4), 482–492.

Koehler, D. A. (2007). The effectiveness of voluntary environmental programs – A policy at a crossroads? *Policy Studies Journal, 35*(4), 689–722.

Krass, D., Nedorezov, T., & Ovchinnikov, A. (2013). Environmental taxes and the choice of green technology. *Production and Operations Management, 22*(5), 1035–1055.

Kroes, J., Subramanian, R., & Subramanyam, R. (2012). Operational compliance levers, environmental performance, and firm performance under cap and trade regulation. *Manufacturing & Service Operations Management, 14*(2), 186–201.

Larson, C. E., Olson, L. J., & Sharma, S. (2001). Optimal inventory policies when the demand distribution is not known. *Journal of Economic Theory, 101*(1), 281–300.

Mc Kinsey. (2019). *New age of the consumer*. US Survey.

Minor, D., & Morgan, J. (2011). CSR as reputation insurance: *Primum non nocere*. *California Management Review, 53*(3), 40–59.

Perera, S., Sethi, S. P. (2022). A survey of stochastic inventory models with fixed costs: Optimality of (*s*, *S*)-type policies. *Production and Operations Management* (Forthcoming).

Porteous, A. H., Rammohan, S. V., & Lee, H. L. (2015). Carrots or sticks? Improving social and environmental compliance at suppliers through incentives and penalties. *Production and Operations Management, 24*(9), 1402–1413.

Porteus, E. L. (1990). Stochastic inventory theory. *Handbooks in OR & MS, 2*, 605–652.

Potoski, M., & Prakash, A. (2005). Green clubs and voluntary governance: ISO 14001 and firms' regulatory compliance. *American Journal of Political Science, 49*(2), 235–248.

Ravichandran, N. (1995). Stochastic analysis of a continuous review perishable inventory system with positive lead time and Poisson demand. *European Journal of Operational Research, 84*(2), 444–457.

Sarkis, J., & Zhu, Q. (2018). Environmental sustainability and production: Taking the road less travelled. *International Journal of Production Research, 56*(1–2), 743–759.

Scarf, H. (1960). The optimality of (*S*,) policies in dynamic inventory problems. In K. J. Arrow, S. Karlin, & P. Suppes (Eds.), *Mathematical methods in social sciences* (pp. 196–202). Stanford University Press.

Sethi, S. P., & Cheng, F. (1997). Optimality of (*s*,*S*) policies in inventory models with Markovian demand. *Operations Research, 45*(6), 931–939.

Silver, E. A. (1974). A control system for coordinated inventory replenishment. *International Journal of Production Research, 12*(6), 647–671.

Sim, J. E., El Ouardighi, F., & Kim, B. (2019). Economic and environmental impact of vertical decision sequences under vertical and horizontal competition and integration. *Naval Research Logistics, 66*(2), 133–153.

Song, Y., Ray, S., & Boyaci, T. (2009). Optimal dynamic joint inventory-pricing control for multiplicative demand with fixed order costs and lost sales. *Operations Research, 57*(1), 245–250.

Wu, J. J. (2009). Environmental compliance: The good, the bad, and the super green. *Journal of Environmental Economics and Management, 90*(11), 3363–3381.

Xiao, W., Gaimon, C., Subramanian, R., & Biehl, M. (2019). Investment in environmental process improvement. *Production and Operations Management, 28*(2), 407–420.

Zipkin, P. (2008). On the structure of lost-sales inventory model. *Operations Research, 56*(4), 937–944.

Chapter 8
Recycling of Multi-source Waste in an Aggregate Circular Economy

Raouf Boucekkine, Fouad El Ouardighi, and Konstantin Kogan

Abstract We investigate how the relationship between capital accumulation and pollution is affected by the source of pollution: production or consumption. We are interested in polluting waste that cannot be naturally absorbed, but for which recycling efforts aim to avoid massive pollution accumulation with harmful consequences in the long run. Based on both environmental and social welfare perspectives, we determine how the interaction between growth and polluting waste accumulation is affected by the source of pollution, i.e., either consumption or production, and by the fact that recycling may or may not act as an income generator, i.e., either capital-improving or capital-neutral recycling efforts. Several new results are extracted regarding optimal recycling policy and the shape of the relationship between production and pollution. Besides the latter concern, we show both analytically and numerically that the optimal control of waste through recycling allows to reaching larger (resp., lower) consumption and capital stock levels under consumption-based waste compared to production-based waste while the latter permits to reach lower stocks of waste through lower recycling efforts.

Keywords Circular economy · Multisource waste · Optimal recycling · Capital accumulation

R. Boucekkine
Rennes School of Business, Rennes, France
e-mail: raouf.boucekkine@rennes-sb.com

F. El Ouardighi (✉)
ESSEC Business School, Cergy Pontoise Cedex, France
e-mail: elouardighi@essec.fr

K. Kogan
Bar-Ilan University, Ramat-Gan, Israel
e-mail: kogank@biu.ac.il

F. El Ouardighi (ed.), *Essays on Pollution Control in Economics and Management Science*, Lecture Notes in Operations Research,
https://doi.org/10.1007/978-3-031-78227-5_8

Introduction

Because it is considered a credible model of sustainable development, potentially enabling efficient production at minimal rates of use of resources and driving polluting waste and emissions down, the circular economy is at the heart of numerous ongoing research projects in operation research and ecological economics. In the related operation research and management domain also referred to as closed-loop supply chain (see Battini et al. (2017), for an overview), the emphasis is the design of reverse logistics to capture values of products consumed or used by businesses, with the ultimate goal to reduce the environmental damage. A more recent trend is the development of circular business models, and even circular business model innovation, as recently highlighted in Geissdoerfer et al. (2020). Lastly, the range of applications of the circular economy or closed-loop supply chains is quite wide: it ranges from the agriculture (Barros et al., 2020 to the manufacturing (Acerbi & Taisch, 2020) sectors through much smaller scale problems such as waste management case studies (Hrabec et al., 2020) or sustainable inventory models (see Suhandi and Chen (2023), in the case of the pharmaceutical industry).

Needless to say, a key tool in the circular economy is recycling. This work examines the role of recycling in a circular economy with multiple sources of waste. The multiplicity of waste sources complicates enormously the optimal design of the corresponding recycling strategy. Within the literature surveyed above, comparatively fewer papers deal with this issue. A few related applied papers incidentally address this issue (see, for example, Nakamura (1999), in the ecological economics literature, and Sheu (2007), in operations research) but the theoretical literature is almost silent on it. Boucekkine and El Ouardighi et al. (2016) built up a circular economy macro-model with two different sources of waste, consumption versus production, but they have not studied the distinct economic implications of each source of waste and the inherent policy implications. However, broadly speaking, polluting waste can emanate either from a single source, for example, solely from manufacturing activities (e.g., mineral wastes such as cement or glass wastes) or consumption activities (e.g., cellulose acetate-based materials such as cigarette butts, soda cans or polythene), or from hybrid sources, i.e., both manufacturing and consumption activities. Plastics are the best illustration of multisource waste. Incidentally, they have become indeed a global pollution issue since the formation of sprawling garbage patches in the Pacific (Eriksen et al., 2013) and Atlantic (Law et al., 2010) oceans. Most recent data show that plastic smog estimated to be over 170 trillion plastic particles float in the world's oceans (Eriksen et al., 2023).

The debate of the safest and most efficient treatment of the global issue posed by plastic waste is indeed vivid (see Verma et al. (2016), for an early view of the options, including burning in open spaces or incineration plants). Significant progress has been recently made in the design of pyrolysis systems based on bio-refineries to convert waste such as plastic and biomass waste into energy. For instance, from used tires, the patented recycling technology, Pyrum thermolysis,[1]

[1] See https://www.pyrum.net/en/about-us/technology

makes it possible to extract 50% oil, 38% coke, and 12% gas, which can be used to ensure energy self-sufficiency. The potential market for the Pyrum technology is considerable since 17 million tons of used tires are generated each year worldwide. In addition to its environmental neutrality, this technology is highly profitable with a return on investment reaching 25%. Further income generation from recycling is key in our theory, as it will be clear below.

We are indeed particularly interested in polluting waste for which natural absorption takes an extremely long time, i.e., several centuries, (e.g., iron, aluminum, mineral residues) or that is nonbiodegradable (e.g., plastic, computer hardware). In both cases, no natural abatement that prevents solid waste accumulation can be reasonably assumed. In this setup, recycling efforts are required to avoid massive accumulation with harmful long-term consequences. We therefore focus on the design of recycling efforts in our analysis of the optimal trade-off between economic performance and polluting waste. Depending on whether the polluting waste emanates from a stock of productive capital or consumption flows, the long-term effects on the stocks of productive capital and polluting waste, and therefore their recycling policy implications, might differ significantly. In this paper, we investigate how the trade-off between capital growth and pollution accumulation is affected by the source of pollution—productive capital or consumption—and identify which source of pollution should be mitigated first to enhance environmental, economic, and welfare performances.

A second aspect is crucial in our theory: while recycling is not without cost and requires an effort (whatever the waste source) which should be accounted for in the social welfare arithmetics, it has potentially a double social benefit. One is obvious and works through the reduction of the stock of waste. The second is more hypothetical: recycling may also generate income. In our paper, we distinguish between two cases: one where recycling efforts benefit capital accumulation, i.e., capital-improving recycling (CIR) efforts, and one where no additional revenue is drawn from recycling efforts, i.e., capital-neutral recycling (CNR) efforts. Capital-neutral recycling reflects a depreciation of the qualitative properties of recycled volume (e.g., Martin, 1982), while CIR is an effective recycling process. An example of CIR can be found in the European plastics industry, where 60 million tons of plastics diverted from landfills are equivalent to over 66 billion euros (Plastics Europe, 2015). More generally, CIR has received strong support in the literature (e.g., Doonan et al., 2005; Pati et al., 2006; Jacobs et al., 2010; Simpson, 2012). Two comments are worth doing at this stage. Of course, net income from recycling may be optimized, several operations research papers are devoted to this question [see, for example, Derigs and Friederichs (2009)]. In ecological economics, the distinction between the original consumption good and the recycled consumption good is precisely done to model the extra income (or welfare) arising from recycling [since the seminal work of Lusky (1976)]. Here we consider a reduced form specification and assume that the net income generated from recycling may or may not benefit capital accumulation, the unique engine of production growth in our stylized circular economy model. Second, and related to the first point, we do not integrate the fact that recycling may enlarge the set of consumption goods and the aggregate level of consumption. Again, this is due to our focus on the much less obvious recycling capital accumulation mechanism and its long-term implications.

With all these elements accounted for, we will tackle three distinct sets of original questions:

- How do consumption and production-based polluting waste respectively affect the short and long-term performance of the economy as measured by consumption, capital stock, recycling effort, or the stock of waste? What is the role played by income generation through recycling in the latter outcomes?
- How do consumption and production-based polluting waste respectively affect the interaction between capital and polluting waste accumulation from both environmental and social welfare perspectives?
- How do CIR efforts influence the interaction between capital and polluting waste accumulation from both environmental and social welfare perspectives?

Incidentally, we will be also able to explore a fourth question, touching on the debate on the so-called Environmental Kuznets Curve (EKC) literature, as we will be able to compute the optimal shapes of the relationship between pollution and income (or production) for a large variety of scenarios. The EKC conjectures that pollution is an inverse U-shaped function of economic development indicators such as income or capital stock: nations undergo a phase of expansion of both capital (or income) and pollution followed by decreasing pollution and further expansion of capital. Beyond a certain turning point, capital growth coincides with a cleaner environment [see, for example, Stokey (1998)]. Both the theoretical [see, for example, Boucekkine et al. (2013)] and the empirical literature (see de Bruyn & Opschoor, 1997; Sengupta, 1997) are inconclusive regarding the occurrence of the EKC. We will examine the original mechanisms of our recycling model, which includes multiple sources of pollution and income generation schemes, and their respective contributions to our understanding of the relationship between pollution and revenue throughout the development process. To our knowledge, the EKC literature has not yet explored the impact of recycling activities on the latter shapes.[2]

We will build on the circular economy aggregate model developed by Boucekkine and El Ouardighi et al. (2016), and expand upon it to explore more deeply the four sets of problems described above. The current study differs from the former in several respects. Primarily, our model is fully linear-quadratic while in the 2016 paper, the utility function is logarithmic in line with related growth theory papers (e.g., Stokey, 1998). More importantly, and besides differences in a few technical assumptions like the latter, which allow us to push the theoretical analysis

[2] Indeed, the EKC literature is huge (e.g., Boucekkine et al., 2013), and there are thousands of cases studied with different ingredients (abatement, recycling, multisectoral economy, circular or not, etc.), which each features a different case for non-monotonicity with the corresponding underlying mechanisms. In our paper, we use "EKC-like" for any capital-waste trajectory that is not monotonic. Of course, our model need not have the same associated dynamics per variable that we find in the classical case, precisely because we rely on different variables and different income generation mechanisms.

clearly further and at a lower algebraic cost in the current paper, we keep the distinction between consumption-based versus production-based waste and the simple reduced form specification of recycling net income generation. It is worth pointing out that Boucekkine and El Ouardighi's paper is not at all concerned with the exploration of the distinctive dynamical and long-term implications of consumption-based versus production-based waste recycling, which are in contrast the focus of the current work. Indeed, they have only explored the local stability properties of the steady state equilibrium in the presence of both sources of waste.

There are other stylized circular economy macro-models in the ecological economics literature, and our setting entails some essential departures with respect to this literature. We have already explained above the crucial difference with respect to the seminal model of Lusky, the central role of capital accumulation in our model. We also depart from other multisectoral models in the literature [like those of Martin (1982) or Fodha and Magris (2015)] in that we use a simple one sector-one good circular economy, rich enough to replicate the multiple trade-offs one can find in the latter more complex frames. However, what makes definitively the originality of our paper is that it is the first one that studies theoretically the distinctive implications of recycling in the short and long run depending on the source of the waste. We do that as explained above by considering different interaction channels between capital accumulation and recycling, particularly through contrasting the CIR vs the CNR configurations.

We show that under consumption-based waste the central planner reaches larger consumption and capital stock levels than under production-based waste. The main rationale behind this result is that when waste is generated through production, the negative environmental externality limits both capital accumulation and consumption. This brake on capital accumulation is inactive under consumption-based waste.

The paper is structured as follows. Second section develops the central planner model where polluting waste is a by-product of either consumption or production. In third section, we derive the qualitative properties of the model under both consumption and production-based waste. Fourth section compares and contrasts the results, and fifth section concludes the paper.

Model

Following standard assumptions, we consider a society that continuously exploits productive capital, denoted by $K(t) \geq 0$, where t stands for time, to linearly generate a flow of revenue, $aK(t)$, where $a > 0$ is the constant marginal revenue generated by the productive capital stock. The revenue generated allows a certain consumption level, denoted by $c(t) \geq 0$.

Polluting waste, denoted by $w(t) \geq 0$, is caused by either the production or the consumption process (Boucekkine & El Ouardighi, 2016). In the case where polluting waste emanates from the exploitation of productive capital (production-based waste), we assume that $w(t) = \kappa aK(t) \geq 0$, where $\kappa > 0$ is the marginal wasting impact of production. In the case where polluting waste emanates from consumption (consumption-based waste), we assume that $w(t) = \beta c(t)$, where $\beta > 0$ is the marginal wasting impact of consumption.

To reduce the stock of polluting waste, denoted by $W(t) \geq 0$, the economy may, in all cases, invest in recycling efforts $v(t) \geq 0$ over time. We assume that the waste-generating processes and recycling operations are mutually independent so that the recycling efforts are nonproportional to the waste emissions. This assumption allows for unbounded recycling efforts, i.e., $v(t) \lesseqgtr w(t)$, to account for the possibility of reduction of past waste emissions. Due to the generally long period of natural absorption of the waste, we assume that there is no natural abatement of pollution waste. In economic terms, this implies that the social planner cannot benefit from any natural amenities.

Finally, a fixed proportion of recycled waste is supposed to possibly generate additional revenues, $\varphi v(t)$, and therefore to positively influence capital growth, $\varphi \geq 0$ being the marginal revenue from recycling. However, if $\varphi = 0$, recycling has a neutral impact on capital growth due to an ineffective recycling process. For simplicity, we also assume zero capital depreciation. Indeed, as one can see just below, adding a nonzero depreciation of capital, δ, will essentially lead to the replacement of the productivity parameter, a, by $a - \delta$ in the law of capital accumulation just below. In a realistic case, $a \gg \delta$. So, this assumption is totally innocuous.

Based on the previous assumptions, the endogenous capital growth process is given as:

$$\dot{K} = aK(t) - c(t) + \varphi v(t), \qquad K(0) = K_0 \geq 0$$

where a positive difference between the total revenues from capital and recycled waste, and current consumption results in investment in productive capital, while a negative difference leads to disinvestment. The initial capital endowment is $K_0 \geq 0$.

It is noteworthy that the modeling of income generation through recycling is exclusively aimed at capturing the link between recycling and capital growth. It turns out that the unique engine of growth in this model is capital accumulation because the production function is linear in the capital stock. Whether recycling may be capital-reducing is not an issue here for two reasons. First, as explained in the introduction, we are referring to primary recycling activities where part of the waste is channeled into a recycling sector. This sector may generate profits, notably through cost reduction (Doonan et al., 2005; Pati et al., 2006; Jacobs et al., 2010; Simpson, 2012); we assume that these profits go to investment to capture the link between revenue growth and recycling, and we model the size of this growth effect linearly through the term $\varphi v(t)$, φ measuring the marginal contribution, that is, the *marginal effectiveness*, of recycling on economic growth. It could be the case, of course, that part of the waste (particularly production waste) comes from capital goods in the workplace. For example, some of these goods may be scrapped for environmental reasons: typically, the oldest capital goods are the most polluting (see Boucekkine et al. (2014), for an explicit modeling of the latter environmental nexus).

In such a case, the loss in value due to the scrapping has to be considered in addition to recycling profits in the law of motion of capital. We do assume here that the net effect is positive. Note that this is the most acceptable assumption as the scrapped goods are the oldest, the dirtiest, and also the least productive. Finally, the model works even though we relax the positivity of the parameter φ, we only stick to the most acceptable case.

The general law of motion of polluting waste is given by:

$$\dot{W} = \kappa a K(t) + \beta c(t) - v(t), \qquad W(0) = W_0 \geq 0$$

Notice that we do not introduce natural depreciation of waste: this is because, as outlined strongly in the introduction, we are particularly interested in the waste for which natural absorption takes an extremely long time, or that is nonbiodegradable, implying a self-degradation rate close to zero. A similar assumption can be found in El Ouardighi et al. (2016, 2021), though over the finite time horizon.

We now define the objective function. We assume that the social utility is the difference between the utility drawn from current consumption and the costs incurred from the waste stock and the recycling efforts, respectively. The instantaneous utility from current consumption is a linear quadratic function, that is, $c(t)(\theta - c(t)/2)$, where $\theta > 0$ is the utility-maximizing consumption level. The justification for this choice, notably with respect to Boucekkine and El Ouardighi et al. (2016), is detailed in the introduction.

Further, the waste stock entails negative externalities such as environmental destruction and biomass extinction (Barnes, 2002), and related health consequences. These negative externalities are internalized by the central planner and valued as an increasing convex function of the waste stock, that is, $eW(t)^2/2$, where $e > 0$ is a disutility coefficient associated with the waste stock.

Lastly, regardless of whether they are related to consumption or production-based waste, recycling efforts generally involve separation and transformation efforts. The disutility engendered by these efforts is expressed as an increasing quadratic function denoted by $fv(t)^2/2$, with $f > 0$. To gain more insights and without loss of generality regarding the qualitative properties of the model, f is normalized to 1, so that e, the disutility coefficient of waste, could be interpreted relative to 1, as the disutility coefficient of recycling.

If $r > 0$ denotes the discounting rate, we assume that the social planner is sufficiently patient that the net marginal revenue generated by the productive capital stock is strictly positive (i.e., $a - r > 0$) to allow sustained capital accumulation.

Assuming an infinite planning horizon, the social planner's optimal control problem is:

$$U = \int_0^\infty e^{-rt}\left[c(t)\left(\theta - \frac{c(t)}{2}\right) - \frac{eW(t)^2}{2} - \frac{v(t)^2}{2}\right]\mathrm{d}t \tag{8.1}$$

subject to:

$$\dot{K} = aK(t) - c(t) + \varphi v(t), \qquad K(0) = K_0 \geq 0 \tag{8.2}$$

$$\dot{W} = \kappa a K(t) + \beta c(t) - v(t), \qquad W(0) = W_0 \geq 0 \tag{8.3}$$

under the control constraints $c(t) \geq 0$ and $v(t) \geq 0$, and the state constraints $K(t) \geq 0$ and $W(t) \geq 0$.

Analytical Properties of the Optimal Paths

We successively determine the optimal consumption and recycling policies related to consumption and production-based waste generation, and characterize the nature of convergence to the steady state, if any. For each source of waste generation, we assess the long-run effects of recycling efforts on capital growth. It might look odd that steady states could co-exist with an AK production function, which usually serves as the reduced form of endogenous (balanced) growth. Nonetheless, as demonstrated already by Stokey (1998), the internalization of the pollution externality by the central planner eventually leads to not letting capital grow indefinitely, thus choosing convergence to an optimal steady state despite the AK production technology. This is also true here (with recycling "replacing" abatement in the Stokey frame). Clearly, replacing the AK production function with a Cobb-Douglas one, besides complicating a lot the algebra, will not change the concept of admissible long-term equilibria as decreasing returns inherent in Cobb-Douglas production functions disable the emergence of balanced growth paths asymptotically.

Consumption-Based Waste Generation

In this case, we assume that $\kappa = 0$ and $\beta > 0$ in the law of motion of polluting waste (Eq. 8.3). Skipping the time index for convenience, the current-value Hamiltonian is (Grass et al., 2008):

$$H = c\left(\theta - \frac{c}{2}\right) - \frac{eW^2}{2} - \frac{v^2}{2} + \lambda(aK - c + \varphi v) + \mu(\beta c - v) \tag{8.4}$$

where $\lambda \equiv \lambda(t)$ and $\mu \equiv \mu(t)$ are costate variables, that, if the state constraints are not binding, evolve according to:

$$\dot{\lambda} = (r - a)\lambda \tag{8.5}$$

$$\dot{\mu} = r\mu + eW \tag{8.6}$$

under the transversality conditions: $\lim_{t \to \infty} e^{-rt}\lambda(t) = 0$ and $\lim_{t \to \infty} e^{-rt}\mu(t) = 0$.

Although Eqs. 8.5 and 8.6 are given for an interior solution in terms of state variables, we will also discuss important cases where the states can become negative. In such cases, nonnegativity will be imposed, $K(t) \geq 0$ and $W(t) \geq 0$, and the costate equations may include jumps (when the constraints are binding), which we denote by $d\eta(t)$. Whenever jumps are possible in both states, $\eta(t)$ will have subscript 1 to identify a jump due to $K = 0$ and subscript 2 due to $W = 0$. Formally, the jumps are characterized by nonnegativity $d\eta_1(t) \geq 0$, $d\eta_2(t) \geq 0$ and complementary slackness, $\int_0^\infty Kd\eta_1 = 0$ and $\int_0^\infty Wd\eta_2 = 0$.

Assuming interior solutions, the necessary conditions for optimality are:

$$H_c = \theta - c - \lambda + \mu\beta = 0 \Rightarrow c = \theta - \lambda + \mu\beta \tag{8.7}$$

$$H_v = -v + \lambda\varphi - \mu = 0 \Rightarrow v = \lambda\varphi - \mu \tag{8.8}$$

From Eq. 8.5, we get $\lambda(t) = Ge^{(r-a)t}$, where G is a constant of integration. Given that $a - r > 0$, this solution satisfies the transversality condition regardless of whether G is different from zero or not. In all cases, λ is zero at the steady state. This implies that, in the long run, the capital stock is disregarded and that the controls depend only on the implicit value of the waste stock.

Lemma 1. For $\kappa = 0$ and >0, the necessary conditions are sufficient for optimality.

Proof See Appendix A1.

The sufficient conditions used are true only for interior solutions (Seierstadt & Sydsaeter, 1987, p. 187).

Lemma 2. Let the state constraint $W(t) \geq 0$ be nonbinding. Then, it always holds that $\mu < 0$. Further, let the state constraint $K(t) \geq 0$ be nonbinding, then $\lambda > 0$ if $\lambda(0) > 0$ and $\lambda < 0$ when $\lambda(0) < 0$. In addition, we have: $\lim_{t \to +\infty} \lambda = 0$.

Proof See Appendix A2.

Conversely, under binding state constraints (Hartl et al., 1995), both costates λ and μ can exhibit negative jumps which, in this case, are given by the differentiable function:

$$\dot{\eta} = r - a + \frac{(\varphi + \beta)(r\mu + eW)}{(1 + \varphi^2)\lambda} \tag{8.9}$$

We summarize the possible binding cases as follows.

Lemma 3. If $W(0) > 0$ and $K(0) > 0$, neither state constraint $K \geq 0$ exclusively nor state constraint $W \geq 0$ exclusively will become binding over the planning horizon. Further, $K = 0$ and $W = 0$ can never be maintained simultaneously over an interval of time with positive controls $v > 0$ and $c > 0$. If $W(0) = 0$ and $K(0) > 0$, then it is optimal to maintain only this state at a zero level, $W = 0$ and $\dot{W} = 0$, over a finite interval of time. Similarly, if $W(0) > 0$ and $K(0) = 0$, then it is optimal to maintain only this state at a zero level, $K = 0$ and $\dot{K} = 0$, over a finite interval of time.

Proof See Appendix A3.

We now turn to the analysis of an interior solution. Plugging the value of c and v from Eqs. 8.7 and 8.8 in Eqs. 8.2 and 8.3 for interior solutions, respectively, the equations:

$$\dot{K} = aK - (\theta - \lambda + \mu\beta) + \varphi(\lambda\varphi - \mu) \tag{8.10}$$

$$\dot{W} = \beta(\theta - \lambda + \mu\beta) - \lambda\varphi + \mu \tag{8.11}$$

along with Eqs. 8.5 and 8.6, form the canonical system in the state-costate space.

Proposition 1. Under consumption-based waste generation, if $\beta\varphi < 1$, the steady state is unique, given by:

$$\begin{pmatrix} K^S & W^S & c^S & v^S \end{pmatrix}^T = \begin{bmatrix} \frac{\theta(1-\beta\varphi)}{a(1+\beta^2)} & \frac{r\theta\beta}{e(1+\beta^2)} & \frac{\theta}{1+\beta^2} & \frac{\theta\beta}{1+\beta^2} \end{bmatrix} \tag{8.12}$$

where the superscript 'S' stands for steady state.

Proof See Appendix A4.

The steady state in Eq. 8.12 exists despite the condition for sustained long-term growth $a > r$. This is due to two ingredients of the model. First, the central planner accounts for the existence of a negative environmental externality. This result was first identified by Stokey (1998). Second, preferences are linear-quadratic: even without the environmental externality, that is only keeping the consumption term in instantaneous utility, the economy would not grow indefinitely. The parametric condition $\beta\varphi < 1$ is worth commenting on. Notice if this condition fails to hold, stationary capital becomes negative. This comes from the accumulation of Eqs. 8.2 and 8.3 in the steady state which jointly imply:

$$aK^S = c^S - \varphi v^S = c^S(1 - \beta\varphi)$$

It can be shown that if $\beta\varphi < 1$ is violated, the unique equilibrium compatible with the nonnegativity constraints is the trivial equilibrium where all variables are nil. As explained before, we disregard this type of equilibrium, which is an artefact of linear-quadratic preferences. Moreover the larger φ or β, the lower the capital stock. This looks at first glance odd at least in what concerns the impact of parameter φ. Indeed, if this parameter increases *ceteris paribus*, recycling generates income and both v and K should trivially go up in the first reaction. However, as income and production are raised, consumption will also increase by a mere income effect, which tends to curb capital accumulation. The overall effect on capital either in the short or long run is therefore ambiguous. This ambiguity shows up clearly in the long run ($aK^S = c^S - \varphi v^S$). In our analytical case, the overall effect is negative and capital goes down when φ increases. The same kind of reasoning can be applied to interpret the impact of a rising φ on stationary waste, recycling, or consumption.

Concerning the stock of waste, by its law of motion, a larger consumption would in principle, as above, increase the recycling effort, therefore driving it down. But again as above, the resulting second-round income effect would push consumption up, resulting in more waste. It appears that the two effects offset each other.

One can draw some more comparative statics from the expressions of the steady state variables' values given here above. Interestingly enough, an increase in β unambiguously decreases long-term consumption but its effect on both recycling and total waste depends on the position of β with respect to 1. In the normal case where $\beta < 1$ (that is, one marginal unit of consumption generates less than one unit of waste), a further increase in β raises the long-term recycling effort but also the stock of waste. While stationary consumption decreases when β rises, the flow of waste βc^S does increase in the long run.

As to the parameters of the utility function, one notes that a greater consumption utility coefficient (θ) increases all the steady state values. This is due to the linear-quadratic specification of preferences. Qualitatively, an increase in θ boosts the marginal utility of consumption, which increases the consumption and the stock of waste. This in turn stimulates recycling. The total impact on capital depends on income generation through recycling and consumption rise. In our case, it is positive. Finally, a greater waste disutility coefficient (e) lowers the steady-state waste stock and does not affect the steady-state capital stock. Note that if the waste disutility coefficient is lower than the recycling disutility coefficient, i.e., $e < 1$, the steady-state waste stock increases. Regarding the discounting rate (r), it has no influence on the steady-state capital stock but has a positive effect on the steady-state waste stock. This property is mainly due to the linear-quadratic preferences: abstracting away from waste and recycling, the AK model with quadratic preferences will deliver long-term consumption and capital levels that only depend on productivity a and the bliss point θ. Adding waste and recycling does not modify this property. However, as the discounting rate increases, leading to a larger preference for the present, consumption will increase along the transition, inducing an increment in the stock of waste in the long run.

We now investigate the structure of the associated stable manifolds.

Proposition 2. Under consumption-based waste generation, the steady state exhibits a local two-dimensional stable manifold. The convergence from a close neighborhood to a steady state is monotonic.

Proof See Appendix A5.

Therefore, for a sufficiently patient social planner, a saddle point exists. In addition, the control paths can be characterized as follows.

Lemma 4. Under consumption-based waste generation, an interior optimal solution satisfies the system:

$$\ddot{c} = r\dot{c} + a(r-a)Ge^{(r-a)t} + \beta e(\beta c - v) \tag{8.13}$$

$$\ddot{v} = r\dot{v} - a\varphi(r-a)Ge^{(r-a)t} - e(\beta c - v) \tag{8.14}$$

Proof See Appendix A6.

This general case is solvable numerically. For $\beta = 1$, we get:

$$c(t) = \frac{1}{2}\left\{\theta + C_3 e^{\frac{1}{2}\left(r-\sqrt{r^2+4e}\right)t} - \left[1-\varphi + \frac{a(1+\varphi)(r-a)}{a(r-a)+e}\right] G e^{(r-a)t}\right\}$$

$$v(t) = \frac{1}{2}\left\{\theta - C_3 e^{\frac{1}{2}\left(r-\sqrt{r^2+4e}\right)t} - \left[1-\varphi - \frac{a(1+\varphi)(r-a)}{a(r-a)+e}\right] G e^{(r-a)t}\right\}$$

whose constants of integration can be found explicitly. Letting $\varphi < 1$, then for $G > 0$ ($G < 0$), the sum $c(t) + v(t)$ increases (decreases) monotonically and concavely (convexly) to converge to θ when $t \to \infty$. Furthermore, the closer φ to one the smaller the change in $c(t) + v(t)$, i.e., the weaker the short-term effect of φ on the evolution of $c(t) + v(t)$. Note, the difference $c(t) - v(t)$ decreases to zero when $t \to \infty$ and φ do not affect it at all.

Remark[3] The system (Eqs. 8.10–8.11–8.5–8.6) has a two-dimensional stable manifold. Any combination of this manifold results in a path of state and costate variables which ends in the steady state for variables W, c or v. However, in this case K can potentially grow indefinitely and asymptotically at a rate a. As a result, a second type of path, which may be optimal, can emerge. Along these paths, the system converges to an asymptotic balanced growth path (in the sense that all variables grow asymptotically at constant rates, equal to zero for variables c, v and W). In our comparison of the consumption vs production-based waste generation cases, we shall disregard these alternative paths for a couple of reasons: economic irrelevance and mathematical non-robustness. Indeed, they are economically irrelevant because they lead to a major violation of Kaldor stylized facts: the ratio $c/Y = c/aK$ tends to zero. Obviously, this property cannot arise generally in a model with CRRA preferences: for example, it cannot happen in the non-circular economy model à la Stokey (1998), already quoted above. The spurious nature of these alternative paths can be captured even more directly, without moving to CRRA preferences or any other fundamental adjustment: it is enough to add a condition ensuring the Kaldor fact holds at large, for example by constraining the limit of c/Y to be larger than or equal to given a strictly positive constant at large (which could be arbitrarily small).

Production-Based Waste Generation

In this case, we assume that $\kappa > 0$ and $\beta = 0$ in the law of motion of polluting waste (Eq. 8.3). Skipping the time index for convenience, the current-value Hamiltonian is:

$$H = c\left(\theta - \frac{c}{2}\right) - \frac{eW^2}{2} - \frac{v^2}{2} + \lambda(aK - c + \varphi v) + \mu(\kappa aK - v) \tag{8.15}$$

[3] We warmly thank a Referee for raising this point.

where $\lambda \equiv \lambda(t)$ and $\mu \equiv \mu(t)$ are costate variables, that, if the state constraints are not binding, evolve according to:

$$\dot{\lambda} = (r - a)\lambda - \mu\kappa a \tag{8.16}$$

$$\dot{\mu} = r\mu + eW \tag{8.17}$$

under the transversality conditions: $\lim_{t\to\infty} e^{-rt}\lambda(t) = 0$ and $\lim_{t\to\infty} e^{-rt}\mu(t) = 0$.

Assuming interior solutions, the necessary conditions for optimality are:

$$H_c = \theta - c - \lambda = 0 \Rightarrow c = \theta - \lambda \tag{8.18}$$

$$H_v = -v + \lambda\varphi - \mu = 0 \Rightarrow v = \lambda\varphi - \mu \tag{8.19}$$

From Eq. 8.18, the greater the implicit price of the capital stock, the lower the consumption rate. In Eq. 8.19, the greater the implicit price of the capital stock and the implicit cost of the waste stock, the greater the recycling effort. These results are intuitive.

Comparing necessary conditions (Eqs. 8.18–8.19 and 8.7–8.8) and their related costate equations, it is obvious that under consumption-based waste, the consumption is instantaneously diminished by the implicit cost of waste engendered by the marginal consumption-based waste, while under production-based waste, it is intertemporally diminished by the implicit cost of waste engendered by the marginal waste due to the marginal revenue of capital.

Lemma 5. The necessary conditions are sufficient for optimality.

Proof See Appendix A7.

Here also, the sufficient conditions used are true only for interior solutions.

Lemma 6. Let the state constraint $W \geq 0$ be nonbinding. Then, it always holds that $\mu < 0$.

Proof See Appendix A8.

Lemma 7. It is not optimal to have $W = 0$ and $K = 0$ along an interval of time.

Proof See Appendix A9.

The above lemma shows that, unlike with consumption-based waste generation, it is not optimal to maintain both states K and W at zero over any time interval.

Plugging the value of c and v from Eqs. 8.18 and 8.19 in Eqs. 8.2 and 8.3, respectively, the equations:

$$\dot{K} = aK - (\theta - \lambda) + \varphi(\lambda\varphi - \mu) \tag{8.20}$$

$$\dot{W} = \kappa aK - \lambda\varphi + \mu \tag{8.21}$$

along with Eqs. 8.16 and 8.17, form the canonical system in the state-costate space.

Proposition 3. Under production-based waste generation, for $r < a$, the steady state is unique, given by:

$$\left(K^S\ W^S\ c^S\ v^S\right)^T = \left[\frac{\theta[(1+\varphi\kappa)a-r]}{a\{(1+\varphi\kappa)[(1+\varphi\kappa)a-r]+\kappa^2 a\}}\ \frac{r\kappa\theta(a-r)}{e\{(1+\varphi\kappa)[(1+\varphi\kappa)a-r]+\kappa^2 a\}}\right.$$
$$\left.\frac{\theta(1+\varphi\kappa)[(1+\varphi\kappa)a-r]}{(1+\varphi\kappa)[(1+\varphi\kappa)a-r]+\kappa^2 a}\ \frac{\alpha\theta[(1+\varphi\kappa)a-r]}{(1+\varphi\kappa)[(1+\varphi\kappa)a-r]+\kappa^2 a}\right]. \tag{8.22}$$

Proof See Appendix A10.

As it transpires from the expressions of the steady state values given above, the model with capital-based or equivalently production-based waste involves more complex relationships because capital accumulation is not only the main engine of income in this economy, it is also the generator of waste, which cause the negative environmental externality and fosters recycling. Henceforth the corresponding comparative statics are more complicated. From Eq. 8.22, we observe that the marginal wasting coefficient of production (α) decreases as expected in the long-run capital stock, waste stock, and consumption flow. More interesting comparative statics are with respect to parameters, which are common to the consumption-based waste and production-based waste models. Let us consider the parameter φ for example. Greater marginal revenue from recycling (φ) increases the long-run consumption, which in turn decreases the long-run capital stock. As a consequence, the long-run waste stock and recycling effort are both reduced. When φ goes up, consumption seems to play a long-term stabilizing role in the stock of waste. This is not the case in the model with consumption-based waste where the recycling effort, while increasing with φ in the short run, ends up independent of this parameter in the long run.

The second common parameter example is the discounting rate (r): it has a negative impact on the long-run capital stock, consumption rate and recycling effort but a positive effect on the long-run waste stock. Recall that in the consumption-based waste case, it only affects (positively as in the current case) the stock of waste. In the model with production-based waste, the stationary capital ends up decreasing with the discount rate, as in the traditional Ramsey model.

We now move to the stability analysis and investigate the structure of the stable manifolds.

Proposition 4. Under production-based waste generation, the steady state exhibits a (local) two-dimensional stable manifold. The convergence to the steady state is either monotonic or oscillatory.

Proof See Appendix A11.

We now highlight an interesting property of the optimal growth model with production-based waste. We show hereafter that under certain conditions, it may give rise to an oscillatory convergence to the steady state, which is not granted in the case of consumption-based waste (Proposition 2). According to Proposition 4, convergence to the saddle point is either monotonic or oscillatory, depending on the magnitude of the wasting impact of the revenue-generation process and the marginal impact of the recycling effort on capital accumulation. Therefore, oscillating convergence to the steady state can be caused either by highly wasteful production, which reflects the use of dirty technology (Stokey, 1998), or by highly CIR, which mirrors the absence of any depreciation of scrap recovery (Martin, 1982), that is, a highly effective recycling process.

A Preliminary Comparison Result

In the next section, we shall go a step further and determine the main differences between consumption and production-based polluting waste, on the one hand, and CNR and CIR efforts, on the other hand, on capital growth and polluting waste accumulation in terms of both the steady state and the transient path, using both analytical and numerical results. Before, and based on our analytical results for consumption and production-based polluting waste, we make an important comparison.

Lemma 8. For $\varphi = 0$ and equal marginal wasting impact of production and consumption, it holds that:

$$c^S_{|\kappa=0,\beta>0} > c^S_{|\kappa>0,\beta=0} \tag{8.23}$$

$$v^S_{|\kappa=0,\beta>0} > v^S_{|\kappa>0,\beta=0} \tag{8.24}$$

$$K^S_{|\kappa=0,\beta>0} > K^S_{|\kappa>0,\beta=0} \tag{8.25}$$

$$W^S_{|\kappa=0,\beta>0} > W^S_{|\kappa>0,\beta=0} \tag{8.26}$$

Proof See Appendix A12.

Therefore, for CNR efforts, i.e., $\varphi = 0$, and equivalent marginal wasting impact in the case of either consumption- or production-based waste, i.e., $\kappa = \beta > 0$ for either $\kappa = 0$ and $\beta > 0$ or $\kappa > 0$ and $\beta = 0$, the steady-state consumption level is greater if the polluting waste is a by-product of consumption rather than of production. Because $c^S_{|\kappa=0,\beta>0} < \theta$ and $c^S_{|\kappa>0,\beta=0} < \theta$, the steady-state consumption is lower than the utility-maximizing consumption level under both production and consumption-based waste. For similar conditions, it follows that the steady-state capital stock is greater under consumption-based waste than under production-based waste. The two latter properties are important and thus deserve a careful interpretation. Why could it be that under consumption-based waste the central planner reaches larger consumption and capital stock levels than under production-based waste? When only consumption generates waste, two effects are at work: on one hand, consumption is directly driven down (with respect to the case $\beta = 0$) as it decreases utility via waste; but on the other hand, a drop in consumption increases savings and capital accumulation, which in turn pushes consumption up (income effect). When waste is generated through production, there is an explicit brake on capital accumulation induced by the environmental negative externality, which limits both capital accumulation and consumption. It turns out that this mechanism (brake on capital accumulation) is inactive under consumption-based waste. As a result, both the long-term consumption and capital stock are higher in the latter case.

Note also that due to $\beta[(a - r) + \kappa^2 a] > (1 + \beta^2)\kappa(a - r)$ when $\kappa = \beta$, the steady-state values of the waste stock and the recycling efforts are lower under production-based waste than under consumption-based waste. This comes as a natural consequence of the abovementioned properties for equal marginal waste generation coefficients ($\kappa = \beta$): as both consumption and capital stock are lower under production-based waste, the resulting stock of waste, and thus the optimal recycling effort (under $\varphi = 0$), are lower.

It is impossible to come out with complete analytical comparison results for any κ, β and φ. However, numerical experiments are conducted hereafter to complement this analytical section.

Numerical Analysis

We now get back to the main research questions stated in the introduction and answer them. We shall particularly investigate how consumption *versus* production-based polluting waste affects the interaction between capital growth and waste accumulation (thus the emergence of an EKC path), and the role of the CIR versus the CNR assumption in the mechanisms involved. To this end, we use numerical experiments.

Numerical Strategy and Calibration

To compute the solution paths, we use a dynamic programming approach (see Appendix A13). The model is calibrated as depicted in Table 8.1 just below:

The values in Table 8.1 reflect a situation where the marginal revenue from the capital stock ($a = 0.1$) is greater than the discounting rate ($r = 0.05$). The range of values selected for κ, β and φ aims to contrast some important configurations, that is, production-based *vs*. consumption-based waste (i.e., $\kappa = 0.5$ *vs*. $\beta = 0.5$), on the one hand, and CIR *vs*. CNR efforts (i.e., $\varphi = 0.1$ *vs*. $\varphi = 0$) on the other hand. In addition, two special cases inducing oscillatory convergence to the steady state are considered, that is, highly wasteful production processes (i.e., $\alpha = 4$) and highly CIR (i.e., $\varphi = 10$). Finally, we assume equivalent parameter values for the disutility coefficient

Table 8.1 Parameter values

r	a	κ	β	φ	θ	e
0.05	0.1	(0,0.5,4)	(0,0.5)	(0, 0.1,10)	1	1

Table 8.2 State and control steady-state values under focal configurations

	Consumption-based waste $\kappa = 0, \beta = 0.5$		Production-based waste $\kappa = 0.5, \beta = 0$		Production and consumption-based waste $\kappa = 0.5, \beta = 0.5$	
	CIR ($\varphi = 0.1$)	CNR ($\varphi = 0$)	CIR ($\varphi = 0.1$)	CNR ($\varphi = 0$)	CIR ($\varphi = 0.1$)	CNR ($\varphi = 0$)
K^S	7.6	8	6.65	6.67	3.94	4
W^S	0.02	0.02	0.015	0.017	0.019	0.02
c^S	0.8	0.8	0.698	0.67	0.43	0.4
v^S	0.4	0.4	0.332	0.33	0.41	0.4

of waste and the disutility coefficient of recycling (i.e., $e = 1$) so that the tradeoff between waste and recycling does not stem from arbitrary decisions concerning parameter values. To assess the sensitivity of the solutions to the initial conditions, we assume in the cases with monotonic convergence two initial values for each state variable, that is, low and high (i.e., lower and higher than the steady-state value), with $K_0 = (6, 10)$ and $W_0 = (0,0.3)$.

The steady state values are reported in Table 8.2. All things being equal, the long-run capital stock and consumption are greater under consumption-based than production-based waste. Despite greater recycling efforts, the long-run waste is greater under consumption-based than production-based waste. On the other hand, the steady-state capital stock is greater under capital-neutral than CIR while the steady-state waste stock is similar under consumption-based waste and greater under production-based waste.

Claim 1. Although CIR limits capital growth more than CNR does, it promotes a slightly cleaner environment only under production-based waste (Table 8.2).

To unburden the presentation and ease the comparison between the different mechanisms involved, we shall study the induced optimal paths and welfare outcomes first under consumption-based waste, and then under production-based waste.

Optimal Time Paths and Implied Welfare Under Consumption-Based Waste

Transient Dynamics

Figure 8.1 illustrates the paths converging to the steady state from various initial conditions under consumption-based waste for CIR (Fig. 8.1a) and CNR (Fig. 8.1b) efforts. The transient paths are drawn for initially low *versus* high capital and waste stocks, that is, $(K_0, W_0) = \{(6, 0), (10, 0), (6,0.3), (10,0.3)\}$.

The optimal path is composed of two arcs corresponding to sequential sets of actions, that is:

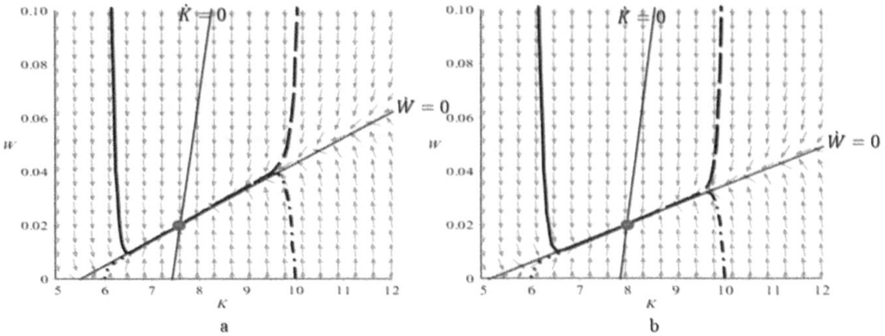

Fig. 8.1 Phase diagrams in the state space for consumption-based waste. (**a**) CIR effort. (**b**) CNR effort

- From both initially high (low) state values, the optimal policy consists of first decreasing (increasing) the waste stock rapidly and the capital stock slowly, and then decreasing (increasing) both the capital and waste stocks at a similar pace until the steady state is reached.
- From initially high (low) capital stock and high (low) waste stock, the optimal policy is first to decrease (increase) the waste stock rapidly and the capital stock slowly, and then to decrease (increase) the capital and waste stocks at a similar pace toward the steady state.

The second arc, which arises beyond the turning point and characterizes the longest portion of the planning horizon, depicts a mutual complementarity between the capital stock and the waste stock. That is, under consumption-based waste, capital and pollution most often either increase or decrease together. This result lets us identify initial conditions for which the first arc, located before the turning point, would exhibit (transient) mutual substitutability between the capital and waste stocks, i.e., an EKC-like arc.

We find that such initial conditions exist (Fig. 8.2): for both CIR and CNR, capital growth allows a cleaner environment from initially intermediate capital stock (i.e., near its steady-state value) and very large waste stock, e.g., $(K_0, W_0) = (8, 2)$, but only for a limited time interval. The substitutability arc between the two stocks is more marked for CIR than for CNR, as in the subsequent complementarity arc, which allows greater contraction of the capital stock but with a much smaller decrease of the stock of waste.

Claim 2. An EKC-like arc exists for consumption-based waste, during which capital accumulation allows for a cleaner environment for both capital-improving and capital-neutral recycling, but with ephemeral and reversible capital growth (Fig. 8.2).

The optimal consumption and recycling effort policies' time paths corresponding to the various paths in Fig. 8.1 are shown in Fig. 8.3. The patterns of optimal consumption and recycling efforts obtained for the various initial conditions are summarized in Table 8.3. The initial recycling efforts are at their highest level for

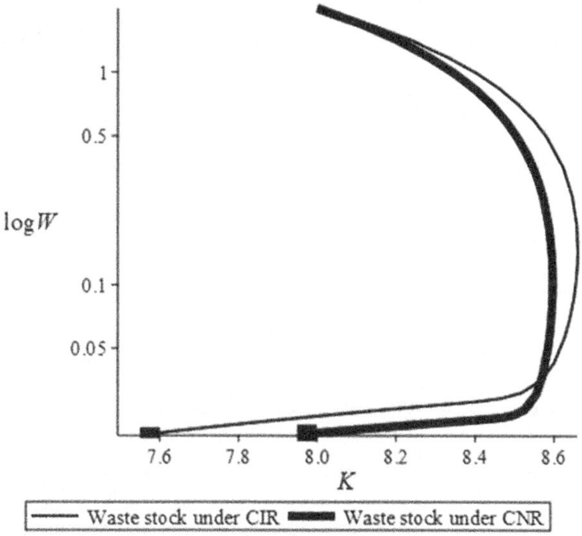

Fig. 8.2 EKC-like state path for consumption-based waste under CIR and CNR with a logarithmic scale for waste stock

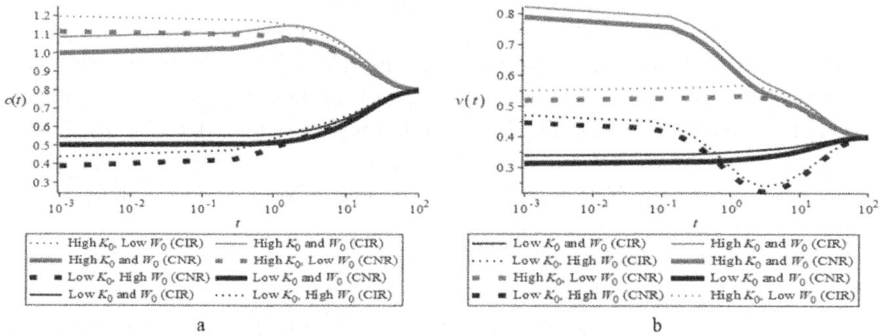

Fig. 8.3 Control paths for consumption-based waste under CIR and CNR (log. time scale). (**a**) Consumption. (**b**) Recycling effort

Table 8.3 Optimal patterns of consumption and recycling efforts for consumption-based waste

	Low initial capital stock	High initial capital stock
Low initial waste stock	Initially low, increasing consumption and recycling efforts	Initially high, decreasing consumption and initially intermediate, decreasing recycling efforts
High initial waste stock	Initially low, increasing consumption and initially intermediate, decreasing then increasing (undershooting) recycling efforts	Initially high, increasing then decreasing (overshooting) consumption and initially high, decreasing recycling efforts

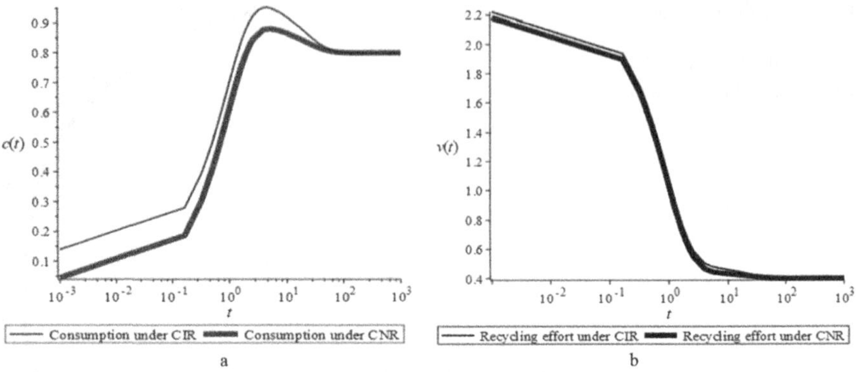

Fig. 8.4 Control paths related to EKC-like path for consumption-based waste under CIR and CNR (log. time scale). (**a**) Consumption. (**b**) Recycling effort

both initially high stocks of capital and waste, while consumption is clearly lower than for initially high capital stock and low waste stock. The reason is that more recycling and less consumption are initially needed to quickly reduce the initial waste stock. Note however that the initial capital stock has a stronger influence than the initial waste stock on both consumption level and recycling efforts.

An interesting feature nevertheless associated with a high initial waste stock is a non-monotonic behavior, i.e., undershooting of recycling efforts linked to initially low capital stock and overshooting of consumption due to initially high capital stock. This behavior is particularly marked in the case of initially low capital stock and high waste stock for which consumption and recycling efforts are first briefly mutual substitutes and then become mutual complements.

In the specific case of initially intermediate capital stock and very large waste stock, e.g., $(K_0, W_0) = (8, 2)$, which allows for an EKC-like path, the optimal policy also exhibits overshooting, predominantly with substitutability between the control variables, i.e., consumption should be initially low and increasing while recycling efforts should be initially high and decreasing over time, as suggested in Fig. 8.4. Finally, in all cases, when recycling does not add to capital growth, consumption is lower than in the converse case, which thus makes recycling efforts less necessary.

Welfare Implications

In Fig. 8.5, the welfare function is computed as a function of the length of the time horizon to emphasize the magnitude of its transient evolution. Due to the effect of discounting, the instantaneous welfare tends to zero as the length of the time horizon tends to infinity.

Apparently, *lower initial waste stock results in higher welfare*. Compared with CIR, CNR is welfare-improving in the case of higher initial capital stock because lower consumption under CNR than CIR entails less disutility from pollution waste

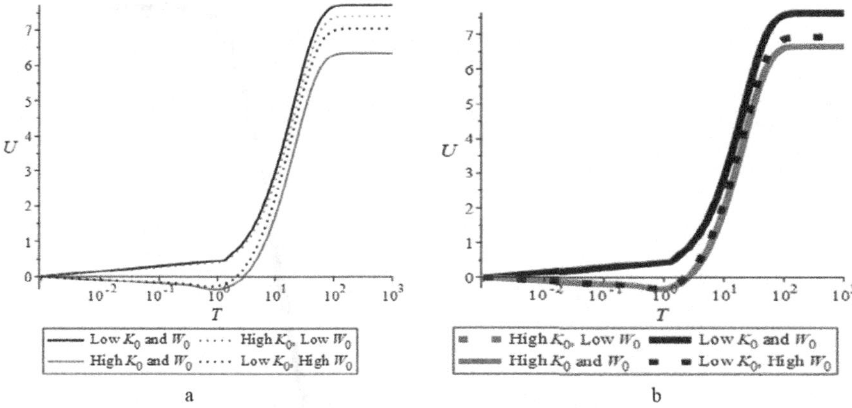

Fig. 8.5 Social welfare for consumption-based waste under (**a**) CIR and (**b**) CNR (log. time scale)

and recycling efforts. In contrast, welfare is similar between CIR and CNR in the case of low initial capital stock due to the initially low and increasing consumption path induced.

Claim 3. Social welfare is greater with both poorer and cleaner rather than both wealthier and dirtier initial conditions (Fig. 8.5).

Welfare is also very similar between CIR and CNR for very high initial waste stock, as in the case of the EKC-like path for which consumption is also initially low and increasing (Fig. 8.6). In all cases, *a greater initial waste stock is more harmful to the transient welfare.*

Optimal Time Paths and Implied Welfare Under Production-Based Waste

Transient Dynamics

We now turn to the case of production-based waste. The phase-portrait diagrams in Fig. 8.7 show the paths converging to the steady state from initially low *versus* large capital and waste stocks, that is, $(K_0, W_0) = \{(6, 0), (10, 0), (6{,}0.3), (10{,}0.3)\}$ for CIR and CNR (resp., Fig. 8.7a and Fig. 8.7b).

Apart from lower steady-state values of the state variables and a slightly weaker complementarity relationship between the two stocks, the patterns do not seem to differ much from those of consumption-based waste. However, for a high initial capital endowment, the contraction of capital is clearly greater for consumption-based waste than for production-based waste, with a similar steady-state stock of waste.

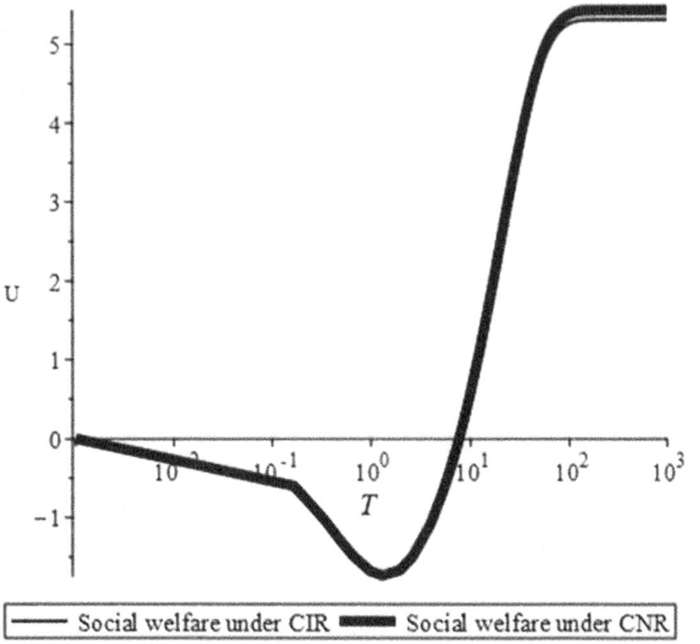

Fig. 8.6 Social welfare for EKC-like state path for consumption-based waste under CIR and CNR (log. time scale)

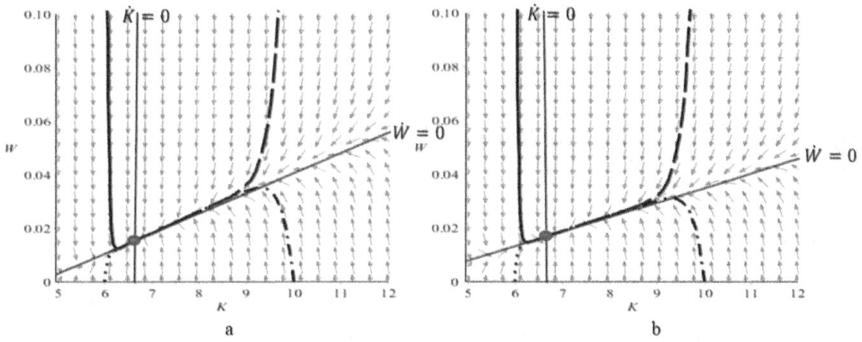

Fig. 8.7 Phase diagram in the state space for production-based waste. (**a**) CIR effort. (**b**) CNR effort

Claim 4. Consumption-based waste is less capital-destructive than production-based waste, notably under capital-neutral recycling (Figs. 8.1 and 8.7).

However, the initial conditions under which mutual substitutability exists between the stocks of capital and waste, i.e., EKC-like path, are different from those related to consumption-based waste (Fig. 8.8). That is, lower initial capital

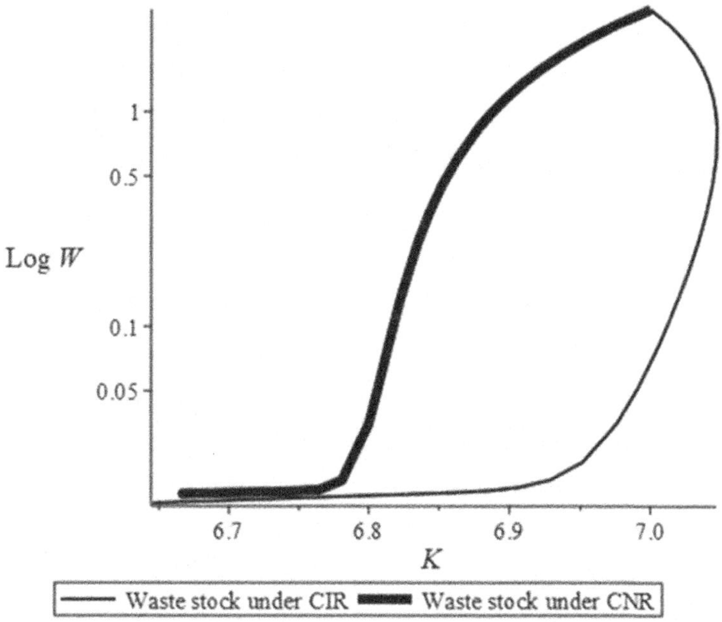

Fig. 8.8 EKC-like state path for production-based waste under CIR and CNR efforts for waste stock (log scale)

stock and larger waste stock than for consumption-based waste, i.e., $(K_0, W_0) = (7, 3)$, are now required for capital growth to allow a transiently cleaner environment but over a smaller interval. However, the arc of substitutability between the stocks of capital and waste does not exist for CNR even for much higher initial waste stock. Thus, complementarity always prevails under production-based waste with CNR, which leads to less severe contraction of capital but also less effective depollution than under CIR.

Claim 5. An EKC-like arc exists for production-based waste, but over a smaller interval and under more restrictive conditions than for consumption-based waste, that is, lower initial wealth and larger waste stock and only for capital-improving recycling (Fig. 8.8).

The optimal consumption and recycling effort policies related to the different paths in Fig. 8.7 are shown in Fig. 8.9. Here again, higher consumption results from higher initial capital stock. Consumption is also positively affected by the initial waste stock, albeit to a lesser extent. This is in contrast with the consumption-based waste case because *consumption here is a non-costly, welfare-improving means to reduce the waste stock via a capital decrease*. Another important difference from the case of consumption-based waste is that recycling is positively influenced by the waste stock. A third important difference from the case of consumption-based waste lies in the fact that while consumption is lower under CNR than CIR efforts, the recycling efforts are now very similar.

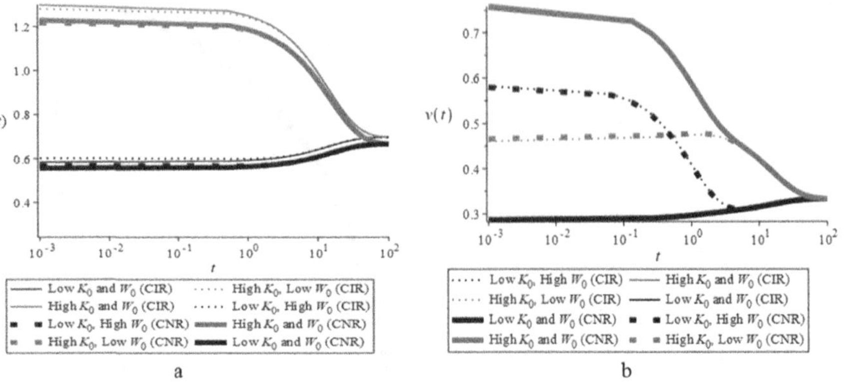

Fig. 8.9 Control paths for production-based waste under CIR and CNR efforts (log. time scale). (**a**) Consumption. (**b**) Recycling effort

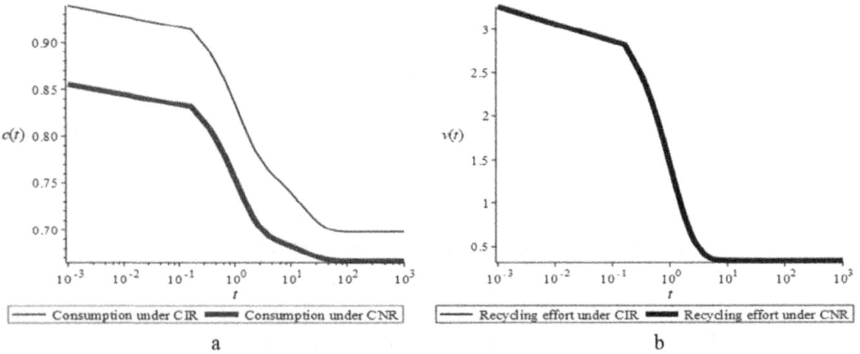

Fig. 8.10 Control paths related to EKC-like path for production-based waste under CIR and CNR (log. time scale). (**a**) Consumption. (**b**) Recycling effort

Claim 6. In contrast with the case of consumption-based waste, consumption is much greater under CIR than CNR for production-based waste (Fig. 8.9 *vs.* Fig. 8.3).

In the specific case of the EKC-like state path, the optimal policy is characterized by complementarity between the control variables, i.e., both consumption and recycling efforts should be initially high and decreasing over time, as suggested in Fig. 8.10. When recycling is CN, consumption is significantly lower than in the converse case, whereas the magnitude of recycling efforts remains unaffected.

Claim 7. The optimal policy related to an EKC-like path for production-based waste exhibits complementarity between the control variables rather than substitutability as observed for consumption-based waste (Fig. 8.10).

Welfare Implications

Figure 8.11 reports the social welfare associated with production-based waste, where higher welfare results from lower initial waste stock and to some extent from higher initial capital stock. Despite higher consumption, initial conditions with both higher capital and waste stocks lead to lower welfare due to greater disutility resulting from higher recycling efforts. Again, the differences in welfare between lower and higher initial capital stock are more marked under CIR than CNR efforts, due to the greater reduction in consumption in the latter case between CIR and CNR for initially higher than lower capital stock. Finally, social welfare under CNR is similar to that under CIR from initially high capital but lower than that from initially low capital, mainly because slightly greater recycling efforts under CNR than CIR entail smaller disutility.

Comparing Figs. 8.5 and 8.11, respectively, we observe that, for CIR, the welfare is greater under production-based waste than under consumption-based waste from initially high capital stock and similar welfare of production-based waste and consumption-based waste from low capital stock. In contrast, for CNR, the welfare is still greater under production-based waste than under consumption-based waste for initially high capital stock, but lower for initially low capital stock.

Claim 8. In the case of production-based waste, poorer but cleaner initial conditions provide greater social welfare than wealthier but dirtier initial conditions (Fig. 8.11).

In the case of the EKC-like state path, the transient welfare is negative with quite low values (Fig. 8.12), due to the initially high disutility generated by both high waste stock and recycling efforts. Afterwards, welfare becomes positive with a slightly greater value under CIR than under CNR because of a greater consumption level.

Overall, production-based waste is more welfare-improving than consumption-based waste if the initial capital endowment is high. The reason is that although

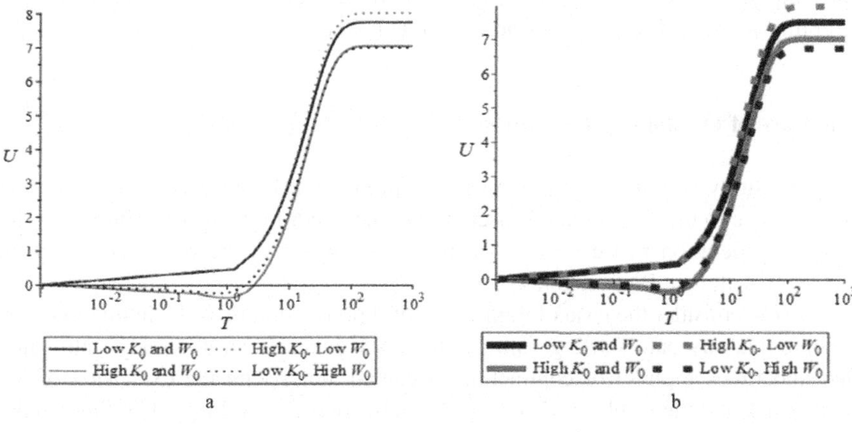

Fig. 8.11 Social welfare for production-based waste under (**a**) CIR and (**b**) CNR (log. time scale)

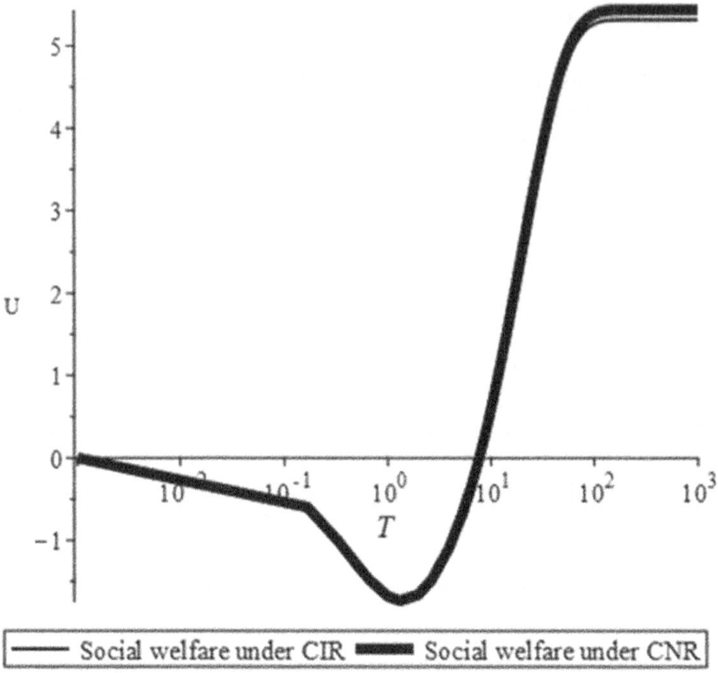

Fig. 8.12 EKC-like social welfare path for production-based waste under CIR and CNR (log. time scale)

consumption-based waste exhibits a higher long-run consumption level for a high initial capital endowment, it has lower consumption in the short run and tends to incur more disutility from polluting waste and recycling efforts than production-based waste over the whole-time horizon. For a low initial capital endowment, production-based waste is still associated with greater transient consumption than consumption-based waste is, but the transient differences in terms of disutility from polluting waste and recycling efforts are now lower.

The Case of Oscillatory Dynamics

We now study two special cases related to production-based waste, for which the convergence to the steady state is oscillatory, due either to the wasting impact of the revenue generation process or to the marginal impact of the recycling effort on capital growth (see Proposition 4).

We first consider the case of highly wasteful production ($\alpha = 4$). In this case, we observe that any path starting with capital stock greater than its steady-state value first undergoes a phase of contraction of capital along with the expansion of the waste stock and then a phase of contraction of the two stocks (Fig. 8.13). Beyond the turning point, the capital decline results in a cleaner environment. Therefore, the

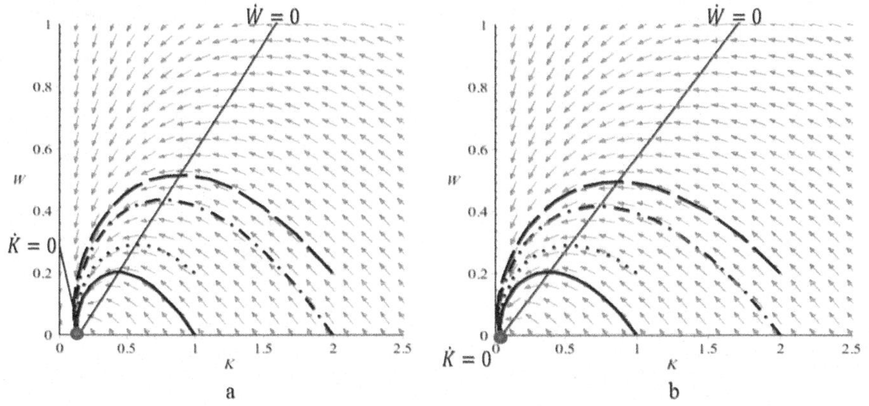

Fig. 8.13 Phase diagram in the state space for highly wasteful production $(K_0, W_0) = \{(1,0), (2,0), (1,0.2), (2,0.2)\}$. (**a**) CIR effort. (**b**) CNR effort

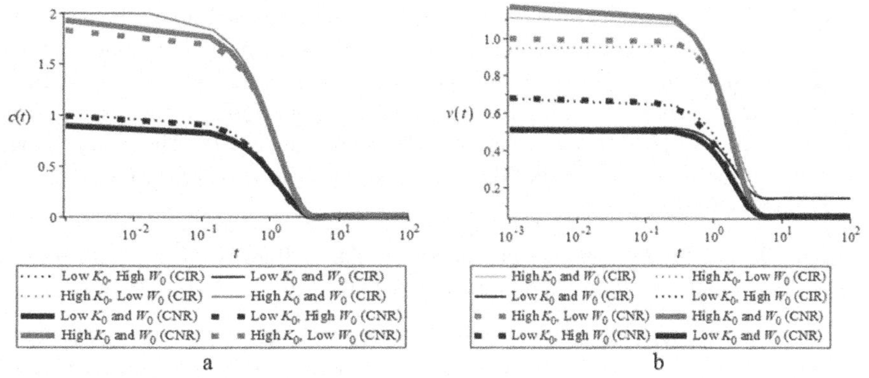

Fig. 8.14 Control paths for highly wasteful production under CIR and CNR (log. time scale). (**a**) Consumption. (**b**) Recycling effort

dirtiest production technologies result in a total reversal of the EKC. Under an ineffective recycling process, i.e., CNR, the total reversal of the EKC is even more severe in terms of capital contraction.

Claim 9. Highly wasteful production is heavily capital-destructive (Fig. 8.13).

In the context of a total reversal of the EKC, the optimal policy is also characterized by complementarity between the control variables (Fig. 8.14). When recycling does not add to capital, consumption is lower than in the converse case only if the initial stock of capital is sufficiently high. Under the latter condition, however, recycling efforts are greater under CNR than in the converse case. The rationale behind this result is that consumption under CIR is a substitute for recycling to reduce waste.

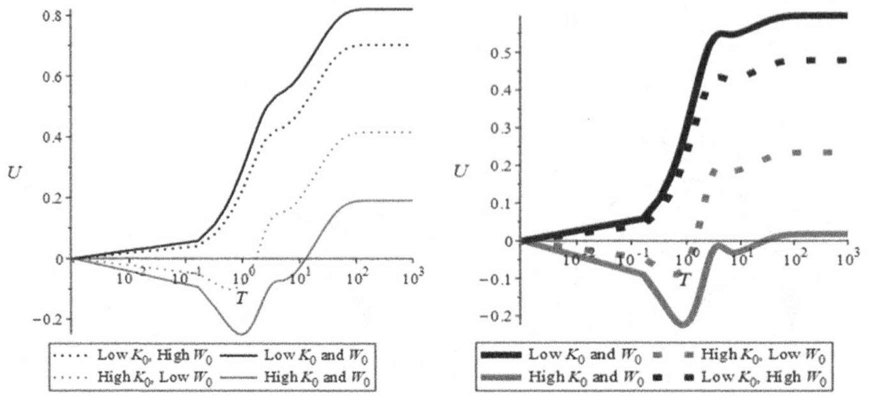

a b

Fig. 8.15 Social welfare for highly wasteful production under (**a**) CIR and (**b**) CNR (log. time scale)

In Fig. 8.15, a higher welfare results from lower initial capital stock and to a lesser extent from lower initial waste stock. Again, despite higher consumption, initial conditions involving both higher stocks of capital and waste lead to lower welfare due to greater disutility resulting from higher recycling efforts. In contrast with the previous cases, welfare is now lower under CNR than under CIR. The reason is that recycling efforts result in smaller disutility under CIR due to greater consumption used to reduce waste.

Claim 10. For highly wasteful production, social welfare is greater from poorer rather than wealthier initial conditions (Fig. 8.15).

We finally analyze the case of highly CIR ($\varphi = 10$). Here, two main phases are observed, that is, a phase of contraction of capital along with expansion of the waste stock and then a phase of expansion of capital along with contraction of the waste stock (Fig. 8.16). Beyond the turning point, capital growth allows a cleaner environment. Therefore, *a highly effective recycling process leads to a partial reversal of the EKC*.

Claim 11. Highly capital-improving recycling is partly capital-destructive (Fig. 8.16).

In the case of partial reversal of the EKC, the optimal policy is also based on complementarity between the control variables (Fig. 8.17). Consumption and recycling are both positively affected by the initial capital stock and are affected to a lesser extent by the initial waste stock.

In Fig. 8.18, higher welfare is derived from lower initial capital stock and to a lesser extent from lower initial waste stock. The ranking in terms of decreasing welfare is consistent with that of consumption, which in turn is allowed by the highly effective recycling efforts.

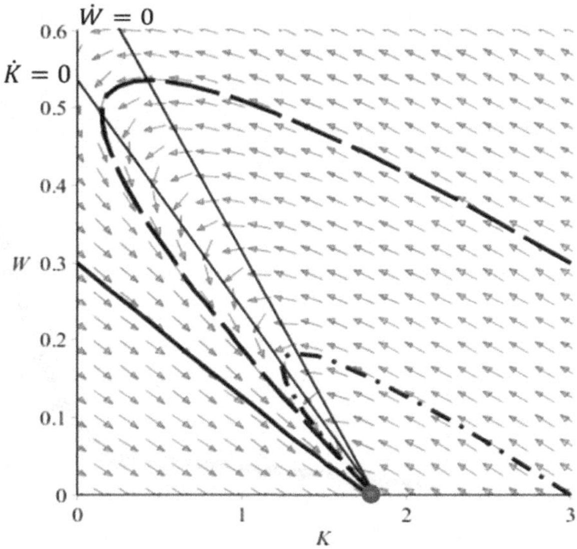

Fig. 8.16 Phase diagram in the state space for production-based waste under highly CIR (K_0, W_0) = {(0,0.3), (3, 0), (3,0.3)}

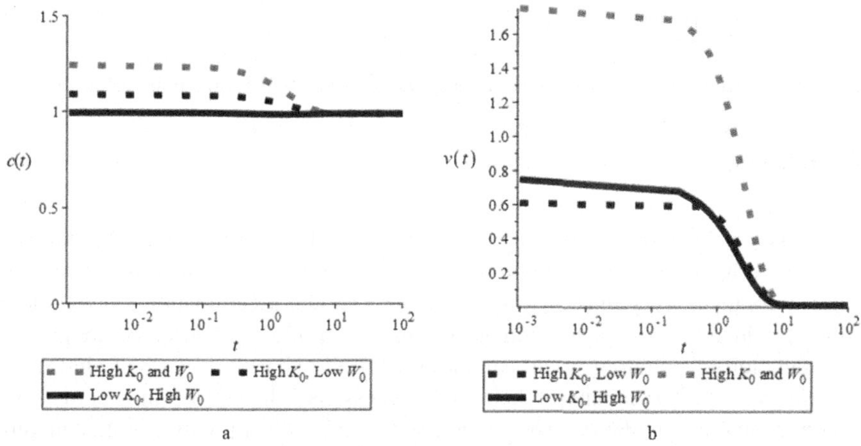

Fig. 8.17 Control time paths for production-based waste under highly CIR (log. time scale). (**a**) Consumption. (**b**) Recycling effort

Claim 12. For highly effective production-based recycling, social welfare is greater with either both poorer and dirtier or both wealthier and cleaner initial conditions rather than both wealthier and dirtier initial conditions (Fig. 8.15).

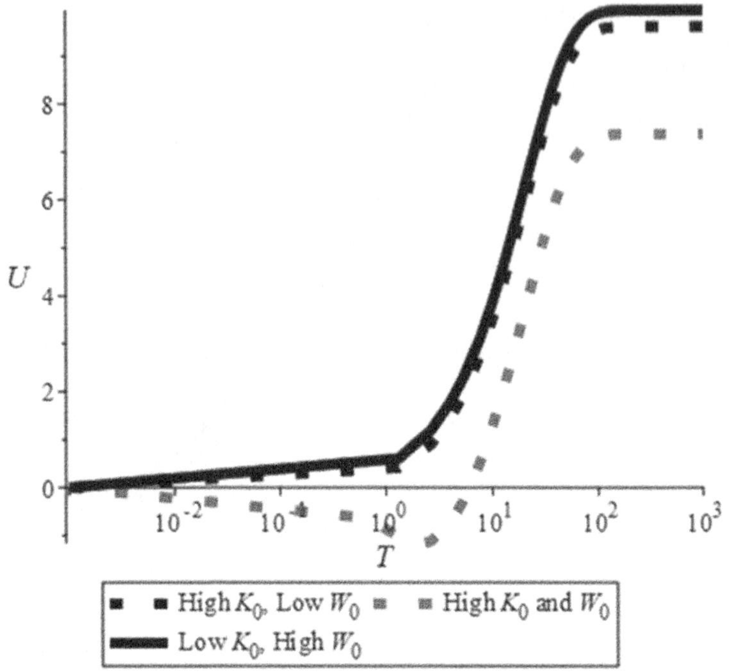

Fig. 8.18 Social welfare for production-based waste under highly CIR (log. time scale)

Conclusion

This paper seeks to determine which source of polluting waste among the stock of productive capital and consumption flows is the most detrimental in terms of welfare and environmental sustainability. We consider a capital accumulation model where either production or consumption is a source of polluting waste. Although the accumulation of polluting waste cannot be naturally abated, it can be reduced by recycling efforts. To account for their effectiveness, recycling efforts may or may not benefit capital accumulation. The economic tradeoff, which involves instantaneous utility from consumption and instantaneous disutility due to polluting waste and recycling over an infinite time horizon, allows us to identify the optimal recycling policy depending on the source of pollution and also to disclose the conditions of the emergence of an Environmental Kuznets Curve path.

Our mathematical analysis shows that consumption and production as sources of polluting waste have quite different consequences on the interaction between capital and pollution accumulation. The main differences can be summarized as follows:

- Convergence from a close neighborhood to the unique interior, locally stable steady state is monotonic under consumption-based waste, but it can also be oscillatory under production-based waste due either to very dirty technology of production or to highly effective recycling.

- To compare the long-term consequences of the optimal control of consumption *versus* production-based waste *via* recycling, we examine the implications of the benchmark case where one additional unit of consumption yields the same additional waste as a supplementary unit of production and recycling does not generate additional income. In this framework, larger consumption and capital stock stationary values are reached under consumption-based waste than under production-based waste. This is because in the latter there is an explicit brake on capital accumulation induced by the environmental negative externality, which limits both capital accumulation and therefore consumption. Such a mechanism does not exist under consumption-based waste.
- In the same analytical set-up, it is shown that lower stocks of waste are obtained with lower recycling efforts in the production-based waste case.

From extensive numerical analysis, we also gained further important insights:

- The analytical results above are shown to be robust under a set of alternative parameterizations where the assumptions adopted in the benchmark case are successively dropped.
- In countries with intermediate wealth and very large waste stock, an EKC-like arc exists under consumption-based waste, during which capital accumulation allows for a cleaner environment for both capital-improving and capital-neutral recycling, but with ephemeral and reversible capital growth. In contrast, under production-based waste, a lower initial wealth and larger waste stock than for consumption-based waste are required for an EKC-like arc to arise, but this happens only for capital-improving recycling and over a smaller interval.
- Under production-based waste involving very dirty technology, a total reversal of the EKC-like arc is observed. Under capital-neutral recycling, the reversal of the EKC is even more severe in terms of capital contraction. Finally, under production-based waste with highly capital-improving recycling, a partial reversal of the EKC-like arc is obtained.

While most economic studies disregard the consequences of the differences between two pollution sources, namely production and consumption, on social welfare and environmental sustainability, our study suggests that, all things being equal, wealthy countries should focus on minimizing the wasteful impact of consumption rather than production because it is less capital-destructive, especially in the case of ineffective recycling. In contrast, national regulations should focus on minimizing the wasteful impact of production in the case of ineffective recycling in poor countries.

Our results can be further generalized by extending the model to the case of two noncooperative countries with different levels of effectiveness in their recycling process. An important issue related to this context is how the countries' noncooperative strategies affect both welfare and environmental sustainability.

Disclosure statement The authors ensure that no conflict of interest is directly or indirectly related to this study. The authors have no financial or proprietary interests in any material discussed in this article.

Appendix

A1. The Legendre-Clebsch condition of joint concavity of the Hamiltonian with respect to the control variables is satisfied, because the Hessian $\begin{bmatrix} H_{cc} & H_{cv} \\ H_{vc} & H_{vv} \end{bmatrix} = \begin{bmatrix} -1 & 0 \\ 0 & -1 \end{bmatrix}$ is negative definite. This guarantees a maximum of the Hamiltonian. Plugging the expressions of c and v respectively from (8.7) and (8.8) in (8.4) gives the maximized Hamiltonian:

$$H^0 = \frac{(\theta - \lambda + \mu\beta)^2}{2} + \frac{(\lambda\varphi - \mu)^2}{2} + \lambda aK - \frac{eW^2}{2}$$

from which the Hessian $\begin{bmatrix} H_{KK} & H_{KW} \\ H_{WK} & H_{WW} \end{bmatrix} = \begin{bmatrix} 0 & 0 \\ 0 & -e \end{bmatrix}$ is negative semi-definite. This ensures that the necessary conditions are sufficient for optimality and thus the existence of a globally optimal solution. □

A2. The proof is by contradiction. Assume a time point t such that the costate $\mu \geq 0$ and the transversality condition holds. Then, equation (8.6) implies that μ can grow only at a rate greater than r. Therefore, the transversality condition will never be met. Such a point cannot exist, i.e., $\mu < 0$. The result for λ is derived in a similar way. Because we assume that $r < a$, starting from any $\lambda(0)$, λ always tends to zero. □

A3. Differentiating (8.9), we find that $d\beta c - dv = 0$ if:

$$\begin{aligned} -d\lambda(\beta + \varphi) + d\mu\left(1 + \beta^2\right) &= -(r - a)\lambda(\beta + \varphi)dt + (r\mu dt + eWdt - d\eta)\left(1 + \beta^2\right) \\ &= 0 \end{aligned}$$

and, with respect to $W = 0$, we have:

$$\dot{\eta} = r\mu - \frac{(\beta + \varphi)(r - a)\lambda}{1 + \beta^2} \tag{A3.1}$$

Since $r < a$, $\dot{\eta} \geq 0$ when $\lambda \geq 0$ and $\mu \geq 0$, which with respect to (8.9) requires:

$$\frac{\beta\theta}{\beta + \varphi} \leq \lambda \tag{A3.2}$$

which holds only for positive λ. Such a solution is clearly feasible, if $c > 0$ and $v > 0$ and implies λ remains positive and tends to zero while μ grows over time and satisfies the transversality condition by a growing negative jump. Next, we verify when the controls are feasible by substituting μ with respect to (8.9). Specifically, $c \geq 0$ if:

$$c = \theta - \lambda + \beta\mu = \theta - \lambda + \frac{\beta[(\beta+\varphi)\lambda - \beta\theta]}{1+\beta^2}$$
$$= \theta\left(1 - \frac{\beta^2}{1+\beta^2}\right) + \left(\frac{\beta+\varphi}{1+\beta^2} - 1\right)\lambda \geq 0$$

or if:

$$\left(1+\beta^2 - \beta - \varphi\right)\lambda \leq 0 \tag{A3.3}$$

and $v = \varphi\lambda - \mu = \varphi\lambda - \frac{(\beta+\varphi)\lambda - \beta\theta}{1+\beta^2} = \frac{\varphi(1+\beta^2)\lambda - (\beta+\varphi)\lambda + \beta\theta}{1+\beta^2}$, that is:

$$\left[\varphi\left(1+\beta^2\right) - (\beta+\varphi)\right]\lambda \geq -\beta\theta \tag{A3.4}$$

Let $\varphi(1 + \beta^2) - (\beta + \varphi) \leq 0$, then (8.4) always holds and (8.4) along with (8.2) results in:

$$\frac{\beta\theta}{\beta+\varphi} \leq \lambda \leq \frac{\beta\theta}{\beta+\varphi-\varphi\left(1+\beta^2\right)} \tag{A3.5}$$

Since $r < a$, λ always decreases to zero and the left-hand side will not hold, i.e., $\beta c - v > 0$. Therefore, W increases. Letting $1 + \beta^2 - \beta - \varphi > 0$, then if $\varphi(1 + \beta^2) - (\beta + \varphi) \leq 0$ still holds, we get the same result, W increases. If $\varphi(1 + \beta^2) - (\beta + \varphi) > 0$, then (8.4) holds. Consequently, (8.3) along with (8.2) results in:

$$\frac{\beta\theta}{\beta+\varphi} \leq \lambda \leq \frac{\theta}{1+\beta^2-\beta-\varphi} \tag{A3.6}$$

Again, W will eventually increase. Thus, $W = 0$ can occur only during a finite interval. Finally, while $W = 0$, K may decrease to zero if $\varphi < \frac{1}{\beta}$.

Regarding the fact that $K(0) = 0$ can be maintained when $W(0) > 0$ only over a finite time interval, this requires a jump as given by (8.9). The jump is due to requiring $\dot{K} = 0$, that is:

$$\begin{aligned}\varphi v - c = & -(\theta - \lambda + \mu\beta) + \varphi(\lambda\varphi - \mu) = -\theta + \left(1+\varphi^2\right)\lambda - (\varphi+\beta)\mu \\ = & \ 0\end{aligned} \tag{A3.7}$$

which, when differentiating to ensure $dc - d\varphi v = 0$, results in:

$$\begin{aligned} d\lambda(1+\varphi^2) - d\mu(\varphi+\beta) &= (rdt - adt - d\eta)\lambda(1+\varphi^2) + (r\mu + eW) \\ &\quad \times (\varphi+\beta)dt \\ &= 0 \end{aligned} \quad \text{(A3.8)}$$

Finally, to have $K = 0$ and $W = 0$ concurrently along with $v > 0$ and $c > 0$ over an interval of time, we need $\dot{K} = 0$ and $\dot{W} = 0$, i.e., $\varphi = 1/\beta$. Assuming that $\varphi \neq 1/\beta$, this would apparently not be possible. Let $\beta\varphi > 1$. If both $W = 0$ and $K = 0$ at a point in time, then from that point on, the nil solution, i.e., no recycling and no consumption, is optimal over the entire time horizon. Note that $\dot{K} = aK - c + \varphi v < 0$ and $\dot{W} = \beta c - v < 0$, if $\beta c < v < \frac{c - aK}{\varphi}$. This implies that if K is small and $\beta < 1$, then it is possible that $\dot{W} < 0$ and $\dot{K} < 0$ occurs starting from some point in time so that both K and W will become equal to zero. Note that, in practice, this solution corresponds to a negative capital state. □

A4. Equating the RHS of (9)-(10)-(5)-(6) to 0 and solving by identification and substitution, we get $\lambda^S = 0$ and $\mu^S = -\frac{\beta\theta}{1+\beta^2}$ and K^S and W^S as given in (8.11). Plugging these expressions in (8.7) and (8.8), respectively, and simplifying yields c^S and v^S in (8.11). From (8.11), it can be shown that the limiting transversality conditions are satisfied for the saddle-paths because:

$$\lim_{t\to\infty} e^{-rt}\lambda(t)K(t) = 0$$

$$\lim_{t\to\infty} e^{-rt}\mu(t)W(t) = \lim_{t\to\infty} \frac{-r\beta^2\theta^2 e^{-rt}}{e(1+\beta^2)^2} = 0$$

This ensures the uniqueness of the globally optimal solution. □

A5. To analyze the stability of the canonical system (10)-(11)-(5)-(6), we write the Jacobian:

$$J = \begin{bmatrix} a & 0 & 1+\varphi^2 & -\beta-\varphi \\ 0 & 0 & -\beta-\varphi & 1+\beta^2 \\ 0 & 0 & r-a & 0 \\ 0 & e & 0 & r \end{bmatrix}$$

Given that λ and μ are evaluated at their steady state value, we compute the determinant:

$$|J| = ae(a-r)(1+\beta^2)$$

which has a positive value for $r < a$. Using Dockner's formula (Dockner, 1985), we determine the sum of the principal minors of J of order 2 minus the squared discounting rate, that is:

$$\Psi = -a(a-r) - e\left(1+\beta^2\right)$$

The necessary and sufficient conditions that ensure that two eigenvalues have negative real parts and two have positive real parts, which corresponds to the case of a two-dimensional stable manifold, are $|J| > 0$ and $\Psi < 0$. The sign of Ψ is negative whenever $r < a$, which implies that a two-dimensional stable manifold (saddle-point) exists in the case of a sufficiently patient social planner. To determine whether the optimal path is monotonic or oscillatory, we compute the expression (Dockner, 1985):

$$\Omega = \Psi^2 - 4|J| = \left[a(a-r) - e\left(1+\beta^2\right)\right]^2$$

A positive (negative) sign of Ω indicates that convergence to the saddle-point is monotonic (spiraling) near the steady state. Because $\Omega > 0$, the convergence to the saddle-point is monotonic near the steady state. □

A6. Differentiating (8.7) with respect to time, we get:

$$\dot{c} = -\dot{\lambda} + \beta\dot{\mu} = -(r-a)\lambda + \beta(r\mu + eW) \tag{A7.1}$$

which, when differentiating again, leads to:

$$\ddot{c} = -(r-a)^2\lambda + \beta\left[r(r\mu + eW) + e\dot{W}\right] \tag{A7.2}$$

(A7.2)

Substituting $\beta(r\mu + eW)$ from (A5.1) and $\dot{W}$from the state equation (8.3), we have:

$$\ddot{c} = r\dot{c} + a(r-a)\lambda + \beta e(\beta c - v)$$

Similarly, from (8.8), we obtain:

$$\dot{v} = \varphi(r-a)\lambda - (r\mu + eW)$$

which, differentiated again, gives:

$$\ddot{v} = \varphi(r-a)^2\lambda - \left[r(r\mu + eW) + e\dot{W}\right] = \varphi(r-a)^2\lambda + \left[r\varphi(r-a)\lambda - e(\beta c - v)\right]$$

Thus, accounting for $\lambda(t) = Ge^{(r-a)t}$, we get (8.13)–(8.14). The transient path converging to a steady state can then be found with the boundary conditions:

$$\dot{v}(0) = \varphi(r-a)G - (r\mu(0) + eW(0)), \dot{c} = -(r-a)G + \beta(r\mu(0) + eW(0)),$$
$$\ddot{c}(0) = r\dot{c}(0) + a(r-a)G + \beta e(\beta c(0) - v(0)), c(0) = \theta - G + \beta\mu(0),$$

$$\ddot{v}(0) = r\dot{v}(0) - a\varphi(r-a)G - e(\beta c(0) - v(0)), v(0) = \varphi G - \mu(0),$$

$$\ddot{c} = -(r-a)^2\lambda(0) + \beta\left[r(r\mu(0) + eW(0)) + e\dot{W}(0)\right], \dot{W}(0) = \beta c(0) - v(0),$$

$$\ddot{v}(0) = \varphi(r-a)^2\lambda(0) - \left[r(r\mu(0) + eW(0)) + e\dot{W}(0)\right].\square$$

A7. The Legendre-Clebsch condition of concavity of the Hamiltonian with respect to the control variables is satisfied, as $\begin{bmatrix} H_{cc} & H_{cv} \\ H_{vc} & H_{vv} \end{bmatrix} = \begin{bmatrix} -1 & 0 \\ 0 & -1 \end{bmatrix}$ is negative definite. This guarantees a maximum of the Hamiltonian. Plugging the respective expressions of c and v in (8.4) gives the maximized Hamiltonian:

$$H^0 = \frac{(\theta-\lambda)^2}{2} + \frac{(\lambda\varphi-\mu)^2}{2} + (\lambda + \kappa\mu)aK - \frac{eW^2}{2}$$

from which the Hessian matrix: $\begin{bmatrix} H_{KK} & H_{KW} \\ H_{WK} & H_{WW} \end{bmatrix} = \begin{bmatrix} 0 & 0 \\ 0 & -e \end{bmatrix}$ is negative semi-definite. This ensures that the necessary conditions are also sufficient for optimality. □

A8. The proof is identical to that for Lemma 2. □

A9. Let $W = 0$ and $K = 0$, $v = 0$ and $c = 0$ due to $\varphi\lambda - \mu \leq 0$ and $c = \theta - \lambda \leq 0$, that is:

$$\theta \leq \lambda \leq \mu/\varphi \qquad \text{(A10.1)}$$

(A10.1)

Then, both $\lambda > 0$ and $\mu > 0$ to meet (8.11) and if:

$$\mu/\varphi > \theta \qquad \text{(A10.2)}$$

(A10.2)

then (8.11) can be feasible. Accordingly, to maintain $W = 0$ and $K = 0$, we need to keep $v = 0$ and $c = 0$ which, with respect to (8.18) and (8.19), leads to:

$$dv = \varphi d\lambda - d\mu = \varphi[(r-a)\lambda dt - \mu\kappa a dt - d\eta_1] - (r\mu dt - d\eta_2) \leq 0$$

$$dc = -d\lambda = -(r-a)\lambda dt + \mu\kappa a dt + d\eta_1 \leq 0$$

Solving this system of inequalities we find:

$$(r-a)\lambda - \mu\kappa a - \frac{r\mu - \dot{\eta}_2}{\varphi} \leq \dot{\eta}_1 \quad \text{(A10.3)}$$

$$0 \leq \dot{\eta}_2 \leq r\mu \quad \begin{matrix}\text{(A10.3)}\\ \text{(A10.4)}\end{matrix}$$

(A10.4)

where $(r-a)\lambda - \mu\kappa a - \frac{r\mu - \dot{\eta}_2}{\varphi} < 0$ due to (8.14). Thus, $\dot{\eta}_1 \geq 0$ and $0 \leq \dot{\eta}_2 \leq r\mu$ hold. Further, from (8.6) we observe that, starting from a positive value, the growth of μ can be fully offset by the maximal jump $\dot{\eta}_2 = r\mu$, so that, $\dot{\mu} = r\mu - r\mu = 0$ which ensures the transversality condition. Further, although according to (8.13) the optimality conditions can be met with a zero jump, $\dot{\eta}_1 = 0$, λ always decreases. Starting from a positive λ that satisfies (8.13), although μ is constant, $\dot{\mu} = 0$, (satisfying (8.12)), we observe that decreasing λ will inevitably violate (8.11). However, for an optimal solution to be feasible in terms of the state constraints, we need the capital to grow by reducing consumption, which is already zero. However, an increase in recycling will immediately lead to a negative waste. Thus, there is no feasible solution to complete the paths. □

A10. Equating the RHS of (20)-(21)-(16)-(17) to 0 and solving by identification and substitution, we get:

$$\lambda^S = \frac{\kappa^2\theta a}{(1+\varphi\kappa)[(1+\varphi\kappa)a - r] + \kappa^2 a}$$

$$\mu^S = -\frac{\kappa\theta(a-r)}{(1+\varphi\kappa)[(1+\varphi\kappa)a - r] + \kappa^2 a}$$

along with K^S and W^S as given in (8.11). Note that $r < a$ implies that $r < (1 + \kappa a)a$, which ensures a feasible steady state. Plugging the expressions in (8.18) and (8.19), respectively, and simplifying, yields c^S and v^S in (8.27). From (8.22), the limiting transversality conditions are satisfied for the saddle-paths because:

$$\lim_{t\to\infty} e^{-rt}\lambda(t)K(t) = \lim_{t\to\infty} \frac{\kappa^2\theta^2 a[(1+\varphi\kappa)a - r]e^{-rt}}{a\{(1+\varphi\kappa)[(1+\varphi\kappa)a - r] + \kappa^2 a\}^2} = 0$$

$$\lim_{t\to\infty} e^{-rt}\mu(t)W(t) = \lim_{t\to\infty} \frac{-r\kappa^2\theta^2(a-r)^2 e^{-rt}}{e\{(1+\varphi\kappa)[(1+\varphi\kappa)a - r] + \kappa^2 a\}^2} = 0$$

This ensures the uniqueness of the globally optimal solution. □

A11. The Jacobian matrix of the canonical system (21)-(22)-(16)-(17) is:

$$J = \begin{bmatrix} a & 0 & 1+\varphi^2 & -\varphi \\ \kappa a & 0 & -\varphi & 1 \\ 0 & 0 & r-a & -\kappa a \\ 0 & e & 0 & r \end{bmatrix}$$

Given that λ and μ are evaluated at their steady state value, we compute the determinant:

$$|J| = ae\left\{(1+\varphi\kappa)[(1+\varphi\kappa)a - r] + \kappa^2 a\right\}$$

which is positive for $r < a$, and the sum of the principal minors of J of order 2 minus the squared discounting rate:

$$\Psi = -a(a-r) - e$$

which is negative. Both signs imply that a two-dimensional stable manifold (saddle-point) exists. To check whether the optimal path is monotonic or cyclical, we compute:

$$\Omega = [a(a-r)+e]^2 - 4ae\left\{(1+\varphi\kappa)[(1+\varphi\kappa)a - r] + \kappa^2 a\right\}$$

Because the sign of Ω is ambiguous, a limit value analysis highlights the role of a, e, φ, and α in the sign of Ω for a given $r < a$ (see Table 8.4). The results suggest that Ω can be either positive or negative depending on the parameters value. For any given a and e, convergence is monotonic near the steady state. However, $\lim_{\kappa\to 0^+} \Omega > 0$ and $\lim_{\kappa\to\infty} \Omega = -\infty$, as well as $\partial\Omega/\partial\alpha < 0$, which implies that there is a threshold $\tilde{\kappa} > 0$ such that for any $\kappa > \tilde{\kappa}$ ($\kappa < \tilde{\kappa}$), we have $\Omega < 0$ ($\Omega > 0$).

Table 8.4 Limit value analysis

$\lim_{a\to 0^+} \Omega = e^2$	$\lim_{a\to\infty} \Omega = \infty$
$\lim_{e\to 0^+} \Omega = a^2(a-r)^2$	$\lim_{e\to\infty} \Omega = \infty$
$\lim_{\varphi\to 0^+} \Omega = a^2\left[(a-r)^2 - 2e(1+2\kappa^2)\right] + e(2ra+e)$	$\lim_{\varphi\to\infty} \Omega = -\infty$
$\lim_{\kappa\to 0^+} \Omega = \left[a(a-r)^2 - e\right]^2$	$\lim_{\kappa\to\infty} \Omega = -\infty$

A similar result is obtained for φ, i.e., for a high value of $\tilde{\varphi} > 0$ such that for any $\varphi > \tilde{\varphi}$, we get $\Omega < 0$. □

A12. Comparing $c^S_{|\kappa=0}$ from (8.12) and $c^S_{|\beta=0}$ from (8.27), for $\varphi = 0$ and $\kappa = \beta > 0$, $\kappa^2 a > \beta^2(a - r)$, proves (8.22)–(8.26). □

A13. To compute the numerical solution of the consumption and production-based waste generation optimal control problems, we solve the related Hamilton-Jacobi-Bellman (HJB) equations, that is:

$$rV^1 = c\left(\theta - \frac{c}{2}\right) - \frac{eW^2}{2} - \frac{v^2}{2} + V^1_K(aK - c + \varphi v) + V^1_W(\beta c - v)$$

for consumption-based waste generation, and:

$$rV^2 = c\left(\theta - \frac{c}{2}\right) - \frac{eW^2}{2} - \frac{v^2}{2} + V^2_K(aK - c + \varphi v) + V^2_W(\kappa a K - v)$$

for production-based waste generation, where $V^i(K, W)$, $i = 1, 2$, are the value functions respectively associated with the consumption and production-based waste generation optimal control problems. Assuming interior solutions, the necessary conditions are:

$$c = \theta - V^1_K + \beta V^1_W, v = \varphi V^1_K - V^1_W$$

for consumption-based waste generation, and:

$$c = \theta - V^2_K, v = \varphi V^2_K - V^2_W$$

for production-based waste generation. Using the necessary conditions, respectively, we get:

$$rV^1 = \frac{\left(\theta + \beta V^1_W - V^1_K\right)^2}{2} - \frac{\left(\varphi V^1_K - V^1_W\right)^2}{2} - \frac{eW^2}{2} + aV^1_K K$$

for consumption-based waste generation, and:

$$rV^2 = \frac{(\theta - V_K^2)^2}{2} - \frac{(\varphi V_K^2 - V_W^2)^2}{2} - \frac{eW^2}{2} + a\left(V_K^2 + \kappa V_W^2\right)K$$

for production-based waste generation. For the value functions, we consider the following conjectures:

$$V^1 = \psi_1 + \psi_2 K + \psi_3 W + \psi_4 K^2 + \psi_5 W^2 + \psi_6 KW$$

for consumption-based waste generation, where $\psi_1, \ldots, \psi_6$ are real parameters, and:

$$V^2 = \xi_1 + \xi_2 K + \xi_3 W + \xi_4 K^2 + \xi_5 W^2 + \xi_6 KW$$

for production-based waste generation, where $\xi_1, \ldots, \xi_6$ are real parameters.

Depending on whether the waste generation results from consumption or production, one solves a system of 6 algebraic equations in 6 unknowns in each case, that is:

$$r\xi_1 - \frac{(\theta + \xi_2)^2}{2} - \frac{(\varphi\xi_2 - \xi_3)^2}{2} - \beta\left(\theta - \xi_2 + \frac{\beta\xi_3}{2}\right)\xi_3 = 0$$

$$r\xi_3 - 2\left[\beta\theta - (\beta + \varphi)\xi_2 + \left(1 + \beta^2\right)\xi_3\right]\xi_5 + \left[\theta - \left(1 + \varphi^2\right)\xi_2 + (\beta + \varphi)\xi_3\right]\xi_6 = 0$$

$$(r - a)\xi_2 + 2\left[\theta - \left(1 + \varphi^2\right)\xi_2 + (\beta + \varphi)\xi_3\right]\xi_4 - \left[\beta\theta + (\beta + \varphi)\xi_2 + \left(1 + \beta^2\right)\xi_3\right]\xi_6 = 0$$

$$(r - a)\xi_6 + 2\left[2(\beta + \varphi)\xi_5 - \left(1 + \varphi^2\right)\xi_6\right]\xi_4 - \left[2\left(1 + \beta^2\right)\xi_5 - (\beta + \varphi)\xi_6\right]\xi_6 = 0$$

$$r\xi_4 - 2\left[a + \left(1 + \varphi^2\right)\xi_4 - (\beta + \varphi)\xi_6\right]\xi_4 - \left(1 + \beta^2\right)\frac{{\xi_6}^2}{2} = 0$$

$$\left[r - 2\left(1 + \beta^2\right)\xi_5\right]\xi_5 + \left[2(\beta + \varphi)\xi_5 - \left(1 + \varphi^2\right)\frac{\xi_6}{2}\right]\xi_6 + \frac{e}{2} = 0$$

for consumption-based waste generation, and:

$$r\psi_1 - \frac{(\theta + \psi_2)^2}{2} - \frac{(\varphi\psi_2 + \psi_3)^2}{2} = 0$$

$$r\psi_3 - \left(1 + \varphi^2\right)\psi_2\psi_6 + (\theta + \varphi\psi_3)\psi_6 - 2(\psi_3 - \varphi\psi_2)\psi_5 = 0$$

$$\left[r-a-2\left(1+\varphi^2\right)\psi_4+\varphi\psi_6\right]\psi_2-(\kappa a+\psi_6)\psi_3+2(\theta+\varphi\psi_3)\psi_4=0$$

$$-2(\kappa a-2\varphi\psi_4)\psi_5+\left[r-a-2\left(1+\varphi^2\right)\psi_4-2\psi_5+\varphi\psi_6\right]\psi_6=0$$

$$\left[r-2\left(1+a+\varphi^2\psi_4\right)\right]\psi_4-\left(\kappa a-2\varphi\psi_4-\frac{\psi_6}{2}\right)\psi_6=0$$

$$\left[r-2(\psi_5-\varphi\psi_6)\right]\psi_5-\frac{(1+\varphi^2){\psi_6}^2}{2}+\frac{e}{2}=0$$

for production-based waste generation. In general, each system of algebraic equations above may admit zero, one or more solutions. An admissible solution for each system, if any, should satisfy the global asymptotic stability criterion. The necessary conditions for consumption-based waste generation rewrite:

$$c=\theta-\psi_2+\beta\psi_3-(2\psi_4-\beta\psi_6)K-(\psi_6-2\beta\psi_5)W$$

$$v=\varphi\psi_2-\psi_3+(2\varphi\psi_4-\psi_6)K+(\varphi\psi_6-2\psi_5)W$$

and the necessary conditions related to production-based waste generation become:

$$c=\theta-\xi_2-2\xi_4K-\xi_6W$$

$$v=\varphi\xi_2-\xi_3+(2\varphi\xi_4-\xi_6)K+(\varphi\xi_6-2\xi_5)W$$

Plugging the optimal expressions into the corresponding system of state equations, we get:

$$\begin{aligned}\dot{K}=&-\theta+\left(1+\varphi^2\right)\psi_2-(\beta+\varphi)\psi_3+\left[a+2\left(1+\varphi^2\right)\psi_4-(\beta+\varphi)\psi_6\right]K\\&-\left[2(\beta+\varphi)\psi_5-\left(1+\varphi^2\right)\psi_6\right]W\end{aligned}$$

$$\begin{aligned}\dot{W}=&\,\beta\theta-(\beta+\varphi)\psi_2+\left(1+\beta^2\right)\psi_3-\left[2(\beta+\varphi)\psi_4-\left(1+\beta^2\right)\psi_6\right]K\\&+\left[2\left(1+\beta^2\right)\psi_5-(\beta+\varphi)\psi_6\right]W\end{aligned}$$

for consumption-based waste generation, and:

$$\begin{aligned}\dot{K} &= -\theta + (1+2\varphi^2)\xi_2 - \varphi\xi_3 + \left[a + 2(1+2\varphi^2)\xi_4 - \varphi\xi_6\right]K \\ &\quad + \left[(1+2\varphi^2)\xi_6 - 2\varphi\xi_5\right]W \\ \dot{W} &= -\varphi\xi_2 + \xi_3 + (\kappa a - 2\varphi\xi_4 + \xi_6)K + (2\xi_5 - \varphi\xi_6)W\end{aligned}$$

for production-based waste generation. Imposing a steady state for each state equations system yields:

$$K^S = \frac{\theta\left[2(2\beta^2+\beta\varphi+1)\psi_5 - (\beta\varphi^2+2\beta+\varphi)\psi_6\right] - (1-\beta\varphi)^2(2\psi_2\psi_5 - \psi_3\psi_6)}{(1-\beta\varphi)^2(4\psi_4\psi_5 - {\psi_6}^2) + a\left[2(1+\beta^2)\psi_5 - (\beta+\varphi)\psi_6\right]}$$

$$W^S = \frac{\begin{array}{c} a[\beta\theta + (\beta+\varphi)\psi_2] - a(1+\beta^2)\psi_3 + \theta\left[2(\beta\varphi^2+2\beta+\varphi)\psi_4 - (2\beta^2+\beta\varphi+1)\psi_6\right] \\ +(1-\beta\varphi)^2(\psi_2\psi_6 - 2\psi_3\psi_4)\end{array}}{(1-\beta\varphi)^2(4\psi_4\psi_5 - {\psi_6}^2) + a\left[2(1+\beta^2)\psi_5 - (\beta+\varphi)\psi_6\right]}$$

for consumption-based waste generation, and:

$$K^S = \frac{(1+\varphi^2)(2\xi_2\xi_5 - \xi_3\xi_6) + \theta(\varphi\xi_6 - 2\xi_5)}{a[(\kappa+\varphi+2\kappa\varphi^2)\xi_6 - 2(1+\kappa\varphi)\xi_5] - (1+\varphi^2)(4\xi_4\xi_5 - {\xi_6}^2)}$$

$$W^S = \frac{a[\kappa\theta - (2\kappa\varphi^2+\kappa+\varphi)\xi_2 + (1+\kappa\varphi)\xi_3] - \theta(2\varphi\xi_4 - \xi_6) - (1+\varphi^2)(\xi_2\xi_6 - 2\xi_3\xi_4)}{a[(\kappa+\varphi+2\kappa\varphi^2)\xi_6 - 2(1+\kappa\varphi)\xi_5] - (1+\varphi^2)(4\xi_4\xi_5 - {\xi_6}^2)}$$

for production-based waste generation. Clearly, the positivity of the steady-state values should hold. Also, the steady-state solution given for each system of state equations, if any, should be globally asymptotically stable. The Jacobian matrix for consumption-based waste generation is:

$$J^1 = \begin{bmatrix} a + 2(1+2\varphi^2)\psi_4 - (\beta+\varphi)\psi_6 & -2(\beta+\varphi)\psi_5 - (1-\varphi^2)\psi_6 \\ -2(\beta+\varphi)\psi_4 + (1+\beta^2)\psi_6 & 2(1+\beta^2)\psi_5 - (\beta+\varphi)\psi_6 \end{bmatrix}$$

whose trace and determinant are respectively:

$$Tr\, J^1 = a + 2(1+2\varphi^2)\psi_4 + 2(1+\beta^2)\psi_5 - 2(\beta+\varphi)\psi_6$$

$$\left|J^1\right| = \left[a + 2\left(1 + 2\varphi^2\right)\psi_4 - (\beta + \varphi)\psi_6\right]\left[2\left(1 + \beta^2\right)\psi_5 - (\beta + \varphi)\psi_6\right] - \left[2(\beta + \varphi)\psi_5 + \left(1 - \varphi^2\right)\psi_6\right]\left[2(\beta + \varphi)\psi_4 - \left(1 + \beta^2\right)\psi_6\right]$$

On the other hand, the Jacobian matrix for production-based waste generation is:

$$J^2 = \begin{bmatrix} a+2\left(1+2\varphi^2\right)\xi_4 - \varphi\xi_6 & \left(1+2\varphi^2\right)\xi_6 - 2\varphi\xi_5 \\ \kappa a - 2\varphi\xi_4+\xi_6 & 2\xi_5 - \varphi\xi_6 \end{bmatrix}$$

whose trace and determinant are respectively:

$$Tr\, J^2 = a + 2\left(1 + 2\varphi^2\right)\xi_4 + 2(\xi_5 - \varphi\xi_6)$$

$$\left|J^2\right| = \left[a + 2\left(1 + 2\varphi^2\right)\xi_4 - \varphi\xi_6\right](2\xi_5 - \varphi\xi_6) - \left[\left(1 + 2\varphi^2\right)\xi_6 - 2\varphi\xi_5\right] \times (\kappa a - 2\varphi\xi_4 + \xi_6)$$

Based on the global asymptotic stability conditions, that is, $Tr\, J^i < 0$ and $|J^i| > 0$, $i = 1, 2$, the solution of $\psi_1, \ldots, \psi_6$ for consumption-based waste generation, and $\xi_1, \ldots, \xi_6$ for production-based waste generation can be selected.

References

Acerbi, F., & Taisch, M. (2020). A literature review on circular economy adoption in the manufacturing sector. *Journal of Cleaner Production, 273*, Article 123086.

Barnes, D. K. A. (2002). Invasions by marine life on plastic debris. *Nature, 416*, 808–809.

Barros, M. V., Salvador, R., de Francisco, A. C., & Piekarski, C. M. (2020). Mapping of research lines on circular economy practices in agriculture: From waste to energy. *Renewables and Sustainable Energy Review, 131*, Article 109958.

Battini, D., Bogataj, M., & Choudhary, A. (2017). Introduction to the special issue on closed loop supply chain (CLSC): Economics, modelling, management and control. *International Journal of Production Economics, 183*, 319–321.

Boucekkine, R., & El Ouardighi, F. (2016). Optimal growth with pollution waste and recycling. In G. Feichtinger (Ed.), *Dynamic perspectives on managerial decision making*. Springer.

Boucekkine, R., Pommeret, A., & Prieur, F. (2013). Technological *vs.* ecological switch and the Environmental Kuznets Curve. *American Journal of Agricultural Economics, 95*(2), 252–260.

Boucekkine, R., Hritonenko, N., & Yatsenko, Y. (2014). Optimal investment in heterogeneous capital and technology under restricted natural resource. *Journal of Optimization Theory and Applications, 163*, 310–331.

de Bruyn, S., & Opschoor, J. (1997). Developments in the throughput-income relationship: Theoretical and empirical observations. *Ecological Economics, 20*(3), 255–268.

Derigs, U., & Friederichs, S. (2009). On the application of a transportation model for revenue optimization in waste management: A case study. *Central European Journal of Operations Research, 17*, 81–93.

Dockner, E. (1985). Local stability in optimal control problems with two state variables. In G. Feichtinger (Ed.), *Optimal control theory and economic analysis* (Vol. 2). North Holland.

Doonan, J., Lanoie, P., & Laplante, B. (2005). Determinants of environmental performance in the Canadian pulp and paper industry: An assessment from inside the industry. *Ecological Economics, 55*(1), 73–84.

El Ouardighi, F., Sim, J. E., & Kim, B. (2016). Pollution accumulation and abatement policy in a supply chain. *European Journal of Operational Research, 248*(3), 982–996.

El Ouardighi, F., Sim, J. E., & Kim, B. (2021). Pollution accumulation and abatement policies in two supply chains under vertical and horizontal competition and strategy types. *Omega, 98*(102108), 1–19.

Eriksen, M., Maximenko, N., Thiel, M., Cummins, A., Lattin, G., Wilson, S., Hafner, J., Zellers, A., & Rifman, S. (2013). Plastic pollution in the South Pacific subtropical gyre. *Marine Pollution Bulletin, 68*(1–2), 71–76.

Eriksen, M., Cowger, W., Erdle, L. M., Coffin, S., Villarrubia-Gómez, P., Moore, C. J., et al. (2023). A growing plastic smog, now estimated to be over 170 trillion plastic particles afloat in the world's oceans—Urgent solutions required. *PLoS ONE, 18*(3), e0281596. https://doi.org/10.1371/journal.pone.0281596

Fodha, M., & Magris, F. (2015). Recycling waste and endogenous fluctuations in an OLG model. *International Journal of Economic Theory, 11*(4), 405–427.

Geissdoerfer, M., Pieroni, M. P. P., Pigosso, D. C. A., & Soufani, K. (2020). Circular business models: A review. *Journal of Cleaner Production, 277*, Article 123741.

Grass, D., Caulkins, J. P., Feichtinger, G., Tragler, G., & Behrens, D. A. (2008). *Optimal control of nonlinear processes with applications in drugs, corruption, and terror*. Springer.

Hartl, R. H., Sethi, S. P., & Vickson, R. G. (1995). A survey of the maximum principles for optimal control problems with state constraints. *SIAM Review, 37*(2), 181–218.

Hrabec, D., Kudela, J., Somplak, R., Nevrly, V., & Popela, P. (2020). Circular economy implementation in waste management network design problem: A case study. *Central European Journal of Operations Research, 28*, 1441–1458.

Jacobs, B., Singhal, V., & Subramanian, R. (2010). An empirical investigation of environmental performance and the market value of the firm. *Journal of Operations Management, 28*(5), 430–441.

Law, K., Moret-Ferguson, S., Maximenko, N., Proskurowski, G., Peacock, E., Hafner, J., & Reddy, C. M. (2010). Plastic accumulation in the North Atlantic subtropical gyre. *Science, 329*(5996), 1185–1188.

Lusky, R. (1976). A model of recycling and pollution control. *Canadian Journal of Economics, 9*(1), 91–101.

Martin, R. E. (1982). Monopoly power and the recycling of raw materials. *The Journal of Industrial Economics, 30*(4), 104–419.

Nakamura, S. (1999). An interindustry approach to analyzing economic and environmental effects of the recycling of waste. *Ecological Economics, 28*, 133–145.

Pati, R. K., Vrat, P., & Kumar, P. (2006). Economic analysis of paper recycling vis-à-vis wood as raw material. *International Journal of Production Economics., 103*(2), 489–508.

PlasticsEurope. (2015). *Plastics – The Facts 2014/2015. An analysis of european plastics production, demand and waste data*. Belgium.

Seierstad, A., & Sydsæter, K. (1987). *Optimal control theory with economic applications*. North Holland.

Sengupta, R. P. (1997). CO2 emission—income relationship: policy approach for climate control. *Pacific Asia Journal of Energy, 7*(2), 207–229.

Sheu, J. B. (2007). A coordinated reverse logistics system for regional management of multi-source hazardous wastes. *Computers & Operation Research, 34*(5), 1442–1462.

Simpson, D. (2012). Institutional pressure and waste reduction: The role of investments in waste reduction resources. *International Journal of Production Economics, 139*(1), 330–339.
Stokey, N. (1998). Are there limits to growth? *International Economic Review, 39*(1), 1–31.
Suhandi, V., & Chen, P. S. (2023). Closed-loop supply chain inventory model in the pharmaceutical industry toward a circular economy. *Journal of Cleaner Production, 383*, Article 135474.
Verma, R., Vinoda, K. S., Papireddy, M., & Gowda, A. N. S. (2016). Toxic pollutants from plastic waste – A review. *Procedia Environmental Sciences, 35*, 701–708.

Index

F. El Ouardighi (ed.), *Essays on Pollution Control in Economics and Management Science*, Lecture Notes in Operations Research,
https://doi.org/10.1007/978-3-031-78227-5

S

T

The manufacturer's authorised representative in the EU is Springer Nature Customer Service Centre GmbH, Europaplatz 3, 69115 Heidelberg, Germany. If you have any concerns regarding our products, please contact ProductSafety@springernature.com

Printed and bound by CPI Group (UK) Ltd, Croydon, CR0 4YY
30/09/2026
02225573-0008